A PRACTICAL GUIDE TO ASSAY DEVELOPMENT AND HIGH-THROUGHPUT SCREENING IN DRUG DISCOVERY

CRITICAL REVIEWS IN COMBINATORIAL CHEMISTRY

Series Editors

BING YAN

School of Pharmaceutical Sciences
Shandong University, China

ANTHONY W. CZARNIK

Department of Chemistry
University of Nevada–Reno, U.S.A.

A series of monographs in molecular diversity and combinatorial chemistry, high-throughput discovery, and associated technologies.

Combinatorial and High-Throughput Discovery and Optimization of Catalysts and Materials
Edited by Radislav A. Potyrailo and Wilhelm F. Maier

Combinatorial Synthesis of Natural Product-Based Libraries
Edited by Armen M. Boldi

High-Throughput Lead Optimization in Drug Discovery
Edited by Tushar Kshirsagar

High-Throughput Analysis in the Pharmaceutical Industry
Edited by Perry G. Wang

A Practical Guide to Assay Development and High-Throughput Screening in Drug Discovery
Edited by Taosheng Chen

A PRACTICAL GUIDE TO ASSAY DEVELOPMENT AND HIGH-THROUGHPUT SCREENING IN DRUG DISCOVERY

Edited by

Taosheng Chen

CRC Press
Taylor & Francis Group
Boca Raton   London   New York

CRC Press is an imprint of the
Taylor & Francis Group, an **informa** business

CRC Press
Taylor & Francis Group
6000 Broken Sound Parkway NW, Suite 300
Boca Raton, FL 33487-2742

First issued in paperback 2019

ISBN-13: 978-1-4200-7050-7 (hbk)
ISBN-13: 978-0-367-38470-8 (pbk)

Library of Congress Cataloging-in-Publication Data

A practical guide to assay development and high-throughput screening in drug discovery / editor, Taosheng Chen.
    p. cm. -- (Critical reviews in combinatorial chemistry)
Includes bibliographical references and index.
ISBN 978-1-4200-7050-7 (alk. paper)
1. High throughput screening (Drug development) 2. Combinatorial chemistry. I. Chen, Taosheng.

RS419.5.P73 2010
615'.19--dc22   2009022719

Visit the Taylor & Francis Web site at
http://www.taylorandfrancis.com

and the CRC Press Web site at
http://www.crcpress.com

Contents

Preface

High-throughput screening (HTS) has become one of the most important approaches for modern drug discovery. A successful HTS campaign relies on the development of suitable biologically relevant assays that are used either for primary screens (assays used to investigate and identify active compounds) or for secondary assays (investigation of mechanisms of action of active compounds identified by primary screens). In addition to the development of suitable assays, integration of appropriate technology and effective management of the essential HTS infrastructure are also critical for the success of a project. This book discusses the important areas related to assay development and HTS in drug discovery.

Assay development and HTS involve multiple disciplines such as biology, chemistry, informatics, instrumentation and automation, technology development, and project management. Although an integrated process for assay development and HTS has been developed in the pharmaceutical and biotechnology industries, academic institutions have only recently begun to implement similar processes to facilitate basic and translational research. In both industrial and academic settings, scientists involved in assay development and HTS typically have biology or chemistry backgrounds, but to be successful they must hone their skills as well as their experience in integrating interdisciplinary knowledge. Gaining this expertise and experience may take years, and even longer for those involved with facilities starting assay development and HTS efforts from the ground up. This book is authored by experts in the assay development and HTS fields who have experienced challenging situations while starting projects and later bringing them to fruition.

The first five chapters of this book discuss assay developments for important target classes such as protein kinases and phosphatases (Glickman, Chapter 1), proteases (Woelcke and Hassiepen, Chapter 2), nuclear receptors (Bai and Johnson, Chapter 3), G protein-coupled receptors and ion channels (McGivern, Qian, and Lee, Chapter 4), and heat shock proteins (Gao and Fong, Chapter 5). Next, assay developments for cell viability and apoptosis (Riss, Moravec, and Niles, Chapter 6), and infectious diseases (Xu, Chapter 7) are discussed. The last three chapters of the assay development portion of the book discuss the application of emerging technologies and systems, including image-based high-content screening (Feng and Wilson, Chapter 8), RNA interference (Wolters and MacKeigan, Chapter 9), and primary cells (Corazza and Wade, Chapter 10) in assay development and HTS. The last four chapters discuss the essential components of the integrated HTS process, such as screening automation (Zheng and Chen, Chapter 11), compound library management (Nie, Chapter 12), and screening informatics (Schürer and Tsinoremas, Chapter 14), and a unique area that requires specific expertise: screening of natural products from botanical sources (Manly, Smillie, Hester, Khan, and Coudurier, Chapter 13).

This book draws knowledge from experts actively involved in assay development and HTS. I hope that by reading through the information in the chapters, you will appreciate the efforts invested and the accomplishments achieved in this field. I also hope that you will appreciate the interdisciplinary nature of assay development and HTS. More importantly, I hope that this book will motivate further discussions and research to bring further advances to this field.

Taosheng Chen

Acknowledgments

The idea for this book originated from a discussion in early 2007 with Dr. Bing Yan, my colleague at St. Jude Children's Research Hospital. I thank him for his encouragement to initiate this work and his suggestions during the process of preparing this book. I thank all the researchers working on various aspects of drug discovery, both in academia and in industry, especially those who contributed chapters to this book. I thank my academic advisors who provided me with the excellent opportunity of scientific training in their laboratories: Dr. Zuyu Luo of Fudan University in Shanghai, China; Dr. Janet Kurjan at the University of Vermont in Burlington; and Dr. Michael Weber at the University of Virginia in Charlottesville.

Over the years I have enjoyed working with many scientists at the National Cancer Institute, Bristol-Myers Squibb, and St. Jude Children's Research Hospital. I thank Bill Farrar, Michele Agler, Martyn Banks, Michael Sinz, and Kip Guy for many inspirational and valuable discussions. I would also like to thank David Bouck, Jimmy Cui, Anang Shelat, and Vani Shanker at St. Jude Children's Research Hospital; Hu Li at GlaxoSmithKline; Ke Liu at the National Institutes of Health Chemical Genomics Center; and Henry Paulus at the Boston Biomedical Research Institute for reading portions of the book and providing many valuable suggestions.

I thank my wife Yajun, my daughter Nora, and my son Tony, for their loving support and inspiration that made the compilation of this book possible.

Taosheng Chen

About the Editor

Taosheng Chen earned a BS in biophysics from Fudan University in Shanghai, China and an MS in molecular cell biology, also from Fudan University, studying the anticancer properties of parvoviruses with Dr. Zuyu Luo. He received his PhD in cell and molecular biology from the University of Vermont at Burlington, working with Dr. Janet Kurjan on desensitization of the yeast pheromone response pathway, a G protein-coupled receptor–MAP kinase pathway. He conducted postdoctoral studies with Dr. Michael Weber at the University of Virginia, studying the roles of the Ras-to-MAP kinase pathway and the androgen receptor in prostate cancer progression.

Between 1998 and 2000, Dr. Chen worked as a research scientist at SAIC-Frederick, National Cancer Institute. From 2000 to 2006, he served as a senior research investigator in the Lead Discovery and Profiling Department at Bristol-Myers Squibb, where he was responsible for various aspects of drug discovery including assay development, HTS, HCS, new technology evaluation and development, and drug–drug interactions. He has also been certified as a PMP (Project Management Professional) by the Project Management Institute.

Dr. Chen joined the faculty of the Department of Chemical Biology and Therapeutics (CBT) at St. Jude Children's Research Hospital in Memphis in 2006. His laboratory focuses on studying the regulation of transcription factors and their implications in drug metabolism and drug resistance as well as in pediatric cancers. He has also been the director of the St. Jude HTS Center (part of CBT). Its members work with St. Jude researchers to discover and develop new chemical entities as tools for translational research or as therapeutic leads for the treatment of catastrophic pediatric diseases. The center employs state-of-the-art automation and assay technologies and an onsite library of 525,000 compounds. In addition, it is developing whole-genome RNA interference technology. Dr. Chen has authored more than 20 peer-reviewed articles in the areas of cancer biology, drug metabolism, signal transduction, and drug discovery technology.

Contributors

Chang Bai
Pharmaron Inc.
Irvine, California

Catherine Z. Chen
National Human Genome Research Institute
National Institutes of Health
Bethesda, Maryland

Taosheng Chen
Department of Chemical Biology
    and Therapeutics
St. Jude Children's Research Hospital
Memphis, Tennessee

Sabrina Corazza
Enabling Technologies
Axxam SpA
Milan, Italy

Louis Coudurier
Digital Fox Consulting, LLC
Chapel Hill, North Carolina

Yan Feng
Novartis Institutes for BioMedical Research
Cambridge, Massachusetts

Susan Fong
Novartis Institutes for BioMedical Research
Emeryville, California

Zhenhai Gao
Novartis Institutes for BioMedical Research
Emeryville, California

J. Fraser Glickman
The Rockefeller University
New York, New York

Ulrich Hassiepen
Novartis Institutes for BioMedical Research
Basel, Switzerland

John P. Hester
Thad Cochran Research Center
National Center for Natural Products Research
Research Institute of Pharmaceutical Sciences
School of Pharmacy
University of Mississippi
University, Mississippi

Eric N. Johnson
Department of Automated Biotechnology
Merck & Co., Inc.
North Wales, Pennsylvania

Ikhlas Khan
Thad Cochran Research Center
National Center for Natural Products Research
Research Institute of Pharmaceutical Sciences
School of Pharmacy
University of Mississippi
University, Mississippi

Paul H. Lee
Amgen Inc.
Thousand Oaks, California

Jeffrey P. MacKeigan
Laboratory of Systems Biology
Van Andel Research Institute
Grand Rapids, Michigan

Susan P. Manly
Thad Cochran Research Center
National Center for Natural Products Research
Research Institute of Pharmaceutical Sciences
School of Pharmacy
University of Mississippi
University, Mississippi

Joseph G. McGivern
Amgen Inc.
Thousand Oaks, California

Richard A. Moravec
Research and Development
Promega Corporation
Madison, Wisconsin

Dalin Nie
HTS Center and Global Support
CNS Discovery
AstraZeneca
Wilmington, Delaware

Andrew L. Niles
Research and Development
Promega Corporation
Madison, Wisconsin

Yi-xin Qian
Amgen Inc.
Thousand Oaks, California

Terry L. Riss
Research and Development
Promega Corporation
Madison, Wisconsin

Stephan C. Schürer
Department of Pharmacology
Miller School of Medicine
University of Miami
Miami, Florida

and

The Comprehensive Center for Chemical
    Probe Discovery and Optimization
The Scripps Research Institute
Jupiter, Florida

Troy Smillie
Thad Cochran Research Center
National Center for Natural Products Research
Research Institute of Pharmaceutical Sciences
School of Pharmacy
University of Mississippi
University, Mississippi

Nicholas F. Tsinoremas
Department of Medicine
Miller School of Medicine
University of Miami
Miami, Florida

Erik J. Wade
Axxam SpA
Milan, Italy

Christopher J. Wilson
Novartis Institutes for BioMedical
    Research
Cambridge, Massachusetts

Julian Woelcke
Novartis Institutes for BioMedical
    Research
Basel, Switzerland

Natalie M. Wolters
Laboratory of Systems Biology
Van Andel Research Institute
Grand Rapids, Michigan

H. Howard Xu
Department of Biological Sciences
California State University
Los Angeles, California

Wei Zheng
National Human Genome Research Institute
National Institutes of Health
Bethesda, Maryland

1 Assay Development for Protein Kinases and Phosphatases

J. Fraser Glickman

CONTENTS

1.1   INTRODUCTION

Enzymatic phosphate transfers are some of the predominant mechanisms for regulating the growth, differentiation, and metabolism of cells. The post-translational modification of proteins with phosphate leads to dramatic changes in conformation, resulting in the modulation of binding, catalysis, and recruitment of effector molecules that regulate cellular signaling pathways. Examples include the recruitment of SH2 domain containing proteins, the activation of gene transcription pathways, and the activation or deactivation of specific cell surface receptors. Regulation and control by kinases and phosphatases are important in the activation of disease pathways, including inflammation and oncogenesis. Protein kinases and phosphatases are responsible for triggering gross physiological changes and represent important classes of drug targets.

High-throughput screening (HTS) techniques are often employed in the pharmaceutical industry to identify kinase and phosphatase modulators by testing large compound libraries for effects on *in vitro* protein kinase and phosphatase assays. The number and variety of assay technologies used has grown tremendously. New technology has made the screening of compound libraries more efficient. Early assays employed very low throughput techniques such as metabolic labeling and immunoprecipitation or sodium dodecylsulfate polyacrylamide gel electrophoresis (SDS-PAGE) gel autoradiography. The purification of pure kinase and phosphatase enzymes allowed for the development of relatively rapid radioactive filter binding assays. However, these techniques do not enable the testing of the millions of compounds often required in modern HTS. As a result, more sophisticated non-separation techniques have become commonplace. The practical design and implementation of these assays will be the focus of this chapter. We will first give a review of the principles of the various HTS assay technologies available, with particular focus on those technologies that have recently become routine. We will discuss some of the advantages and disadvantages of each technology and will then provide some practical guidelines and principles for assay development. The chapter will conclude with some of the future developments in kinase assay technology and address some of the critical questions for assay design in the future.

1.2   ASSAY "READOUT" TECHNOLOGIES

The keys to protein kinase and protein phosphatase assay development lie in the ability to choose an appropriate "readout" technology; to have ample quantities of enzymes, cell lines, antibodies, and reference compounds; and to optimize the assay for buffer conditions, reagent concentrations, timing, stopping, order of addition, plate type, and assay volume. We will discuss these aspects in separate sections below. The readout technologies summarized in Table 1.1 present many options for assay development and often depend on the laboratory infrastructure, the cost of reagents, the desired substrate, and the secondary assays needed to validate the compounds and determine the structure–activity relationships.

1.2.1   ENZYME ASSAYS

Kinase and phosphatase assay technologies can be divided roughly into those employing whole cells and those employing purified enzymes. Protein kinase enzyme assays require ATP, magnesium (and sometimes manganese) cofactors and a peptide or protein substrate. The requirements for phosphatase enzyme assays are simpler and well described by Montalibet et al. (2005). One must have a method to detect the conversion of substrate by determining either the formation of

TABLE 1.1
HTS Assay Development for Kinases and Phosphatases

Assay Technology (Commercial and Alternate Names)	Technology Principles	Advantages	Disadvantages	References
Fluorescence polarization (anisotropy) version 1 (InVitroGen PolarScreen)	Fluorescently labeled substrate peptide binds to anti-phospho antibody after phosphorylation; change in Brownian motion of peptide–antibody complex results in change in anisotropy measured by polarization of incoming light	High-throughput, only one labeled substrate required	Susceptible to compound interference; peptide must be relatively small; precludes use of protein substrates	Parker (2000); Sills (2002); Newman (2004); Turek-Etienne (2003b)
Fluorescence polarization (anisotropy) version 2 (IMAP)	Fluorophore labeled peptides bind to special detection beads coated with trivalent metal; binding results in change in Brownian motion measured as with FP1 above	Versatile without need for antibody	Susceptible to compound interference; peptide must be relatively small; precludes use of protein substrates	Turek-Etienne (2003a)
Scintillation proximity (FlashPlate, SPA)	Product of reaction is a ^{33}P labeled peptide biotin that can be captured on a detection bead that scintillates from proximity to ^{33}P; dephosphorylation by phosphatases can be detected in signal decrease assay	High-throughput, relatively artifact free in imaging-based systems; universal readout for kinases; versatile	Radioactive waste disposal; may be less sensitive than TR-FRET	Park (1999); Sills (2002); von Ahsen (2006)
Fluorescence resonance energy transfer (quenched fluorescence, InVitroGen Z'-LYTE)	Peptide labeled with fluorescein and coumarin is quenched until cleavage by protease, modification by phosphorylation, or dephosphorylation by kinase or phosphatase produces resistance to proteolytic cleavage	Miniaturizeable, ratiometric readout normalizes for pipetting errors; can be applied to kinases and phosphatases	Coupled assay can be susceptible to protease inhibitors	Rodems (2002)
Immunosorbent assays (enzyme-linked or fluorescent-linked, cell signaling, PathScan)	Antibodies coated onto MTP wells capture kinase or phosphatase substrate; phosphorylation state is detected by anti-phosphopeptide antibody coupled to detector dye; can be read by time-resolved fluorescence (DELFIA) technique	Can be used as sensitive probe for cell lysates in cell-based assays	Lower throughput and wash steps required; must have suitable cell line and antibody pair	Waddleton (2002); Minor (2003); Zhang (2007)

(continued)

TABLE 1.1 (CONTINUED)
HTS Assay Development for Kinases and Phosphatases

Assay Technology (Commercial and Alternate Names)	Technology Principles	Advantages	Disadvantages	References
Kinase-dependent cell growth BaF3	Cell number or metabolism measured by standard cell detection methods such as ATP detection or MTT dye, Alamar blue; cell growth dependent upon specific tyrosine kinase	Very high-throughput cell-based assay	Limited for use with tyrosine kinases; cell line must be made	Warmuth (2007)
Luciferase-based ATP detection (Promega kinase Glo, Perkin-Elmer Easylite Kinase)	ATP-dependent luminescent signal from luciferase conversion of luminal; kinase-dependent depletion of ATP is measured	Versatile and non-radioactive	Signal decrease assay; susceptible to luciferase inhibitors	Koresawa (2004)
Luminescent oxygen channeling (Perkin Elmer AlphaScreen, Surefire)	Anti-phosphotyrosine or -phosphopeptide antibodies bind only to phosphorylated substrate; complex is detected by streptavidin and protein A functionalized beads that when bound together produce channeling of singlet oxygen when stimulated by light; singlet oxygen reacts with acceptor beads to emit photons of lower wavelength than their excitation frequency	Sensitive, high-throughput; can be applied to cell lysates as substitute for ELISA; proximity distances very large relative to energy transfer; emission frequency lower than excitation frequency, thus eliminating potential artifacts by fluorescent compounds; can be applied to whole cell assays	Can be susceptible to interference by compounds that trap singlet oxygen; must work under subdued or specialized lighting	Leoprechting (2004); Warner (2004)
Time-resolved Förster resonance energy transfer (version 1: InVitroGen LanthaScreen, Perkin Elmer LANCE, CysBio KinEase)	Phosphopeptide formation is detected by europium chelate and U light acceptor dye PKA substrates; dephosphorylation by phosphatases can be detected in signal decrease assay	Very sensitive and miniaturizeable; ratiometric readout normalizes for pipetting errors	Requires two specialized antibodies; susceptible to interference; low dynamic range for substrate turnover; binding interaction should be within restricted proximity for optimal efficiency	Moshinsky (2003); Vogel (2006); Von Ahsen (2006); Schroeter (2008)

Assay	Description	Advantages	Disadvantages	References
Time-resolved Förster resonance energy transfer (version 2: BellBrook Labs Transcreener, Adapta)	ADP formation by kinase is detected by displacement of red-shifted TR-FRET system between Alexafluor 647-ADP analog and a europium chelated anti-ADP antibody	High-throughput; miniaturizeable; versatile; ratiometric readout	Signal decrease assay; binding interaction should be in close proximity (7 to 9 nM)	Huss (2007)
In-Cell Western (Li-COR Cytoblot)	Anti-phospho specific antibody labeled with infrared fluor is used to probe fixed cells on MTP; second color can be used for normalization; binding detected via laser scanning technique	Only requires one antibody; has potential for multiplexing to control for total substrate expression in cells	Multiple wash steps; requires cell fixation; lower throughput	Chen (2005)
Prompt fluorescence with small molecule substrates (DiFMUP, AnaSpec Sensolyte)	Action of enzyme on substrate converts substrate into fluorescent compound; excellent substrates available for phosphatases	Versatile; very strong signal	Requires highly purified enzymes, or contaminating phosphatases can be deceptive	Watanabe (1998); Johnston (2007); Montalibet (2005)
Enzyme fragment complementation (DiscoveRx ED-NSIP HitHunter)	Two fragments of reporter protein fusion are joined via biomolecular interaction, thus reconstituting activity of reporter protein; kinases can be assayed in displacement mode using staurospaurine conjugate to one fragment and kinase fused to second fragment of β-galactosidase (ED-NSIP); test compound displaces kinase–staurospaurine interaction, and decreases B-gal activity	High-throughput; sensitive; amplified enzymatic; chemiluminescent signal less susceptible to interference	Coupled assay; may have interference with β-galactosidase binding compounds; compound must be competitive with probe	Eglen (2002); Vainshtein (2002)
Imaging assays (DiscoveRx PathHunter, BioImage)	Can use recombinant green fluorescent or yellow fluorescent fusion proteins to measure kinase-stimulated signalling events such as protein translocation to nucleus or membrane to cytosolic translocations	Biological assay; can present kinase in native form; examines functional consequences of test compound	Difficult to develop; lower throughput	Nickischer (2006); Traskjr (2006)

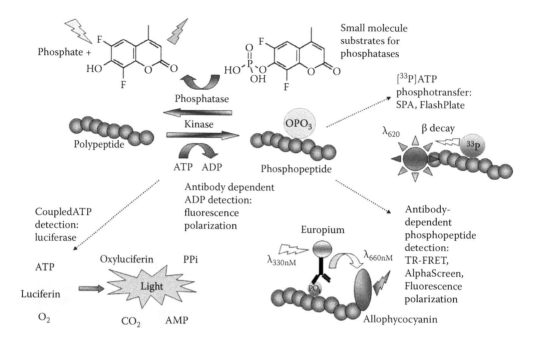

FIGURE 1.1   HTS enzyme assay concepts for kinase and phosphatase screening.

phosphopeptide, or phosphoprotein, or the disappearance of adenosine triphosphate (ATP) or the formation of adenosine diphosphate (ADP).

Phosphatase enzyme essays most commonly employ artificial small molecule substrates that become fluorescent by removal of the phosphate moiety. Other methodologies similar to those employed in kinase assays can also be used, in which the removal of phosphate from a peptide or protein substrate can be detected. Figure 1.1 shows the basic principles. Many commercially available kits and published references describe these methodologies (Table 1.1).

1.2.1.1   Radioactive Assay Technologies

The earliest assays were based on the use of ^{32}P as a label either on the ATP cofactor for kinases or on a peptide substrate for phosphatases. With kinases, the transfer of the ^{32}P from the γ position of ATP to a peptide or protein substrate resulted in a ^{32}P-labeled peptide or protein that would be separated away from the ATP by capture on a filter and subsequent washing. The quantity of phosphoprotein could be quantified by scintillation counting.

The availability of a new version of isotopically labeled ATP, [^{33}P]ATP, provided benefits of safety and longer half-life. The lowered energy was also better suited for scintillation proximity assays (SPAs). The SPA was a major step forward because it eliminated the need for wash steps by capturing the [^{33}P]-labeled peptide on a functionalized scintillating crystal, usually via a biotin–streptavidin interaction.

The specific signal in the SPA is a consequence of a radiolabeled peptide or protein substrate becoming closely bound to the scintillation material. As a result, photons are given off due to a transfer of energy from the decaying ^{33}P particle to the scintillation material. Non-specific signals (non-proximity effects) can result from decay particles emitted from free [^{33}P]ATP molecules interacting with the scintillation material at greater distances.

All SPAs are based upon the phenomenon of scintillation. Scintillation is an energy transfer that results from the interactions of particles of ionizing radiation and the de-localized electrons found in conjugated aromatic hydrocarbons or in inorganic crystals. When the decay particle collides with the scintillation material, electrons are transiently elevated to higher energy levels. Because of the return to the ground state, photons are emitted. Frequently, scintillation materials are doped with fluorophores that capture these photons (usually in the ultraviolet spectrum) and fluoresce at a "red-shifted" wavelength more "tuned" to the peak sensitivity of the detector.

Conventionally, the scintillation materials used in bioassays were liquids composed of aromatic hydrocarbons. These bioassays required a wash step before the addition of the scintillation liquid and counting in a liquid scintillation counter. With SPA technology, crystals of polyvinyltoluene (PVT), yttrium silicate (YS), polystyrene (PS), and yttrium oxide (YOx) are used as scintillants. These materials are functionalized with affinity tags to detect the decay particles directly in the bioassay without wash steps. The newer generation of red-shifted FlashPlates and SPA beads yields emission frequencies around 615 nm and thus may be detected by charge-coupled device (CCD) imagers rather than PMT (photomultiplier tube) readers. The advantages of these "imaging" beads and plates lie in both the ability to simultaneously read all wells in a microtiter plate (MTP) and in the reduction of interference from colored compounds, due to the red-shifted emission wavelength.

Because of the cost of disposing of radioactive reagents and the requirement for special safety infrastructure, the use of SPA is becoming less common, although it presents some distinct advantages (Glickman and Ferrand, 2008). First, it does not require phosphopeptide- or phosphotyrosine-specific antibodies, as the [^{33}P] ATPγ is the only cofactor needed. Second, interference of compounds can occur only at one wavelength, whereas fluorescence presents more potential sources of interference. Third, readout is relatively direct and can be "tuned" over a large range of enzyme, substrate and cofactor concentrations, as compared, for example, with time-resolved fluorescence (or Förster) resonance energy transfer (TR-FRET) and techniques that measure ATP depletion or formation.

SPA techniques are well suited for utilizing a variety of biologically relevant substrate proteins. Phosphatase assays can be constructed by pre-labeling the substrate peptide with ^{33}P and measuring the enzyme-dependent signal decrease over time (Sullivan et al., 1997). Universal substrates that are biotinylated, such as poly-glutamine tyrosine (polyEY), can be used for the tyrosine kinases. Generalized substrates such as myosin basic protein or casein, or specialized peptide substrates must be used for the serine–threonine kinases.

1.2.1.2 Luminescence: Fluorescence and Phosphorescence

Because of the complex and often overlapping principles behind kinase and phosphatase assays, I will review the principles of the various fluorescent and luminescent technologies. A textbook by Joseph R. Lakowitz (1999) titled *Principles of Fluorescence Spectroscopy* is recommended for detailed information on the biophysics of fluorescence. Olive (2004) and Von Ahsen and Boemer (2005) wrote good reviews on the advantages and disadvantages of various luminescent (including fluorescent) technologies for kinase assays.

Luminescence is the emission of light from any substance when energy is absorbed (including radioactive emissions) and is roughly divided into fluorescence and phosphorescence. Fluorescence occurs when singlet state electrons in excited orbitals return to the ground state. The photon emission lifetime is usually between 1 and 10 nanoseconds (ns). Phosphorescence is the emission of light from triplet excited states. Transitions to the ground state are impossible and because of this, the emission lifetimes are slow, ranging between 1 millisecond (ms) and 1 second (s). The fluorescence resulting from transition-metal–ligand complexes (that have mixed singlet and triplet states) has an intermediate lifetime. These substances display lifetimes between 400 ns and several hundred microseconds. This type of fluorescence can be used to time-gate the emission (dissociation-enhanced lanthanide fluorescent assay [DELFIA], LANCE®, TR-FRET, homogeneous time-resolved fluorescence [HTRF], etc.) and will be discussed below in Section 1.2.1.4. The "brightness" of a fluorophore relates to the quantum efficiency. The brighter the fluorophore, the more sensitive the detection will be.

The second factor in using fluorophores for HTS assays is red-shifting (Vedvick and Eliason, 2004). One should generally try to use fluorophores that absorb and emit in the longer wavelengths to minimize yellow range interference common with small molecule test compounds. We will see later that this is significant for the use of time-gated resonance energy transfer to measure kinase activity, commonly called TR-FRET or HTRF. These terms are misnomers when applied to HTS, because the HTS techniques do not "resolve" in the time dimension, but rather the long-lived fluorescence enables the "gating" of the reading time window to eliminate background from prompt fluorescence.

1.2.1.3   Fluorescence Assays

Although fluorescent assays are very useful in HTS, the classic issue with these assays is that they are susceptible to interference from compounds that either absorb light in the excitation or emission range of the assay (known as inner filter effects) or that are themselves fluorescent, resulting in false negatives. At typical compound screening concentrations between 1 and 10 μM, these types of artifacts can become significant.

There are several approaches to minimizing this interference. One is to use longer wavelength (red-shifted) fluorophores. This reduces much compound interference because most organic medicinal compounds tend to absorb at shorter wavelengths. Another approach is to use as "bright" a fluorescent label as possible. Bright fluorophores exhibit high efficiencies of energy capture and release. This means that an absorbent or fluorescent compound will exert a lowered impact on the total signal of the assay. Assays with higher photon counts tend to be less sensitive to fluorescent artifacts from compounds as compared with assays having lower photon counts.

Minimizing the test compound concentration can also minimize these artifacts and one must balance the potential of compound artifacts versus the need to find weaker inhibitors by screening at higher concentrations. Another method to reduce compound interference is to use a "kinetic readout" that requires reading the microtiter plate at several time points, and looking for the effect of the test compound on the slope of a curve representing the change in signal over time. Kinetic readouts can be employed typically in kinase and phosphatase assays exhibiting an enzyme-catalyzed linear increase in product formation over time that correlates with an increase in fluorescent signal over time. The inner filter effects of compounds can be thus eliminated because these would not have an effect on the slope of the reaction progress, whereas true inhibitors would have an effect on this slope.

The best examples of simple fluorescence assays are those used to detect the tyrosine phosphatase activity of purified phosphatase enzymes (Johnston et al., 2007; Watanabe et al., 2007). A very informative review of the fluorescence substrates available for tyrosine phosphatases by Montabilet and Skorey (2005) details good starting protocols as well as a presentation of the various substrates that can be synthesized. There are a variety of different synthetic substrates that form a fluorescent product when a phosphate is removed from an aryl phosphate moiety. A common example is 6,8-difluoro-4-methylumbellyferyl phosphate (DiFMUP), available from Molecular Probes Inc. (Catalog Number D-6567). This compound is not fluorescent until it is dephosphorylated and then it fluoresces intensely at 450 nm when irradiated with 360 nm light. With such substrates, continuous monitoring of the reaction is also possible. Most phosphatases have K_m values for this substrate between 1 and 30 μM and k_{cat} levels between 0.03 and 214 per second. Because these substrates are used at high micromolar concentrations and give strong fluorescent signals, they do not tend to be susceptible to fluorescent compound artifacts even though the assays are run at a relatively short wavelength (450 nm).

The availability of anti-phosphotyrosine antibodies, anti-phosphopeptide antibodies, and antibodies to fluorescent ADP analogs enabled the performance of several homogeneous methods using fluorophores including TR-FRET and fluorescence polarization (FP).

1.2.1.4   Time-Resolved Förster Resonance Energy Transfer (TR-FRET)

These assays are based upon the use of a europium or terbium chelate (a transition metal–ligand complex displaying long-lived fluorescent properties) and labeled anti-phosphopeptide or anti-phosphotyrosine antibodies that can bind to phosphorylated peptides. The antibodies are usually labeled with aromatic fluorescent tags such as Cy5, rhodamine, or fluorescent proteins such as allophyco-cyanin—a light-harvesting protein that absorbs at 650 nm and emits at 660 nm (Moshinsky et al., 2003; Newman and Josiah, 2004; Schroeter et al., 2008). When the anti-phosphotyrosine or anti-phosphopeptide antibodies bind to a labeled phosphorylated peptide, the proximity of the antibody to the labeled peptide results in a transfer of energy.

The energy transfer is a consequence of the emission spectrum of the metal–ligand complex overlapping with the absorption of the labeled peptide. If the donor fluor is within 7 to 9 nm of the acceptor fluor, Förster resonance energy transfer can occur although this distance can vary considerably (Vogel and Vedvik, 2006). The action of a kinase enzyme would increase the concentration of phosphopeptide over time and result in an increased signal in such an assay. The action of a phosphatase would decrease the signal in this assay format.

TR-FRET assays have two main advantages. The first is the time-gated signal detection; the emission is not measured for 100 to 900 µs after the initial excitation frequency is applied, resulting in a reduction in fluorescence background from the MTP's buffers and compounds. Data is acquired by multiple flashes per read, to help improve the sensitivity and reproducibility of the signal. The second advantage is that one can measure the ratio of the emission from the acceptor molecule to the emission from the donor molecule. Because of this "ratiometric" calculation, variations in signal due to variations in pipetting volume can be reduced. Generally one observes less inter-well variation with TR-FRET assays (Glickman et al., 2002).

1.2.1.5 Fluorescence Anisotropy (Polarization)

Anisotropy can be measured when a fluorescent molecule is excited with polarized light. The ratio of emission intensity in each polarization plane, parallel and perpendicular relative to the excitation polarization plane, gives a measure of anisotropy, more commonly and incorrectly referred to in HTS as fluorescence polarization (FP). This anisotropy is proportional to the Brownian rotational motion of the fluorophore.

Changes in anisotropy occur when the fluorescent small molecule binds to a much larger molecule, affecting its rotational velocity. Kinase assays can be set up using anti-phosphopeptide antibodies and labeled phosphopeptides (Newman and Josiah, 2004) or by using a metal ion affinity material to capture labeled phosphopeptides (Turek-Etienne and Kober, 2003). The formation or degradation of the phosphopeptide in an enzymatic reaction causes a change in binding state and consequently a change in anisotropy. Phosphatase assays can be configured to present the enzyme with a labeled phosphopeptide substrate (Parker et al., 2000). The advantage of FP is that it requires labeling of only one small molecule (instead of two as with TR-FRET or AlphaScreen).

FP assays are known to be susceptible to artifacts (Turek-Etienne and Small, 2003). In principle, the assays are ratiometric and should normalize for variations in total excitation energy applied as would occur with inner filter effects, and newer generations of red-shifted fluorophores should help to eliminate interference (Vedvik et al., 2004). However, introducing a test compound with fluorescent or absorbent properties at 5 or 10 µM with the typically sub-micromolar concentrations of fluorophores in an FP assay can significantly skew the measurements. For example, if the compounds are insoluble, they can scatter and depolarize light. A concentration-dependent effect on an FP assay could result from an increase in the amount of insoluble compound.

1.2.1.6 Protease Sensitivity Assays

Kinase substrates can become resistant to the actions of proteases due to their phosphorylations. Thus, the fluorescence quench assays (described in Chapter 2 covering protease assays) can be used to measure kinase activity. The assays can be viewed as "coupled" because they require a second enzyme to convert a product or substrate into a detectable signal. With kinase assays, the formation of phosphopeptide inhibits the protease action on the peptide and the signal remains quenched and therefore decreased (Rodems et al., 2002). Inhibiting the kinase results in increases in protease sensitivity and in signal.

For assaying phosphatases, the situation is reversed and phosphatase activity results in an increase in signal due to an increase in the cleavage of the site. The strength of this assay as used for phosphatases is that it allows the use of physiological substrates as compared to DiFMUP, but with all coupled assays, one must be careful that the inhibitors or activators are not working through activity against the coupling systems.

1.2.1.7   Luminescent Technologies: Oxygen Channeling (AlphaScreen)

AlphaScreen technology, first described in 1994 by Ullman et al., and based on the principle of luminescent oxygen channeling, has become a useful technology for kinase assays. AlphaScreen is a bead-based, non-radioactive, amplified luminescent proximity homogeneous assay in which donor and acceptor pairs of 250 nm diameter reagent-coated polystyrene microbeads are brought into proximity by a biomolecular interaction of anti-phosphotyrosine and anti-peptide antibodies immobilized to these beads. Irradiation of the assay mixture with a high intensity laser at 680 nm induces the conversion of ambient oxygen to a more excited singlet state by a photosensitizer present in the donor bead.

The singlet oxygen molecules can diffuse up to 200 nm and, if an acceptor bead is in proximity, can react with a thioxene derivative present in the bead, generating chemiluminescence at 370 nm that further activates the fluorophores contained in the same bead. The fluorophores subsequently emit light at 520 to 620 nm. The donor bead generates about 60,000 singlet oxygen molecules resulting in an amplified signal. Since the signal is very long-lived, with a half-life in the 1-s range, the detection system can be time-gated, thus eliminating short-lived background (the AlphaScreen signal is measured via a delay between illumination and detection of 20 ms). Furthermore, the detection wavelength is of a shorter wavelength than the excitation wavelength, further reducing the potential for fluorescence interference. The sensitivity of the assay derives from the very low background fluorescence. The larger diffusion distance of the singlet oxygen enables the detection of binding distance up to 200 nm, whereas TR-FRET is limited to 9 nm (Glickman et al., 2002).

Kinase or phosphatase assays based on the AlphaScreen principle are similar to TR-FRET assays in that they usually require a biotinylated substrate peptide and an anti-phosphoserine or tyrosine antibody. These two reagents are "sandwiched" between biotin and protein A-functionalized acceptor and donor beads. A kinase assay would show an enzyme-dependent increase in antibody binding (and thus signal) over time and a phosphatase assay would show an enzyme-dependent decrease in antibody binding over time. In some cases, the phosphorylation of an epitope will block the antibody binding and thus a phosphatase assay in principle can be constructed as a signal increase assay (Von Leoprichting and Kumpf, 2004; Warner et al., 2004).

1.2.1.8   ATP Detection as Kinase Assay

Because kinases convert ATP into ADP, the activity of purified kinase enzymes can be measured in a "coupled" assay that detects ATP depletion over time by using the phosphorescence of luciferase and luciferin (Koresawa and Okabe, 2004; Schroter et al., 2008). Luciferases are enzymes that produce light by utilizing the high energy bonds of ATP to convert luciferin to oxyluciferin. Oxygen is consumed in the process, as follows:

$$\text{Luciferin} + \text{ATP} \rightarrow \text{luciferyl adenylate} + \text{PPi}$$

$$\text{Luciferyl adenylate} + \text{O}_2 \rightarrow \text{oxyluciferin} + \text{AMP} + \text{light}$$

The reaction is very energy efficient: nearly all of the energy input into the reaction is transformed into light. A reduction in ATP results in a reduction in the production of photons. This type of assay has the advantage of not needing specialized antibodies and is applicable to all kinases. It is also easy to run the assays with high ATP concentrations as a way of selecting against ATP-competitive inhibitors (Kashem et al., 2007). The assay has the disadvantage of being sensitive to luciferase inhibitors that may bind to the ATP binding site of luciferase. These compounds may produce false negatives.

Another limitation is that this technique is a signal decrease assay in which the enzyme-mediated reaction proceeds with a decrease in signal over time. As the kinase depletes the ATP over time, less light is produced by the luciferase reaction. An inhibitor compound prevents this decrease. A 50% consumption of ATP will result in roughly a twofold signal-to-background using the 100% inhibited reaction as a control. These assays must be run under conditions of relatively high turnover that would produce the effect of slightly weakening the apparent potency of compounds (Wu et al., 2003). Another consequence of a signal decrease assay is that it can have low sensitivity for particularly low turnover enzymes (i.e., those having low k_{cat} values). Slower enzymes may require longer incubation (several hours) to reach a suitable signal. With other assay formats that measure phosphopeptide formation (especially TR-FRET and AlphaScreen), only a 2% to 10% conversion is required before the assay can be read, since the detecting reagents are product-specific.

1.2.1.9 Direct Binding Assays

In principle and like other enzymes, kinases and phosphatases can be viewed as receptors. It is possible to use the HTS techniques typically used in receptor binding assays to screen for small molecules that displace labeled ligands. Such ligands can be radiolabeled and applied to the SPA technique or can be fluorescently labeled and applied to the FP or TR-FRET techniques. In many of the ligand binding assays, the kinase must have some affinity tags such as biotin or glutathione-S-transferase (GST). The advantage of this type of assay is that one can focus on inhibitors that bind to a desired pocket, depending on the labeled probe used. The disadvantage of this approach is the need to produce a labeled probe and often the labeling may disturb the binding interaction. Consequently, one should have some knowledge about the nature of the ligand–receptor interaction, perhaps through x-ray crystallographic or nuclear magnetic resonance structural information.

1.2.2 CELL-BASED TECHNOLOGIES

In contrast to enzyme assays, cell-based assays present the target in a more physiological milieu. With enzyme assays, it may be difficult to purify and express active kinases and phosphatases in their full-length forms and they may require the use of fusion proteins with kinase activity domains. Cell-based technologies, on the other hand, present the opportunity to express the targets with regulatory domains included. Furthermore, cell-based assays usually detect only cell-permeable inhibitors and have the potential to identify more unusual mechanisms, as described earlier.

However, cell-based technologies present the potential for artifacts resulting from "off-target" effects of compounds. They also tend to be more complicated with respect to automation and reagent addition steps, often requiring a lysis, a plate-to-plate transfer step, and several washes. Cellular assays also tend to be more difficult and demanding in obtaining statistical quality for HTS, related to the accurate and reproducible production of cell batches. One usually needs to use downstream assay methods such as enzyme assays to validate the mechanism of action and guide structure-based lead optimization.

The principles for cell-based assays are based on signal transduction. The assay technologies used can, in one way or another, detect intracellular changes in the phosphorylation states of downstream targets of the kinase or phosphatase of interest (Figure 1.2). Alternatively, they can detect some consequences of kinase or phosphatase inhibition or activation such as cell growth or differentiation or specific gene expression. For this purpose, specialized anti-phosphopeptide site antibodies or kinase-regulated reporter gene systems must be made often. These projects represent a challenge because the antibodies must be of suitable affinity and specificity and because the cell line must

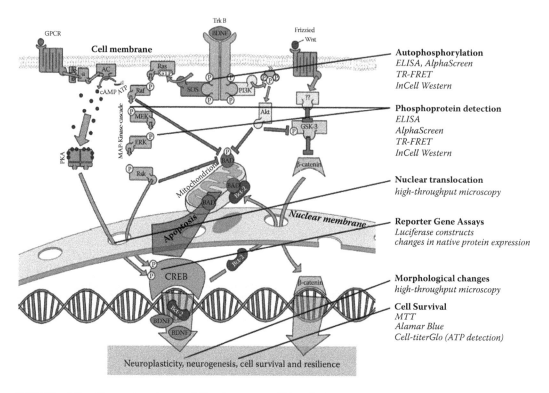

FIGURE 1.2 (See color insert following page 114.) Cell-based assay concepts for kinase and phosphatase HTS.

have high enough expression levels of analytes to be detectable with the desired readout technology. In the following sections, we will review some of the basic cellular techniques for kinase and phosphatase HTS.

1.2.2.1 Cellular Enzyme-Linked Immunoadsorbent Assay (ELISA)

The technique of cellular ELISA is to treat the cells with a test compound, and by modulating kinase or phosphatase activity, cause a change in a particular signaling pathway. The change in phosphorylation state of the signaling components can be detected by lysing the cells and measuring the phosphorylation state of the analytes by capturing them between an anti-analyte antibody and an anti-phosphopeptide or anti-phosphotyrosine antibody (Minor, 2003). The ELISA plates are coated with the anti-substrate antibody, blocked, and then the lysate is added, incubated, and washed. The plate is then probed with a labeled anti-phosphotyrosine or -phosphopeptide antibody that is incubated and washed. The last detection antibody can be an anti-species-specific antibody labeled with europium or with horseradish peroxidase (HRP). Various reagents for detecting HRP activity or europium (DELFIA assays) are available. The plate readers used depend on the label to which the final antibody is coupled and can be fluorescent, chemiluminescent, or time-resolved fluorescent; all versions are relatively sensitive assays.

Early manifestations of this format for measuring the activation of the epidermal growth factor receptor (EGFR) kinase were described by Minor (2003); Zhang et al. (2007) described methods for the detection of dephosphorylation of the insulin receptor by protein tyrosine phosphatase 1b (PTP1b). For example, cell lines expressing the EGFR tyrosine kinase can be given EGF, lysed, and then the EGFR phosphorylation state can be probed by capture with a microtiter plate coated with anti-EGFR antibody. The microtiter plate can then be probed with an anti-phosphotyrosine antibody and an EGF-dependent increase in receptor phosphorylation can be observed. An inhibitor of EGFR kinase would block this EGF-dependent autophosphorylation signal.

Waddleton et al. (2002) described a time-resolved fluorescence-based ELISA assay for measuring insulin-dependent insulin receptor phosphorylation in KRC-7 cells. An anti-insulin receptor biotinylated antibody was bound to streptavidin-coated MTPs and the cell lysate containing the insulin receptor was applied, washed, and probed with anti-phosphotyrosine antibodies labeled with europium. After washing, the europium was released through a commercially available solvent containing compounds that promote a strong long-lived fluorescent signal. These DELFIA (time-resolved fluorescence or TRF) assays are extremely sensitive because the signal reading is time gated to remove background fluorescence from the plates. The many wash and transfer steps required make these assays difficult to adapt to HTS. The advent of various non-separation techniques have in many cases eliminated the need for plate washing and these are described below.

1.2.2.2   Measuring Lysates: Non-Separation-Based Cellular Assays

The newer technologies that allow elimination of the wash step in an ELISA assay in many cases are based on the techniques of AlphaScreen and TR-FRET, the principles of which are described in the earlier enzyme assays section. They all work on a similar proximity principle in which the cellular phosphoprotein is captured between an anti-phosphopeptide and anti-peptide antibody that gives a signal either through TR-FRET or AlphaScreen (luminescent oxygen channeling). These techniques can produce excellent readouts, eliminating many wash steps. However, the antibodies used must be exquisitely selective because they have the potential to bind to a large variety of intracellular proteins. The lack of washing may lower the potential sensitivity of the assay. A typical protocol with such assays would require no plate-to-plate transfer and no wash step. Thus the protocol might be: (1) seed the cells onto an MTP, (2) add compound and incubate, (3) add concentrated lysis solution, (4) add detector reagents, and (5) read in a plate reader.

1.2.2.3   In-Cell Western Assay

The newly available Odyssey and Aerius infrared imaging systems make it possible to probe whole cells with two-color infrared fluorescently labeled antibodies (anti-phosphopeptide, for example) that are used to detect changes in intracellular kinase signaling (Chen et al., 2005). The advantage of using infrared-labeled probes lies in the increased sensitivity and dynamic range and consequent reduction in the use of reagents.

These assays resemble a hybrid of an immunohistochemistry assay and ELISA. Whole cells are fixed, for example with 3.7% formaldehyde, to MTPs permeabilized by repetitive washing with 0.1% Triton X-100, blocked with a protein solution, probed with primary antibodies (phospho-specific, and non-phospho-specific), washed, and subsequently the secondary antibodies labeled with infrared fluorescent tags are added. After washing, these assays are read in a reader (such as the Odyssey or Aerius) designed for high sensitivity detection of two colors. The two colors are useful because one color can be used to accommodate a stain assigned as a total protein or cell number control or as an antibody to total protein, which allows for normalization. These assays may only require a single antibody versus the dual antibody "sandwich" required for ELISA.

1.2.2.4   Translocation Assays and High-Throughput Microscopy

The advent of advanced high-throughput microscopic systems such as EvoTec's Opera, General Electric's INCell Analyzer 3000, Cellomics' Array Scan, Acumen's Explorer, and Molecular Devices' ImageXpress makes it possible to collect and analyze image information on large numbers of samples in standard MTP format. These technologies can be applied to cellular signal transduction involving kinases. Examples were presented by Nickischer et al. (2006) and Trask et al. (2006) in measuring the subcellular translocation events associated with MAP kinase pathway activation.

The principle of the assay system is that stimulation of a transmembrane receptor results in kinase- and phosphatase-mediated changes in the phosphorylation states of intracellular signaling proteins that produce conformational changes and also changes in binding to various proteins that mediate nuclear transport. Recombinant signaling proteins fused with the green fluorescent protein can be expressed and the receptor-mediated transport of these fusion proteins can be scored by digital imaging algorithms. These algorithms can be programmed to find the images of cells, locate nuclei by nuclear staining, and subsequently associate the fluorescence associated with GFP with the fluorescence associated with the nucleus and cytoplasm. Thus, the distribution of nuclear and cytoplasmic GFP fusion proteins can be used as readouts for receptor or pathway activation. Although these assays do not offer a direct way to visualize phosphorylation within cells, they have the potential for reasonably high-throughput and have the ability to measure the structural consequences of pathway modulation.

1.2.2.5 Reporter Gene Assays

Reporter gene assays represent another simple way to measure activation of a pathway involving kinase or phosphatase regulation. In these assays, the kinase-mediated transduction pathway results in the activation of a promoter that can be engineered to drive expression of reporter proteins such as green fluorescent protein, luciferase, or β-lactamase. These assays can be run in an HTS format using standard fluorescent plate readers; however, they generally suffer the weakness of requiring long incubation (8 to 36 hr) before the signal is obtained. The long incubation reflects the time required by the promoter to initiate protein expression. Because of this, these assays can be prone to cytotoxins and non-specific effects. It is also possible that compounds can create artifacts by interacting with the reporter proteins.

1.2.2.6 Kinase Sensors

Some cellular assays also present the ability to measure kinase activation using recombinantly expressed intracellular sensor molecules that fluoresce in response to phosphorylation. These sensors are reviewed by Lawrence et al. (2007). The most interesting versions use yellow and blue fluorescent proteins to create a FRET through phosphorylation-induced folding with an intramolecular SH2 domain. SH2 domains are known to bind tightly to phosphotyrosine molecules and have been successfully shown to serve as phosphotyrosine detectors. In this type of assay, imaging is not necessary because the total fluorescence can be measured in a population of cells using a standard plate reader.

1.2.2.7 BaF3 System

The BaF3 cell-based assay is very useful for HTS targeting the receptor-associated protein tyrosine kinases. The BaF3 cells are a mouse interleukin 3 (IL3)-dependent pro-B cell line. In the "normal" case, the BaF3 cells require interleukin 3 for proliferation, but when an oncogenic tyrosine kinase is overexpressed, the normal IL3 pathway is subverted by the oncogenic kinase and growth is driven by the recombinant protein. Alternatively, receptor tyrosine kinases can be expressed in these cells and one can then observe a proliferation response to the ligand. Cell-permeable inhibitors of the oncogenic fusion protein or receptor tyrosine kinase will block proliferation in the IL3-dependent parental cell line, but not the IL3-independent proliferation. Non-selective or toxic inhibitors will inhibit both pathways and by running the assays against the IL3-dependent cells, one can identify more selective kinase inhibitors.

Proliferation of the BaF3 cells can be monitored rapidly by using standard techniques such as with Alamar blue or the commercially available CellTiter-Glo. These systems measure cell numbers by detecting intracellular ATP concentrations or by measuring intracellular reduction of the Alamar blue dye. The BaF3 system is excellent for looking at larger kinase constructions and receptor tyrosine kinases (Warmuth et al., 2007).

1.3 ASSAY DEVELOPMENT

Much of the cited literature in this chapter describes good protocols for kinase and phosphatase assays. Additionally, the PubChem website published by the U.S. National Institutes of Health includes a large number of protocols associated with HTS data on kinase and phosphatase assays (http://www.ncbi.nlm.nih.gov/sites/entrez, search PubChem assays with keywords *kinases* and *phosphatases*). These protocols provide an excellent starting point for developing new assays. A general strategy for assay development might include the following steps:

1. Choose an appropriate readout technology.
2. Generate cell lines and enzymes, substrates.
3. Design a starting protocol based on prior literature or experimental information on substrate specificity; test the protocol for enzyme-based or ligand-stimulated activity and use reference inhibitors when possible.
4. Gain knowledge of kinetic and mechanistic parameters as guidelines for assay optimization.
5. Set-up "matrix" experiments by varying pH, ionic strength buffers, and various parameters mentioned below to optimize signal-to-background ratios.
6. Choose volumes appropriate for HTS workstations or automated systems.
7. Test a screening protocol in a pilot study and determine assay quality based on the coefficient of variations across the assay wells, the Z prime measurements (Zhang et al.,1999) and dynamic range toward various control or reference compounds.

1.3.1 CONSIDERATIONS AND MECHANISMS

1.3.1.1 Standard Mechanisms of Drug Action

It is of critical importance when designing kinase and phosphatase assays to consider carefully the desired mechanism of action (MOA) of the inhibitor; some of the concepts of inhibition mechanisms when applied to drug discovery are worth discussing briefly. MOA is a very complex topic which is nevertheless well described in terms of drug discovery strategies by David Swinney (2004); examples are given of drugs that work by a variety of mechanisms and fall into three basic categories: (1) competition with substrate or ligand in an equilibrium or steady state, (2) inhibiting at a site distinct from substrate binding, or (3) inhibiting in a non-mass action equilibrium.

A disadvantage of competitive inhibitors [(1) above] is that unless there is a way to degrade the accumulated substrate, this accumulation can reduce the efficacy of a competitive inhibitor by overcoming the competition. This disadvantage can be avoided by inhibiting at sites distal to the catalytic site or by utilizing non-mass action inhibition mechanisms. Non-mass action inhibition mechanisms occur when the initial binding is driven by equilibrium mass action, but is coupled to a process that prevents the system from attaining equilibrium. This coupled process can occur in several ways: (1) covalent modification of the drug target, sometimes because of catalytic action of the enzyme on the inhibitor (mechanism-based inhibitors), (2) slow, tight binding in the picomolar affinity range, or (3) pseudo-irreversible activity with a very low rate of dissociation compared with the substrate.

1.3.1.2 Mechanisms of Action of Kinase Inhibitors

Vogel et al. (2008) presented a good overview of the mechanisms of action of protein kinase inhibitor. These inhibitors can act by binding directly in the ATP binding site competitively, (type I inhibitors), but they tend to be less specific because of the shared characteristics of the ATP binding pockets among various kinases. More specificity can be attained with type II inhibitors that can extend into an allosteric site next to the ATP pocket and is only available in the inactive (non-phosphorylated) forms of the enzymes. Imatinib is an example of this, with a 200-fold increased potency to the inactive form of the enzyme, observed in cell-based versus enzyme assays. Often, these inhibitors bind with slower off-rate and on-rate due to a requirement for conformational changes.

Type III inhibitors bind to sites distal to the ATP binding site and are often inactive in simple kinase enzyme assays. This apparent inactivity arises because these compounds can bind to the kinase and render it a poor substrate for an activating upstream kinase, thus disabling its activation. As a consequence, a cascade or cell-based assay may be required.

1.3.1.3   Overcoming Resistance

One of the major issues in using kinase inhibitors for cancer is drug resistance. Because tumors can evolve, they can become resistant to drug therapy by generating mutations that preclude drug binding. The question then arises as to whether assays can be designed to find compounds that prevent this problem. One approach would be to screen against an enzyme containing the resistance mutations. This would require the development of a collection of compounds targeted to different forms of the kinase. One hypothesis suggests that the strength of using an inhibitor that interacts with the catalytic site is that an enzyme that would evolve to mutate its catalytic site would also lose its function. These types of inhibitors are favored to be less susceptible to resistance; however, they tend to be less selective and thus more prone to side effects. Inhibitors that bind to allosteric pockets without critical functions, although desirable for reasons of mass action (discussed above), are probably more likely susceptible to resistance mutations. Daub et al. (2004) suggest that inhibitors that target the active form of the kinase may preclude their selection for drug-resistant mutations because the "gatekeeper" residues cannot evolve to exclude the inhibitor from the active site.

1.3.1.4   Application of Mechanistic Principles to Assay Design

Important questions related to the desired inhibitor or agonist mechanism such as whether, in the HTS, one desires to find allosteric, competitive, slow-binding inhibitors, or inhibitors of an active or inactive form of the enzyme should be considered. These mechanisms may suggest the appropriate incubation times, substrate concentrations, order of addition, and appropriate recombinant construct to use. Unfortunately, the decisions related to assay set-up are not always straightforward. There are advantages to each type of inhibitor and the choices also depend largely upon the biology of the disease to be treated. As discussed earlier, purely ATP site-competitive inhibitors may present advantages with respect to drug resistance, but disadvantages regarding selectivity, potency, and cellular activity. Inhibitors that compete with the peptide binding site are generally difficult to block with a small molecule since the peptide–enzyme interaction presents a large surface area. Allosteric site inhibitors might have an advantage in potency and selectivity, but also may be resistance-prone. Layered on top of this complexity is the difficulty in expressing a well-defined form of the enzyme target that is physiologically relevant.

One important guideline for assay design is that, whenever possible, a pre-incubation step of compound with empty target enzyme should be included before starting the reaction. This pre-incubation will help to favor the identification of slow, tight binding inhibitors (Glickman and Schmid, 2007). This step may add significant time to a screening protocol and also may be difficult if the enzyme preparation is unstable. Furthermore, it is sometimes necessary to pre-incubate the kinase enzyme with ATP to activate it. This can be accomplished in the "stock" solution of enzyme, before diluting the enzyme (and thus the ATP) into the assay. With protein kinase "hit lists," it is possible to distinguish the ATP-competitive from the non-competitive inhibitors by re-screening under high and low ATP concentrations, and looking for shifts in the IC_{50} or percent inhibition. An ATP-dependent shift in compound potency suggests competition with the ATP site.

In cell-based assays where the inhibitor is present for a long incubation time, such as with the BaF3 assay (discussed above), a slow allosteric inhibitor can be identified. However, with other faster readout cell-based assays, it may be necessary to include a long pre-incubation step to allow the compound to diffuse into the cell and bind to the target.

When working with purified enzymes, it can be useful to perform a close examination of their phosphorylation states and molecular masses. Mass spectrometry is often useful for this purpose. Post-translational modifications or sequence truncations can potentially alter the compound binding sites available and can also change the structure of potential inhibitory sites. For example, with protein kinases, phosphorylations distal from the ATP binding site can inactivate the kinase whereas phosphorylations near the ATP binding site can activate the catalytic activity. Often, practice does not permit control of such situations because the purified systems are often mixtures and cannot be controlled in the commonly used recombinant expression technologies.

To reduce the likelihood of competitive inhibitors, one should run the assays with high concentrations of substrates relative to K_m levels. To favor competitive inhibitors, one should run the assay below the K_m values for the substrates. Thus, for making suitable conclusions for assay design, knowledge of the kinetic and binding constants of receptors and enzymes, such as K_d, k_{cat}, K_m, B_{max}, is useful. Stoichiometric information, such as the number of enzyme molecules per assay, may also be useful because it can serve as a guideline to ensure that the assays are maximally sensitive to the mechanism of action one wants to discover. Problems in assay development often occur when the conditions required for sensitivity to the desired mechanism of action do not yield the best conditions for statistical reproducibility; therefore, compromises and balances of these two opposing factors must be often made.

The percent substrate consumption when the data are collected is also of importance in enzyme assay design. Typically, enzymologists like to ensure that steady-state conditions are maintained in the study of inhibitor constants such as K_i or IC_{50}. However, many assay technologies, combined with the requirements for a robust signal-to-background giving a good Z prime (Zhang et al., 1999), necessitate assay set-up requiring more turnover. This typically causes a trend toward the reduction of compound potency depending on the mechanism of action. Therefore, one must balance the need to have the most sensitive assay toward inhibition by low molecular weight (LMW) compounds and achieve good assay signal-to-background to obtain statistically relevant results. These types of effects have been modeled by Wu et al. (2003) and additionally confirmed by empirical determination. These investigators found that 50% inhibition at low conversion (near zero) can translate into 31% inhibition if the assay is run at 80% conversion. IC_{50} values can shift as much as threefold.

1.3.2 Enzyme Assay Optimization

It is very difficult to give specific guidelines for assay development in HTS due to the complexity of variables involved and because one often must balance cost, speed, sensitivity, statistical robustness, automation requirements, and desired mechanisms. The typical parameters to test are shown in Figure 1.3. Schematically, the assay development process can be thought of as a cycle in which several variables can be tested and the parameters that give a better "reading window" can be fixed, allowing further parameters to be tested under the prior fixed conditions. Often, one observes interactions among the various parameters. For example, the optimal detergent concentration may not be the same for every pH and that is why the same fixed parameters must sometimes be retested under newly identified optimal parameters. Furthermore, the type of optimization experiments depends upon the particular technology used.

In its simplest form, building an assay is a matter of adding several reagent solutions to a microtiter plate (MTP) with a multi-channel pipette, at various incubation times, stopping the reaction if required, and reading the MTP in a plate reader. A typical procedure may involve: (1) adding the enzyme solution to a compound-containing microtiter plate and incubating for 15 min; (2) adding substrates and incubating for 15 to 60 min; (3) adding a stopping reagent such as EDTA (for kinases) or sodium orthovanadate (for phosphatases); (4) adding sensor or detector reagents such as labeled antibodies or coupling enzymes; and (5) measuring in a plate reader. The specific detector reagents, assay reagent volumes, concentrations of reagents, incubation times, buffer conditions, MTP types, and assay stopping reagents are all important parameters that must be tested.

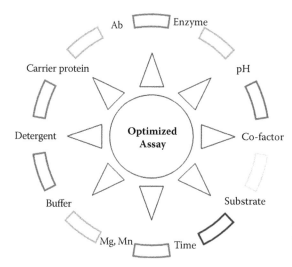

Assay Component	Example Range
Enzyme conc.	1–50 nM
pH value	5–9
Co-factors	0.1–10 uM
Substrate conc.	0.1–10 uM
Time	0.25–3 hours
Divalents	Mg, Mn (mM)
Buffer	20 mM HEPES, TRIS, MES
Detergent	0.05% NP–40, Tween-20, CHAPS, pluronic
Carriers	BSA, gelatin, casein, 0.1%
Antibodies	5–20 nM

FIGURE 1.3   Assay optimization cycle and typical test parameters.

A very important part of assay development is making an appropriate choice of substrates. Where possible, it is desirable to choose a physiological substrate, although this is not always practical. For example, with phosphatases, using fluorescent small molecule substrates to obtain good assay robustness presents good advantages. With kinases, it is often possible to perform a substrate screen in which many random peptide sequences are tested to identify good substrates.

Based on the many factors that must be optimized, statistical design of experiment software (e.g., JMP software from SAS) can be employed to determine the minimal number of variables for drawing statistically significant inferences (Rodgers et al., 2003). Furthermore, design of experiment software can be coupled to automated pipettes to rapidly run multi-variable experiments (e.g., SAGIAN from Beckman Instruments). The disadvantage of using purely statistical parameters in assay optimization is that these systems do not take into account the desired physiological or biochemical mechanism of action in determining optimal reagent concentration. For example, the optimal substrate concentration to use in an enzyme assay may not necessarily be the one that gives the best statistics with respect to reproducibility. The optimal salt concentration required for reproducibility may be far different from physiological concentration. Therefore, one must be careful in using these systems in a way that is consistent with the desired mechanism of action of the lead compound.

1.3.2.1 Enzyme Preparations

One cannot over-emphasize the importance of the enzyme preparation in the ability to develop an assay. First, purity is important because even the slightest contamination with another enzyme can lead one to screen with a measurement of the wrong activity. Specific reference inhibitors can be used to establish that the observed catalytic activity is the correct one. Of course, when working with novel targets, such reference inhibitors may not exist. With recombinant expression systems, one can generate catalytically inactive mutations to establish that the host cell is not the source of a contaminating phosphatase or kinase. In the end, however, the key requirement is to have an extremely pure and highly active enzyme preparation.

If most of the enzyme molecules are inactive or if the enzyme molecules exhibit a very low catalytic efficiency (k_{cat}/K_m), then one will need to have a high concentration of enzyme in the assay to obtain a good signal; this can limit the ability to distinguish between weaker and stronger inhibitors. If one needs to have a 100 nM enzyme concentration in the reaction, then inhibitors with a K_i of 10 nM cannot be distinguished from those with a K_i of 100 nM in a steady-state IC_{50} experiment.

1.3.2.2 Assay Volumes

Assay volumes usually range from 3 μL (for 1536-well MTPs) to 50 μL (384-well MTPs). Within a given total assay volume, smaller volumes of reagents are added. Frequently, we find it convenient to add reagents into the assay in equivalent volumes of assay buffer. As an example, for a 15-μL assay, one might add 5 μL of compound solution, 5 μL of enzyme stock solution, 5 μL of substrate mix, followed by 10 μL of quench solution in a "stop" buffer. For kinase assays, the stop buffer may be EDTA and for phosphatase assays, sodium orthovanadate.

The advantages of using equal volumes are (1) maintaining the volumes at a level that is best for the particular pipette during automation, (2) minimizing the requirements for various specialized instruments, and (3) help in the mixing of the reagents. The disadvantage is that this method introduces transient changes in the final concentrations of reagents during the times between the various additions. In addition, enzymes are not always stable in the large batches of dilute buffer required for HTS. Therefore, it may be preferable to add a low volume of a more concentrated stock solution (10X to 50X) into a higher volume of assay buffer in the well. This step often requires a liquid dispenser able to handle very low volumes of submicroliter liquid.

Assay miniaturization helps to reduce the consumption of very expensive assay reagents. The problems encountered as one attempts to miniaturize an assay relate to the change in surface-to-volume ratio, lowered sensitivity, and low volume dispensing of materials. For instance, as one moves to smaller volumes, the surfaces available for unspecific binding increase relative to the volume. Furthermore, the smaller the volume, the less the amount of product-sensing material can be added; thus the sensitivity of the assay is reduced.

1.3.2.3 Plate Types

Generally, white plates are preferred for phosphorescent assays, such as those employing luciferase, because they reflect outgoing photons. Black plates are preferred for fluorescent assays because they reduce reflection. Transparent plates are generally used for imaging assays. Polystyrene is the material of choice because it can be molded reproducibly such that the plates are sized consistently to fit into automated systems. Polystyrene tends to have some level of non-specific affinity for biomolecular reagents. Therefore, some manufacturers produce proprietary low binding plates. Polypropylene plates have low non-specific binding levels, but are more difficult to shape consistently. However, they are often useful as source plates for reagents due to their tolerance to freezing and low levels of stickiness. The choice of 96, 96 low volume, 384, 384 low volume, or 1536 well trays depends on the assay volume and throughput required. For 96-well plates, volumes range from 80 to 100 μL; for 384-well plates, volumes range from 15 to 100 μL; for 384 low-volume plates, volumes range from 4 to 10 μL; and for 1536-well MTPs, volumes range from 3 to 8 μL.

1.3.2.4 Incubation Times

Incubation times at different steps depend upon the binding kinetics or enzyme kinetics and can range anywhere from 10 min to 15 hr. It is generally preferred to read the reaction at a point when the signal increase over time is linear, but not when the enzymatic reaction is running out of substrates. We also like to add a short pre-incubation step of compound with enzyme before the initiation of the reaction to allow for slow-binding inhibitors that require a conformational change for forming a tight complex with the target. Longer incubation steps can also add significant amounts of plate processing time in HTS automation because incubation time often represents the rate-limiting step in the HTS process. Reaction rates can be increased by increasing the enzyme concentration or temperature and, in this fashion, incubation times can be reduced.

1.3.2.5 Buffers and Solvents

The concentrations of detergents, buffers, carrier proteins, reducing agents, and divalent cations can affect the specific signals and apparent potencies of compounds in concentration response curves (Schroter et al., 2000). In general, it is good to stay as close to physiological conditions as possible, but there are many exceptions. For instance, sometimes carrier proteins are required for enzyme stability and to reduce unspecific binding to reaction vessels. A very good summary of solvent and buffer conditions used in kinase assay optimization is described by von Ahsen and Boehmer (2005) and two good reviews by Montalibet et al. (2005) and McCain and Zhang (2001) for tyrosine phosphatases are available. It is often necessary to provide additives such as protease inhibitors to prevent digestion of the assay components or phosphatase inhibitors such as vanadate to keep the product of a kinase assay intact and protected from contaminating phosphatases. In addition, it is often necessary to quench assays or to add a stopping buffer such as EDTA for kinases or vanadate for phosphatases at the end of the reaction and before reading in an MTP plate reader. A quench step or stopping step is especially important when an automated HTS system does not allow for precise timing or scheduling of the assay protocol.

The presence of "promiscuous" inhibitors or "aggregating" compounds in a chemical library has become a recent area of research. These compounds can cause false positives that can be reduced with certain detergents and can frequently be recognized by steep concentration response curves or enzyme concentration-dependent shifts in the IC_{50} (Feng and Shoichet, 2006; Shoichet, 2006). The exact mechanism of these false positives is not known but it is possible that these compounds induce the formation of compound–enzyme clusters that reduce enzyme activity. Thus, it is important to have some detergent present in enzyme assays and to re-test hits in orthogonal assays to reduce the possibility of identifying promiscuous inhibitors in the HTS.

1.3.2.6 Dimethylsulfoxide (DMSO) Concentration

In most cases, the test compounds are dissolved in DMSO and are added from a source plate into the assay. Thus, the tolerance of the assay for DMSO should be tested, by looking at the activities at various increasing concentrations of DMSO. Generally, enzymatic or biomolecular binding assays are more tolerant of high DMSO concentrations (often up to 5% to 10% DMSO). Cell-based assays usually can tolerate up to 0.5% DMSO.

1.3.2.7 Detector Concentrations

Because of the very large variety of assay detection methods, it is difficult to cover all the parameters needed to optimize for each detection system; however, it is important to mention that all the systems discussed in the assay readout technology section above must be optimized. For example, when using an antibody pair in a TR-FRET assay, it is important to find a concentration of antibodies that can trap the product efficiently at the desired time point in the reaction.

In scintillation proximity assays (SPAs), for example, optimal SPA bead concentration should be determined empirically. In principle, the detector reagents should be present in high enough concentration to capture the analyte stoichiometrically. Excessive concentrations of detectors can be wasteful and expensive and can create higher background signals. Too-low concentrations of detector reagents can compromise the dynamic range and sensitivity of the assay. A control product can sometimes be useful as a calibration standard for these types of optimization experiments. Many commercial assay kit providers provide excellent protocols in optimizing the use of the detector reagents.

1.3.2.8 Pilot Screens

The final step in the assay development process is to run a "pilot" screen in which a small subset of libraries is screened to observe the hit rate, the distribution of the high and low signals, typically employing the Z and Z' concept of Zhang and Chung (1999) to assess quality. The Z factors combine the principles of signal-to-background ratio, coefficient of variation of the background, and coefficient of variation of the high signal into a single parameter that provides a general idea of the screening quality. One should be careful to closely examine the raw data and data trends from screening rather than to rely only on the Z values for quality control.

For pilot studies, the MTPs should be arrayed with reference compounds at various concentrations, to ensure that the screening procedure produces adequate dynamic range and sensitivity to inhibitors or activators. In general, coefficients of variation below 10% and Z values above 0.5 are desirable, along with a reasonable "hit rate," for example, one hit for every 2,000 compounds tested. Cellular assays generally have higher hit rates than enzymatic assays due to the preponderance of off-target hits. Of course, the hit rate depends on the particular library screened and biased libraries may exhibit higher hit rates than random libraries.

1.3.3 CELL-BASED ASSAY OPTIMIZATION

In addition to many of the factors for biochemical assays, many cell-based assays require optimizations of cell numbers and the lysis or fixation step. The steps of a typical cell-based assay protocol may include: (1) plating cells in culture medium onto MTPs and allowing them to adapt overnight; (2) adding compound in a DMSO solution; (3) stimulating cells with a specific hormone, growth factor or other receptor ligand; (4) adding lysis solution; (5) adding detection reagents; and (6) reading in a multi-functional MTP reader. With imaging-based assays, the protocols may include a fixation step followed by probing with labeled antibodies and nuclear stains for cell counting. Cell-based assay procedures are generally more diverse and complex than modern homogeneous or wash-free biochemical assays.

The optimal number of cells per well to add to an assay depends on a few factors, such as the expression levels of the analyte, the growth rate of the cells, and the incubation time with the compound. In most cases, these numbers range from 500 to 2,000 cells per well in a 1536-well format and 1,000 to 20,000 cells per well in a 384-well format. Microscopic imaging-based assays can also be susceptible to the effects of cell-to-cell contact and one should consider the effects upon the imaging algorithms. Many assays require the dispensing of cells in such a way that when the assay is measured in the optical plate reader, the cells do not touch each other so that it remains possible for the image analysis algorithm to identify and score separate cells based on morphology.

Cell lysis is required when using the ELISA, TR-FRET, or AlphaScreen readouts to detect the relative quantity of a specific intracellular phosphorylated protein sequence. A typical 2X to 4X RIPA buffer (commonly used for immunoprecipitation) containing a non-ionic detergent such as NP-40 or Tween20 is used. The lysis solution should contain protease and phosphatase inhibitors so that all further metabolism of the analyte is stopped. Often, an MTP shaking step is required to efficiently lyse the cells.

1.3.4 CONSIDERATIONS FOR MICROTITER PLATE READERS

MTP readers are critical to kinase assay development and design. These instruments are available in a large variety of choices and it is not our intention to bias the reader toward any system. The choice often depends on factors such as budget, existing HTS infrastructure, throughput needs, and project-specific scientific considerations. Plate readers are based on many principles, including:

1. Photomultiplier tube (PMT) detection. The photon counts from each well are measured one at a time by the PMT and the excitation is achieved through an excitation beam or laser. Radioactive assays and luminescent assays do not require excitation beams. Because radioactive assays require long counting times using PMT-based systems, these counters often accomplish throughput by running many PMTs in parallel.
2. Super-cooled charge-coupled device (CCD)-based imagers such as the Perkin Elmer ViewLux or the General Electric LeadSeeker collect all data on the wells simultaneously through a CCD and work like digital cameras. The readers, although expensive, tend to be faster, but less flexible.
3. Microscopic imaging and analysis readers such as EvoTec's Opera, General Electric's INCell Analyzer 3000, Cellomics' Array Scan, Acumen's Explorer, and Molecular Devices' ImageXpress transform the data from each well into a fluorescent microscopic

image. There are many approaches to high-throughput microscopic image analysis and these will be discussed in a later chapter.

4. Miscellaneous confocal scanning systems that can irradiate and detect light emission in small volumes of the assay and compile these data into images or average these data in such a way as to detect only the bound fraction of a labeled probe. Examples of this type of technology include the FMAT and LiCor InCell Western system.

Trends in biochemical screening assays seem to favor the use of multi-function PMT-based readers that allow for various MTP well densities (96, 384, and 1536 well plates), can handle a number of readout formats such as prompt fluorescence, luminescence, fluorescence polarization, time-gated fluorescence, and luminescent oxygen channeling or AlphaScreen. Examples of this type including the Perkin Elmer EnVision, TECAN-Ultra, BMG FluoStar, and LJL Analyst GT can be employed for a variety of the assay technologies described above.

Important features to consider are the ease of software use and the ability to change the optical filters depending upon the peak absorption and emission spectra of the labels used. The dynamic range and sensitivity of the readers are important factors that can affect the assay protocol. A protocol that is suitable for one type of reader may not work as well in a second type of reader due to variations in the optics resulting in differences in dynamic range and sensitivity. Sensitivity is important because it aids in the miniaturization of assays and helps reduce the amounts of expensive reagents required. The ability to detect small quantities of product in an enzyme reaction means that less enzyme will be required.

Dynamic range is important because a large dynamic range allows more flexibility in varying the kinetics and timing of the assay. One can design relatively simple tests for comparing the sensitivity and dynamic ranges of instruments using reference standards in concentration response experiments. The varieties of technologies employed for high-throughput data collection are constantly improving and expanding and perhaps newer readout technologies will provide increased throughput, sensitivity, and dynamic range in the future and pave the way for more information content per well.

1.4   CONCLUSION

The key to success in high-throughput screening is to have a variety of relevant high-throughput assays in place to characterize large numbers of hits from an HTS campaign. Although we would all like a "gold-standard" kinase or phosphatase assay that reveals the true, physiologically relevant potency and selectivity of a compound, we are often faced with combining various types of assays to interpret the mechanisms of inhibition and the observed structure–activity relationships. The various assays alone can give only a partial view of the mechanisms of action of large numbers of compounds, but taken together, more information is provided with regard to potency, mechanism, cell-based activity, and cell permeability, giving a fuller picture and enabling a more informed selection of the best compounds.

Although improving assay sensitivity and throughput of HTS is always needed, the more important technical limitations still lie in expressing kinases and phosphatases in a physiologically relevant form. For example, most crystal structures and enzyme assays focus on truncated forms that do not contain regulatory or transmembrane domains because it is difficult to express natural forms of kinases and phosphatases. Any technological approach that can address these problems will be of great benefit.

REFERENCES

Chen, H. et al. 2005. A cell-based immunocytochemical assay for monitoring kinase signaling pathways and drug efficacy. *Anal. Biochem.* 338, 136–142.

Daub, H., K. Specht, and A. Ullrich. 2004. Strategies to overcome resistance to targeted protein kinase inhibitors. *Nat. Rev. Drug Discov.* 3, 1001–1010.

Feng, B.Y. and B.K. Shoichet. 2006. A detergent-based assay for the detection of promiscuous inhibitors. *Nat. Protoc.* 1, 550–553.

Glickman, J.F. and S. Ferrand. 2008. Scintillation proximity assays in high-throughput screening. *Assay Drug Dev. Technol.* 6, 433–455.

Glickman, J.F. et al. 2002. A comparison of AlphaScreen, TR-FRET, and TRF as assay methods for FXR nuclear receptors. *J. Biomol. Screen.* 7, 3–10.

Glickman, J.F. and A. Schmid. 2007. Farnesyl pyrophosphate synthase: real-time kinetics and inhibition by nitrogen-containing bisphosphonates in a scintillation assay. *Assay Drug Dev. Technol.* 5, 205–214.

Johnston, P.A. et al. 2007. Development and implementation of a 384-well homogeneous fluorescence intensity high-throughput screening assay to identify mitogen-activated protein kinase phosphatase-1 dual-specificity protein phosphatase inhibitors. *Assay Drug Dev. Technol.* 5, 319–332.

Kashem, M.A. et al. 2007. Three mechanistically distinct kinase assays compared: measurement of intrinsic ATPase activity identified the most comprehensive set of ITK inhibitors. *J. Biomol. Screen.* 12, 70–83.

Koresawa, M. and T. Okabe. 2004. High-throughput screening with quantitation of ATP consumption: a universal non-radioisotope, homogeneous assay for protein kinase. *Assay Drug Dev. Technol.* 2, 153–160.

Lakowicz, J.R. 2006. *Principles of Fluorescence Spectroscopy.* 3rd ed. Berlin: Springer.

Lawrence, D.S. and Q. Wang. 2007. Seeing is believing: peptide-based fluorescent sensors of protein tyrosine kinase activity. *Chem. BioChem.* 8, 373–378.

McCain, D.F. and Z.Y. Zhang. 2001. Assays for protein-tyrosine phosphatases. *Meth. Enzymol.* 345, 507–518.

Minor, L.K. 2003. Assays to measure the activation of membrane tyrosine kinase receptors: focus on cellular methods. *Curr. Opin. Drug Discov. Devel.* 6, 760–765.

Montalibet, J., K.I. Skorey, and B.P. Kennedy. 2005. Protein tyrosine phosphatase: enzymatic assays. *Methods* 35, 2–8.

Moshinsky, D.J. et al. 2003. A widely applicable, high-throughput TR-FRET assay for the measurement of kinase autophosphorylation: VEGFR-2 as a prototype. *J. Biomol. Screen.* 8, 447–452.

Newman, M. and S. Josiah. 2004. Utilization of fluorescence polarization and time resolved fluorescence resonance energy transfer assay formats for SAR studies: Src kinase as a model system. *J. Biomol. Screen.* 9, 525–532.

Nickischer, D. et al. 2006. Development and implementation of three mitogen-activated protein kinase (MAPK) signaling pathway imaging assays to provide MAPK module selectivity profiling for kinase inhibitors. *Meth. Enzymol.* 414, 389–418.

Olive, D.M. 2004. Quantitative methods for the analysis of protein phosphorylation in drug development. *Expert Rev. Proteomics* 1, 327–341.

Parker, G.J. et al. 2000. Development of high-throughput screening assays using fluorescence polarization: nuclear receptor-ligand-binding and kinase/phosphatase assays. *J. Biomol. Screen.* 5, 77–88.

Rodems, S.M. et al. 2002. A FRET-based assay platform for ultra-high density drug screening of protein kinases and phosphatases. *Assay Drug Dev. Technol.* 1, 9–19.

Rodgers, G. et al. 2003. Development of displacement binding and GTPγ scintillation proximity assays for the identification of antagonists of the micro-opioid receptor. *Assay Drug Dev. Technol.* 1, 627–636.

Schroter, T. et al. 2008. Comparison of miniaturized time-resolved fluorescence resonance energy transfer and enzyme-coupled luciferase high-throughput screening assays to discover inhibitors of Rho-kinase II (ROCK-II). *J. Biomol. Screen.* 13, 17–28.

Schroter, A., C. Trankle, and K. Mohr. 2000. Modes of allosteric interactions with free and [3H] N-methylscopolamine-occupied muscarinic M2 receptors as deduced from buffer-dependent potency shifts. *Naun. Schmied. Arch. Pharmacol.* 362, 512–519.

Shoichet, B.K. 2006. Interpreting steep dose-response curves in early inhibitor discovery. *J. Med. Chem.* 49, 7274–7277.

Sullivan, E., P. Helsley, and A. Pickard. 1997. Development of a scintillation proximity assay for calcineurin phosphatase activity. *J. Biomol. Screen.* 2, 19–23.

Swinney, D.C. 2004. Biochemical mechanisms of drug action: what does it take for success? *Nat. Rev. Drug Discov.* 3, 801–808.

Trask, O.J., Jr. et al. 2006. Assay development and case history of a 32K-biased library high-content MK2-EGFP translocation screen to identify p38 mitogen-activated protein kinase inhibitors on the Array Scan 3.1 imaging platform. *Meth. Enzymol.* 414, 419–439.

Turek-Etienne, T.C., Kober, T. P. et al. 2003. Development of a fluorescence polarization AKT serine threonine kinase assay using an immobilized metal ion affinity-based technology. *Assay Drug Dev. Technol.* 1, 545–553.

Turek-Etienne, T.C., Small, E.C. et al. 2003. Evaluation of fluorescent compound interference in 4 fluorescence polarization assays: 2 kinases, 1 protease, and 1 phosphatase. *J. Biomol. Screen.* 8, 176–184.

Ullman, E.F. et al. 1994. Luminescent oxygen channeling immunoassay: measurement of particle binding kinetics by chemiluminescence. *Proc. Natl. Acad. Sci. USA* 91, 5426–5430.

Vedvik, K.L. et al. 2004. Overcoming compound interference in fluorescence polarization-based kinase assays using far-red tracers. *Assay Drug Dev. Technol.* 2, 193–203.

Vogel, K.W. and K.L. Vedvik. 2006. Improving lanthanide-based resonance energy transfer detection by increasing donor-acceptor distances. *J. Biomol. Screen.* 11, 439–443.

Vogel, K.W. et al. 2008. Developing assays for kinase drug discovery: where have the advances come from? *Expert Opin. Drug Disc.* 3, 115–129.

Von Ahsen, O. and U. Boemer. 2005. High-throughput screening for kinase inhibitors. *Chem. BioChem.* 6, 481–490.

Von Leoprechting, A. et al. 2004. Miniaturization and validation of a high-throughput serine kinase assay using the AlphaScreen platform. *J. Biomol. Screen.* 9, 719–725.

Waddleton, D. et al. 2002. Development of a time-resolved fluorescent assay for measuring tyrosine-phosphorylated proteins in cells. *Anal. Biochem.* 309, 150–157.

Warmuth, M. et al. 2007. Ba/F3 cells and their use in kinase drug discovery. *Curr. Opin. Oncol.* 19, 55–60.

Warner, G. et al. 2004. AlphaScreen kinase HTS platforms. *Curr. Med. Chem.* 11, 721–730.

Watanabe, T. et al. 1998. Synthesis of fluorescent substrates for protein tyrosine phosphatase assays. *Bioorg. Med. Chem. Lett.* 8, 1301–1302.

Wu, G., Y. Yuan, and C.N. Hodge. 2003. Determining appropriate substrate conversion for enzymatic assays in high-throughput screening. *J. Biomol. Screen.* 8, 694–700.

Zhang, J.H., T.D. Chung, and K.R. Oldenburg. 1999. A simple statistical parameter for use in evaluation and validation of high-throughput screening assays. *J. Biomol. Screen.* 4, 67–73.

Zhang, Y.L. et al. 2007. An enzyme-linked immunosorbent assay to measure insulin receptor dephosphorylation by PTP1B. *Anal. Biochem.* 365, 174–184.

2 Fluorescence-Based Biochemical Protease Assay Formats

Julian Woelcke and Ulrich Hassiepen

CONTENTS

2.1 INTRODUCTION

Proteases are enzymes catalyzing the hydrolysis of peptide bonds. They form one of the largest enzyme families encoded by the human genome, with more than 500 active members. Based on the different catalytic mechanisms of substrate hydrolysis, these enzymes are divided into four major classes: serine/threonine, cysteine, metallo, and aspartic proteases. In serine, cysteine, and threonine proteases, the nucleophile of the catalytic site is a side chain of an amino acid in the protease (covalent catalysis). In metallo and aspartic proteases, the nucleophile is a water molecule activated through the interaction with amino acid side chains in the catalytic site (non-covalent catalysis) (Gerhartz et al., 2002).

From the perspective of the substrate, the proteases are classified as endopeptidases and exopeptidases according to the nomenclature by Barrett and MacDonald (1986). In endopeptidases, the substrate runs through the entire length of the active site and is cleaved at an internal peptide bond in the middle of its sequence (Barret, 2004). In exopeptidases, substrate binding is structurally constrained so that only one or two amino acid residues of the substrate can specifically bind to the protease. The exopeptidase reaction is defined as the cleavage near or at either end of the substrate molecule, commonly directed by the recognition of the free, charged amino (N) or carboxy (C) terminal groups of the substrate by the protease. The aminopeptidases bind and cleave their substrates from the N terminus. The carboxypeptidases bind and cleave their substrates from the C terminus.

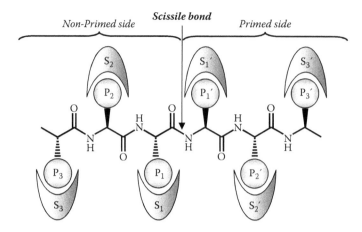

FIGURE 2.1 Standard nomenclature for substrate residues and their corresponding binding sites on the protease. The subsites toward the N terminus of the substrate are called non-primed sites and are numbered S1 to Sn, beginning with S1 at the N terminal side of the scissile bond. The subsites toward the C terminus of the substrate are called primed sites and are numbered S1′ to Sn′ beginning with S1′ at the C terminal side of the scissile bond. The substrate residues that the enzymatic subsites accommodate are numbered P1 to Pn and P1′ to Pn′, respectively.

The surface of a protease that is able to accommodate one single amino acid side chain of the substrate sequence is called a subsite. The subsites toward the N terminus of the substrate are called non-primed sites and are numbered S1 to Sn beginning with S1 at the N terminal side of the scissile bond (Figure 2.1). The subsites toward the C terminus of the substrate are called primed sites and are numbered S1′ to Sn′ beginning with S1′ at the C terminal side of the scissile bond. The substrate residues that the enzymatic subsites accommodate are numbered P1 to Pn and P1′ to Pn′, respectively (Berger and Schechter, 1976). The structures of the catalytic, non-primed, and primed sites play an important role for the specific recognition of a protein or peptide substrate by the protease.

Initially identified as participants in mammalian food digestion in the intestinal tract, proteases were later recognized to play a central role in many physiological and pathological processes such as cell proliferation, blood coagulation, blood pressure control, protein catabolism, neurodegeneration, bacterial and viral diseases, and inflammation and cancer (Turk, 2006). Therefore the strategy to achieve pharmacologically relevant inhibition of proteases has been an attractive drug discovery strategy since the 1950s and paid off with a broad number of protease-inhibiting drugs on the market today.

The most prominent drug target example from the carboxypeptidase group is the angiotensin-converting enzyme (ACE). It is a metalloprotease that acts as carboxy dipeptidase, cleaving two amino acids from the C terminus of angiotensin. ACE plays a central role in the treatment of hypertension with inhibitors on the market for more than 20 years. Prominent examples from the class of the endopeptidases playing an important role as drug targets are the aspartic renin and HIV proteases. Inhibitors of both proteases are on the market. Rasilez treats hypertension; ritonavir and saquinavir treat AIDS. Prominent examples of exopeptidases as drug targets are the dipeptidylpeptidase IV (DPPIV) serine protease for the treatment of type 2 diabetes and the serine proteases of the blood coagulation cascade, especially thrombin and factor Xa for the treatment of thrombosis (Turk, 2006). Several drugs inhibiting these serine proteases are on the market already (DPPIV inhibitors Januvia and Galvus; fXa inhibitor Xarelto) or in clinical development. All three enzymes belong to the aminopeptidase group.

A major starting point for drug discovery in the pharmaceutical industry is the identification of low molecular weight inhibitors by high-throughput screening (HTS) campaigns. Large numbers of compounds, typically a million, are tested for their inhibitory potential on the drug target of interest. HTS is followed by compound validation and optimization activities during which inhibitors are modified to improve potency on the target and selectivity over other members of the same target class or family. Both HTS and follow-up activities require robust and sensitive assays.

In the second section of this chapter, strategies to identify substrates for biochemical protease assays are discussed. Section 2.3 focuses on theoretical and practical aspects of various fluorescence-based readouts for biochemical protease assays. Finally concrete experiments for the determination of enzyme kinetics relevant for the development of robust and sensitive biochemical protease assays are summarized in Section 2.4. Altogether this chapter offers guidelines for the development of biochemical protease assays for the purpose of protease inhibitor-directed drug discovery.

2.2　SUBSTRATE FINDING APPROACHES

In vitro protease activity assays are based on monitoring the cleavage of the substrate, either a short peptide or a whole protein. Peptide substrates are sufficient if the peptide sequence in its denatured form carries the accurate specificity information of the protease's active site. Proteases prefer whole proteins as substrates if native components of structural elements for recognition and/or activation of the protease are required. In these cases, substrate design must be hypothesis-driven, based on the recognition/activation mechanism. The substrate ubiquitin-Rh110-glycine is an example of such an approach. It was used to monitor the activity of the deubiquitinating proteases (DUBs) UCH-L3 and USP2 *in vitro* (Hassiepen et al., 2007).

In the cases where a simple peptide substrate is sufficient, substrate discovery and optimization can be done by a variety of standard approaches: (1) testing standard protease substrates such as oxidized insulin B chain, (2) screening commercially available substrates for other proteases, (3) screening libraries of synthetic, fluorescently labeled peptides generated via combinatorial chemistry, or (4) applying genetic methods such as substrate phage display. The different strategies and methods for the identification and optimization of protease substrates were reviewed and discussed extensively by Richardson (2002) and Diamond (2007).

Positional scanning synthetic combinatorial libraries are better suited for substrate optimization than for substrate finding. They become efficient if basic knowledge of the substrate specificity of the protease under investigation is available. The synthesis and subsequent screening activities may be outsourced to service providers (e.g., JPT, Berlin, Germany). For substrate discovery on a novel protease with unknown substrate specificity, a large number of peptides differing in sequence is desirable for testing. A diversity of approximately 2.7×10^5 different sequences has been reported for a bead-based peptide library generated with a split/mix combinatorial chemistry approach. Seven positions were randomized using all 20 genetically encoded amino acids. However, the number of beads used as the solid support for the peptide synthesis was limited to 2.7×10^5 in the experiment—far off the 20^7 projected for all theoretically possible combinations (St. Hilaire et al., 1999).

A far greater complexity ranging up to 10^8 different peptide sequences (that also do not suffer from the disulfide constraints sometimes reported for combinatorial peptide libraries) may be covered by utilizing substrate phage display (Li et al., 2007). Filamentous bacteriophage virons can be engineered to display foreign peptides instead of their native surface proteins. If the surface protein encoding DNA is partially randomized, each phage will display a different peptide from a huge number of possible sequences. The randomized peptide sequences are additionally affinity tagged to enable phage binding to a solid support. Selection of the optimal substrate sequence is obtained by repeated cycles of protease-mediated phage release and enrichment. The substrate sequence is finally identified by sequencing the phage DNA.

Schilling and Overall (2008) recently reported a solution for the problem of generating a peptide library that contains all relevant sequences of a human proteome, without including those that are biologically irrelevant (not present in humans, such as repeating tracts of rare amino acids). These "proteome-wide peptide libraries" are derived from the human proteome or other species of interest. To generate this new class of peptide library, natural proteomes, for example, cell lysates and secretomes, are endoproteolytically digested with peptidases possessing different canonical cleavage site specificities such as trypsin (R/K in P1), GluC (E/D in P1), or chymotrypsin (W/Y/L/F in P1). Using these three proteases, peptides having different C terminal residues are generated. Thus, it is ensured that the substrate specificities of most proteases under investigation are covered.

The endopeptidases are inactivated after the digest, followed by carboxyamidomethylation of sulfhydryl groups and dimethylation of amino termini and ε-amino groups of lysine residues. This forms a biologically relevant, database-searchable peptide library useful for many applications including the characterization of protease cleavage specificity.

Following incubation of the blocked proteome-wide libraries with the protease of interest, the carboxy cleavage products (prime side) of proteolyzed peptides are selected by a pullout procedure using NHS–biotin reactivity to the cleaved neo-N terminus. These are then analyzed by tandem mass spectrometry and the sequence is identified. Then the amino acid sequences of the non-prime site sequences can be identified through database searches using the MS/MS-identified primed side sequence. Thus, both the non-prime and prime sequences can be discovered in a single analysis workflow. Hence, proteomic identification of protease cleavage site specificity (PICS) can quickly identify a cleavable peptide sequence that can then be used in assay development. It is important to note that no other current technique for determining protease consensus sites yields both the primed and non-primed sides of the cleaved bond in the same experiment.

In contrast to synthetic peptide libraries, proteome-wide peptide libraries possess amino acid compositions reflecting the average amino acid composition of the respective species. Biologically, it is a huge advantage for profiling target molecules to use only those sequences to which the target molecule may be exposed in a proteome. Moreover, complementary libraries can be prepared when different human cellular fractions or different endopeptidases are used for the generation of peptides. Then a huge diversity of peptide sequences can be used for screening, which is not restricted to a particular cellular proteome or peptidase cleavage pattern. PICS was used to find cleavage sites using proteases from every catalytic class and so is broadly applicable to most proteases.

Prior to the start of any experimental substrate finding activity, databases should be mined. A tremendous amount of information about proteases, substrates, inhibitors, and structures can be retrieved from two searchable databases: MEROPS (Rawlings et al., 2006) (http://merops.sanger.ac.uk) and BRENDA (www.brenda-enzymes.de), that serve as good starting points for assay development in many cases. These databases are available to the public and should be consulted as primary sources of information.

2.3   ASSAY FORMATS

Fluorescence-based readouts build by far the most important basis of protease assays for HTS and compound profiling in the drug discovery industry today. These assay formats became increasingly popular in the 1990s (Burbaum and Sigal, 1997; Silverman et al., 1998). The basics of fluorescence are very well described in detail by Lakowicz (2006). This discussion will be restricted to biochemical protease assays in homogeneous formats based on fluorescence readouts that are suitable for HTS and automated compound profiling.

2.3.1   FLUORESCENCE INTENSITY

Most biochemical assays used to determine the potency of inhibitors against (1) aminopeptidases and (2) endopeptidases with minor contributions to the substrate binding efficiency by the S′ site are based on the measurement of fluorescence intensity (FI). The FI readout principle is shown in Figure 2.2. The dynamic range of the FI readout basically scales with the fluorescence quantum yield, that is, the efficiency of fluorescence emission of the dye label. For the FI readout, peptide substrates with fluorogenic groups such as acetylmethoxycoumarin (AMC), 7-amino-4-trifluoromethyl

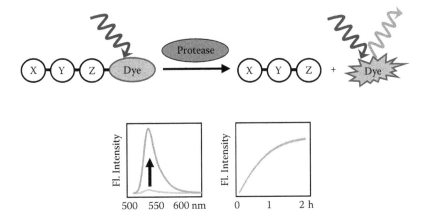

FIGURE 2.2 Fluorescence intensity readout principle. In the intact peptidic substrate (amino acids symbolized by X, Y, and Z) labeled with a fluorophore at the C terminus, the intensity of fluorescence emission (light gray arrow) after excitation (dark gray arrow) is low. Through the cleavage of the substrate between the C terminal amino acid (Z) and the fluorophore by a protease, the intensity of fluorescence emission is strongly enhanced. An increase of fluorescence intensity over time dependent on the enzymatic velocity is observed.

coumarin (AFC), and rhodamine 110 (Rh110) attached to the C terminus are frequently used. Table 2.1 summarizes the key characteristics of this selection of fluorophores.

Dipeptidylpeptidases like DPPIV cleave the peptide bond between the amino acids in P1 and P1′ positions with the label attached to the P1′ position. Aminopeptidases cleave the pseudo peptide bond between the amino acid in P1 position (= C terminus) and the fluorophore (Figure 2.2). This is possible with high catalytic efficiency (k_{cat}/K_M), because the aminopeptidases primarily recognize and bind to the N termini of their substrates. The dye attached to the C terminus does not alter the protease–substrate interaction. The cleavage of the labeled peptide results in an increase in total fluorescence due to the reconstitution of the fluorophore system. Historically, AMC was most frequently used as a fluorogenic group in biochemical protease assays with mono-labeling of the substrate. Three publications on cathepsin G assays serve as examples of 20 years of progress in assay design based on FI as the readout vehicle (Tanaka et al., 1985; Réhault et al., 1999; Attucci et al., 2002).

AMC is still a popular dye today. According to our experience, the IC_{50} values determined for protease inhibitors in AMC-based protease assays may be biased and misinterpreted due to the interference of label and compound autofluorescence characteristics (Figure 2.3). In the worst case, this can lead a drug discovery program in a wrong direction with respect to prioritization of compound classes. At the excitation and emission wavelengths of 350 and 500 nm, respectively, used for AMC, many compounds also display fluorescence characteristics.

TABLE 2.1
Fluorophores Frequently Used for Protease Assays Based on FI Readout

Fluorophore	Excitation Wavelength (nm)	Emission Wavelength (nm)	Reference
AMC	350	500	Okun et al., 2006
ACC	350	450	Harris et al., 2000
AFC	400	505	Gurtu et al., 1997
Rh110	485	535	Grant et al., 2002

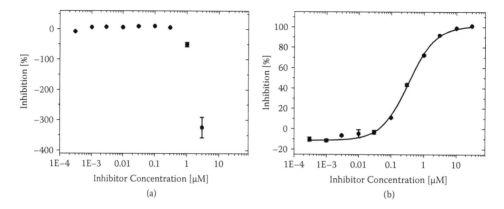

FIGURE 2.3   Dose–response curves for protease inhibitor with autofluorescence characteristics. (a) Result of standard fluorescence intensity-based assay employing an AMC-labeled substrate at a concentration of 2 μM. Profiling data could not be obtained due to the interference of the compound's fluorescence with the readout. (b) Result obtained from assay employing a Rh110-based substrate at a concentration of 0.5 μM. The profiling data (IC_{50} value of 335 nM and Hill coefficient of 1.0) were obtained for the depicted data set.

Red-shifted dyes are much less prone to interfering artifacts and thus result in significantly fewer false-positive and false-negative hits in HTS or compound profiling (Turconi et al., 2001a; Banks et al., 2000). Therefore, fluorophores such as Rh110 are applied more frequently today (Grant et al., 2002; Graziano et al., 2006). In our experience, Rh110 is the most suitable dye for protease activity assays to determine inhibitor potency based on the FI readout. The fluorescence of Rh110 is usually excited at 488 nm and the emission is detected at 535 nm. Commercially available peptides carrying Rh110 labels are symmetrically *bis*-substituted in most cases, e.g., XYZ-Rh110-ZYX with the amino acid sequence XYZ at positions P3 to P1, because the synthesis of these substrates is easier than for asymmetric peptides. Therefore most Rh110-based protease assays are developed using symmetric peptidic substrates.

The application of asymmetric peptides is much less frequently described (Cai et al., 2001). The production of asymmetric Rh110-labeled *bis*-substituted peptides is much more difficult because the synthesis requires significantly more steps than required for symmetrically *bis*-substituted peptides. The critical step is the synthesis of the mono-substituted intermediate product Rh110-ZYX with low yield in most cases (personal communication). However, due to the fact that a strong fluorescence increase is only observed after the fluorophore has been cleaved off from both peptide chains of the symmetrically *bis*-substituted substrate (two-step proteolysis), the kinetics of the enzymatic reaction can be followed more accurately and easily with peptides comprising only one copy of the scissile bond.

To improve the solubility of asymmetric peptides, it is advisable to use the non-primed site recognition sequence of the protease combined with a Rh110 molecule linked to an acidic amino acid (e.g., glutamate) via the γ-carbonyl function, i.e., XYZ-Rh110-γGlu. In some cases, this may lead to a slight increase of the K_M value or a slight reduction of the catalytic efficiency (k_{cat}/K_M) in comparison to the analogous peptidic substrate mono-labeled with AMC and without γGlu. In general, the catalytic efficiency determined with C terminally labeled peptides tends to be lower in comparison to unlabeled substrates because of the location of the fluorophore at the scissile bond, substituting the natural amino acid in position P1′. The fluorophores are bulky aromatic groups differing from the natural amino acids. Moreover, the amide bond between the C terminal amino acid and the fluorophore differs from the usual peptide one.

As a rule, substrate concentrations of 100 to 500 nM for Rh110 and concentrations of 1 to 5 μM for AMC are sufficient for the development of robust assays suitable for HTS and high-throughput

TABLE 2.2
Dynamic Ranges for Various Fluorescence-Based Biochemical Protease Assay Formats

Reference	Substrate Sequence	Protease	Δ Signal
FI-1	Suc-L-L-V-Y ↓ AMC	Chymase	$f_{dyn} \geq 250$
FI-2	Suc-L-L-V-Y ↓ Rh110-γE	Chymase	$f_{dyn} \geq 250$
FRET-1	Ac-R-E(EDANS)-E-V-L-F-Q ↓ G-P-K(DABCYL)-R-NH$_2$	HRV 3C protease	$f_{dyn} \geq 40$
FP-1	BTN-T-T-R-P-G-S-G-L-T-N-I-K-T-E-E-I-S-E-V-N-L ↓ D-A-E-F-R-H-D-K-TAMRA	BACE-1	160 mPU
FP-2	Ac-R-R-K(TAMRA)-L-L-V-Y ↓ H-K(BTN)-OH	Chymase	330 mPU
FLT-1	Ac-E-F-K-P-I-L-W ↓ R-L-G-C(PT14)-E-NH$_2$	Kallikrein 7	5.5 ns
FLT-2	W-P ↓ S-G-T-F-T-K-C(PT14)-NH$_2$	DPP IV	5.8 ns
FLT-3	C(PT14-ME)-G-G ↓ W-OH	Carboxypeptidase A	9 ns
FLT-4	Ac-C(PT14)-V-P-R ↓ A-W-E-NH$_2$	Thrombin	3.7 ns
FLT-5	PT14-D-E-V-D ↓ W-E-NH$_2$	Caspase-3	4.8 ns

All data obtained with Tecan Ultra Evolution MTP reader. The following excitation and emission wavelengths were used: EDANS and AMC: 350 and 500 nm; Rh110: 485 and 535 nm; TAMRA: 535 and 595 nm; PT14: 405 and 450 nm. ↓ = primary cleavage site confirmed by MS. AMC = aminomethylcoumarin. Rh110 = rhodamine 110. γE = glutamic acid attached to Rh110 via its carbonic acid in side chain. EDANS = fluorophore 5-[(2-aminoethyl)amino]naphthalene-1-sulphonic acid. DABCYL = 4-(4-dimethylaminophenylazo)benzoic acid quencher. BTN = biotin. PT14 = acridone-based fluorescence lifetime label.

compound profiling according to our experience. Signal-to-noise ratios are higher with Rh110 in comparison to AMC and AFC due to a much higher extinction coefficient, fluorescence quantum yield, and photostability (Liu et al., 1999). Thus assay systems with low K_M values and low substrate concentrations will usually give best signal-to-noise ratios with Rh110 compared to AMC. The FI-1 and FI-2 examples in Table 2.2 give an impression of the high dynamic range that can be achieved with FI-based assays. Due to the high speed and ease of detection, the FI readout is ideally suited for high-throughput applications. The dyes and synthesis of labeled peptide substrates are affordable and thus allow cost-efficient operations. FI assays can easily be adapted to the 1536-well format. On the other hand FI assay results can suffer from fluorescence interference by the tested compounds, leading to false positive and false negative results.

FI is not frequently used as a readout for carboxypeptidases because the assay principle cannot be applied easily. In C terminally labeled peptide substrates, the primed site part of the carboxypeptidase recognition sequence is missing, resulting in high K_M values, incompatible with the development of robust protease activity assays. The same limitation of the FI assay principle is observed with endopeptidases in which amino acids on the primed site of the substrate (primarily P1′) strongly contribute to the binding energy of the peptide.

Enzymatic assays for aminopeptidases can also be based on the readouts described below, which are applied primarily for endopeptidases and carboxypeptidases. These readouts use peptide substrates that span from the non-primed to the primed side.

2.3.2 FLUORESCENCE RESONANCE ENERGY TRANSFER (FRET)

FRET-based readouts are the most prominent methods used for endopeptidase and carboxypeptidase activity assays. In general, a huge number of biological assays have been developed based on the FRET principle (Van der Meer et al., 1994; Andrews and Demidov, 1999). FRET substrates extend on both sides of the scissile bond. This is important for proteases, where the S′ binding site significantly contributes to the binding affinity of the substrate. For example, the activity of human

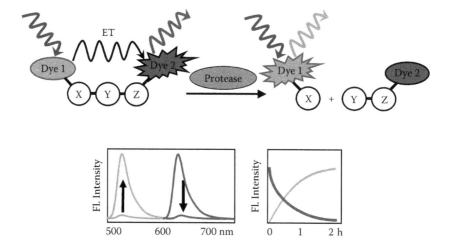

FIGURE 2.4 FRET readout principle. In the intact peptidic substrate (amino acids symbolized by X, Y and Z) labeled with two fluorophores (dye 1 and dye 2) at the opposite sites of the scissile bond. Dye 1 serves as fluorescence donor and dye 2 as fluorescence acceptor. Through an energy transfer (ET) from the donor to the acceptor, the fluorescence emission of dye 2 is observed (dark gray arrow). Thus the emission is significantly shifted to longer wavelengths compared to the emission of dye 1 (light gray arrow). After proteolytic cleavage of the substrate between amino acids X and Y by a protease, the energy transfer is disrupted and only the fluorescence emission of dye 1 is observed. An increase of fluorescence intensity of dye 1 and a decrease of fluorescence intensity of dye 2 over time dependent on the enzymatic velocity is recorded.

neutrophil elastase (HNE) depends greatly on the lengths of the synthetic substrates (Lestienne and Bieth, 1980). The S1′ site of this protease in particular plays an important role in substrate binding (Stein and Strimpler, 1987).

The FRET readout depends on the presence of fluorescence donor and fluorescence acceptor molecules. In protease assays, these two dyes are incorporated into the cleavable substrate. The donor and the acceptor molecule must be attached on either side of the scissile bond but must remain in close proximity to each other to enable an efficient energy transfer from donor to acceptor (Figure 2.4). The mechanism of resonance energy transfer is formulated by classic and quantum mechanical theory. The Förster theory (Förster, 1948; Turro, 1965) describes the transfer mechanism as a weak dipole–dipole resonance coupling between donor and acceptor. According to this theory, the efficiency of the energy transfer is dependent on the quantum yield of the donor in the absence of transfer and the spectral overlap between donor and acceptor. This means that the maximum of the fluorescence emission spectrum of the donor must be at or near the excitation maximum of the fluorescence acceptor to enable a significant energy transfer. The efficiency of energy transfer decreases with the sixth power of the distance.

2.3.2.1 FRET Quench

In most cases of FRET dye pairs for protease assays, the acceptor is a fluorescence dark quencher (Figure 2.5). The fluorescence quencher does not emit fluorescence and the transferred energy is lost through, for example, singlet–triplet transitions or internal conversions. The energy can also be transferred by a collision mechanism arising from direct contact between donor and acceptor when electron shares of donor and acceptor overlap. An effective quenching by the collision mechanism requires a short distance between donor and acceptor. The peptide substrates carrying the two fluorophores are usually flexible molecules for which the predictions of the distance between donor and acceptor and thus of the energy transfer mechanism are difficult.

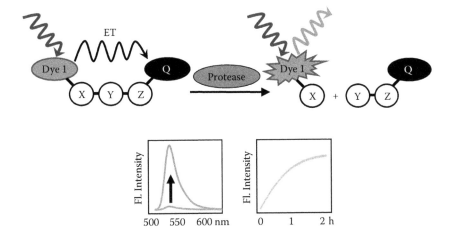

FIGURE 2.5 FRET quench readout principle. In the intact peptidic substrate (amino acids symbolized by X, Y and Z) labeled with a fluorophore (dye) and a quencher (Q) at the opposite sites of the scissile bond, the fluorescence emission is quenched through an energy transfer (ET) from the fluorophore to the quencher. After cleavage of the substrate between amino acids X and Y by a protease, the energy transfer is disrupted and an increase in fluorescence emission is observed (light gray arrow). An increase of fluorescence intensity over time dependent on the enzymatic velocity is recorded.

In the substrate, the fluorescence of the donor is quenched by the quenching label in close proximity. Upon cleavage of the substrate by the protease, this proximity is lost and a strong increase in fluorescence can be recorded. Table 2.3 summarizes a selection of fluorescence donor and acceptor pairs frequently used for protease activity assays based on a FRET quench readout. In principle, the FRET quench effect can be achieved by labeling the substrate with two different dyes with overlapping spectra (hetero-double labeling).

Historically, the *ortho*-aminobenzoyl group (Abz) was frequently used as fluorescence donor because of its small size, high hydrophilicity, and high quantum yield (Gershkovich and Kholodovych,

TABLE 2.3
Examples of Fluorescence Donor and Acceptor Pairs Frequently Used for Protease Assays Based on FRET Quench Readout Principle

Donor	Excitation Wavelength (nm)	Emission Wavelength (nm)	Acceptor	Excitation Wavelength (nm)	Reference
Abz	340	415	pNA	315	Stambolieva et al., 1992
Abz	340	415	Phe(NO$_2$)	280	Nishino et al., 1992
Abz	340	415	EDDnp	360	Carmel et al., 1977
Abz	340	415	Nitrotyrosine	428	Angliker et al., 1995
Abz	340	415	Dnp	360	Mao et al., 2008
EDANS	336	495	DABCYL	472	Wang et al., 1990
Mca	328	393	Dnp	360	Xia et al., 1999
DANSYL	330	520	Phe(NO$_2$)	280	Knight et al., 1992; Chen, 1968
Cy5.5	675	695	NIRQ820	790	Malfroy and Burnier, 1987
Cy3	530	570	Cy5Q	640	www.gehealthcare.com
TAMRA	547	573	QSY7	560	www.invitrogen.com

1996). Abz was combined with a broad variety of non-fluorescent acceptors such as *p*-nitrobenzyl for leucine aminopeptidase (Carmel et al., 1977), pNA for trypsin (Bratanova and Petkov, 1987), 4-nitrophenylalanine [Phe(NO$_2$)] for HIV protease (Toth and Marshall, 1990), and *N*-(2,4-dinitrophenyl) ethylenediamine (EDDnp) for thermolysin and trypsin (Nishino et al., 1992). Lecaille et al. (2003) described a FRET quench assay based on a specific substrate for cathepsin K labeled with Abz and EDDnp. This substrate is not cleaved by the other C1 cysteine cathepsins and serine proteases in contrast to methoxycoumarin (Mca)-based substrates described earlier (Aibe et al., 1996; Xia et al., 1999) and merely covered the non-primed site of the scissile bond. The 5-[(2-aminoethyl)amino] naphthalene-1-sulfonic acid (EDANS) compound is a second example of a fluorescence donor historically used for many FRET quench-based protease assays, e.g., in combination with tryptophan as a quencher in an ECE activity assay (Von Geldren et al., 1991). The FRET-1 example in Table 2.2 shows the typical dynamic range that can be achieved with an EDANS/DABCYL-based assay.

EDANS is still a prominent fluorescence donor in protease assay development today, mainly in combination with 4-(4-dimethyl-aminophenylazo)-benzoic acid (DABCYL) as a non-fluorescence quencher. Zou et al. (2005) describe a recent application of this dye pair in an ADAM33 assay. The combination of Mca as a fluorophore and dinitrophenyl (Dnp) as a quencher shows similar high levels of fluorescence intensity and high quenching efficiency of >90% as EDANS/DABCYL. In contrast to many other FRET dye pairs, both EDANS/DABCYL and Mca/Dnp generate a dynamic range between cleaved and uncleaved substrate, sufficient for the application of such protease assays in HTS and high-throughput compound profiling under initial velocity conditions.

One recent example for the application of the FRET quench principle with Mca and Dnp as a dye pair for the development of robust assays for ADAMTS-4 and ADAMTS-5 was described by Lauer-Fields et al. (2007). High quenching efficiencies of >90% were also observed with DANSYL as a donor and tryptophan and Phe(NO$_2$) as a quencher, respectively, in assays for neutral endopeptidase (Malfroy and Burnier, 1987; Florentin et al., 1984). One disadvantage of the EDANS/DABCYL and Mca/Dnp dye pairs is the excitation maximum in the ultraviolet range. As a result, the potential for the interference of the fluorescence signal with that of the chemical compounds tested in protease assays is high, leading to enhanced false negative rates or misinterpretation of IC$_{50}$ data.

Many drug-like compounds display fluorescence characteristics when excited at wavelengths in the ultraviolet range. In addition, poor solubility of Mca/Dnp can limit the application of this dye pair for the development of robust protease assays in an HTS environment. Other FRET dye pairs with red-shifted fluorescence signals diminish the avoidance of fluorescence artifacts caused by interference with the compounds tested. The near-infrared (NIR) probe pair consisting of Cy5.5 a as fluorescence emitter (excitation and emission at 667 and 690 nm, respectively) and NIRQ820 as a fluorescence quenchner is an example. The successful application of this dye pair in an MMP7 assay was demonstrated (Pham et al., 2004).

Other red-shifted dye pairs include tetramethylrhodamine (TAMRA), MR121 or europium chelates (Truepoint™ by Perkin Elmer) in combination with Cy3/Cy5Q or members of the QSY dark quencher family (QSY7 or QSY21). Mao et al. (2008) recently described an assay based on the Truepoint principle to measure the inhibition of the NS3-4A hepatitis C virus protease. Konstantinidis et al. (2007) compared the interferences of compounds with substrates labeled with EDANS/DABCYL and TAMRA/QSY7 (excitation and emission at 547 and 573 nm), respectively, in a hepatitis C virus NS3-4A assay. As expected, they observed higher compound interference with EDANS/DABCYL. George et al. (2003) compared a Cy3/Cy5Q-labeled substrate with the respective Mca/Dnp-labeled analogue in an MMP3 assay and observed higher robustness and less compound interference with the Cy dye substrate.

The major disadvantages of these alternatives to EDANS/DABCYL and Mca/Dnp are the much greater expense and the frequent generation of dynamic signal ranges that are not sufficient for measurement under initial velocity conditions in our hands. Moreover, a limitation of the Truepoint™ is the minimum length of the applicable peptide substrates of 8 to 15 amino acids (personal communication). A disadvantage of double labeling is that the quenching mechanism often affects the conformational flexibility within the substrate in a way that compensation by increasing enzyme concentrations is required to obtain a sufficiently high fluorescence signal. This increase in enzyme concentration causes a reduction of sensitivity.

In general, a FRET quench readout is simple. A broad range of available fluorescence donors and acceptors allows cost-efficient operations in an industrialized HTS and automated compound profiling environment. On the other hand, the readout can suffer from inner filter effects due to high absorption coefficients of the dyes and fluorescence artifacts by the tested compounds, resulting in enhanced false positive and false negative rates. Moreover, the readout is limited to substrates in which short distances between donor and acceptor dye can be realized without disturbing the interaction of enzyme and substrate. The flexibility of the peptide conformation makes the prediction of the effective distance between the dyes and consequently the prediction of the FRET effect difficult. The distance between donor and acceptor cannot be easily approximated by the mean hydrodynamic radii of the dyes.

2.3.2.2   Time-Resolved (TR) FRET

TR-FRET measurements are interesting alternative readouts to the above described FRET quench. The technology is very popular due to its sensitivity, ease of use, and ease of assay development. The key aspect of this readout technology is the use of donor fluorophores with large Stokes shifts and extremely long fluorescence emission half-lives. The fluorescence emission measured after energy transfer to the acceptor fluorophore has a typical half-life ($t_{1/2}$) in the microsecond to millisecond range. Consequently, the measurement of the fluorescence can be time-gated. This means that the fluorescence is recorded only several hundred nano- to microseconds after excitation by flash illumination, e.g., a laser pulse. Consequently, short-lived background fluorescence significantly decreases, resulting in a high signal-to-background ratio and high assay sensitivity (Pope et al., 1999).

As the fluorescence lifetimes of typical drug-like compounds are in the low nanosecond range (Moger et al., 2006), fluorescence artifacts caused by the compounds can be excluded by time-gated measurements in TR-FRET-based assays. Lanthanide–organic complexes display fluorescence characteristics with long half-lives needed for donor molecules in this readout. Several vendors offer TR-FRET assay formats (CisBio, Perkin Elmer, GE Healthcare, Invitrogen, Covalis and others). Comley (2006) provided a comprehensive comparison of the various products. In the homogeneous time resolved fluorescence (CisBio's HTRF®; Mathis, 1999 and Perkin Elmer's LANCE Ultra™ Karvinen et al., 2002) assays, europium cryptates linked to streptavidin are attached to the protease substrate via a biotin molecule synthesized onto the peptide substrate.

In an assay for carboxypeptidase B, Ferrer et al. (2003) used a monoclonal antibody labeled with Eu^{3+}-cryptate as donor that specifically recognized the C terminus newly generated after enzymatic cleavage. In the LanthaScreen™ (Invitrogen) technology, a terbium complex attached to the protease substrate via a cysteine moiety is applied as long-living donor fluorophore. Allophycocyanin (APC) is used in most TR-FRET assays as an acceptor fluorophore attached to the opposite site of the scissile bond. It is applied because of its high efficiency documented by a high extinction coefficient ($\varepsilon = 7 \times 10^5$ M^{-1} cm^{-1}) and high quantum yield. In the HTRF® assay, modified allophycocyanin XL665 is used and fluorescence emission measured at 665 nm. The APC molecule is usually conjugated to an antibody that recognizes an antigen on the protease substrate, for example, a label like Dnp (Preaudat et al., 2002), FLAG-tag, GST-tag, His-tag or HA-tag (Comley, 2006). Alternatively, APC is conjugated to streptavidin, which binds to a biotin attached to the peptide substrate. The toolbox labeling reagents most frequently used for TR-FRET assay development are generally streptavidin, biotin, and anti-GST (Comley, 2007).

An advantage of this readout is the opportunity to carry out ratiometric measurements, because two "true" fluorophores with different excitation and emission maxima are used. This strongly increases the robustness of the assay. In general, TR-FRET-based assays can easily be miniaturized to the 1536-well format. The disadvantage of APC as a fluorescence acceptor is the size of the antibody–APC complex. XL665 is a large hexameric structure of 105 kDa that may cause steric hindrance in combination with short peptide substrates. Alternatively, Cy5 or Dy647 can serve as acceptors. In the LanthaScreen™ technology, a green fluorescent protein (GFP) or other common fluorophores such as fluorescein are used as a fluorescence acceptor with the advantage of avoiding the use of an antibody–APC complex in the assay system and the option to use physiologically relevant substrates as GFP fusion proteins. Horton et al. (2007) describe an UHCL-3 assay based on the LanthaScreen principle. The substrate is a yellow fluorescent protein (YFP) fusion of ubiquitin with a short C terminal extension containing an engineered cysteine residue labeled with a terbium chelate as a donor.

The high costs of reagents represent one major disadvantage of TR-FRET-based protease assays in general. In a survey among pharmaceutical and biotechnology companies, the most frequent reason not to choose TR-FRET as a readout was reagent price (Comley, 2006). Another disadvantage is that the fluorescence signal decreases with increasing enzyme activity. The maximum signal is intrinsically given for the intact, uncleaved protease substrate and cannot be enhanced by higher substrate concentrations or longer incubation times. Consequently, the dynamic range of the protease assay can be too small under initial velocity conditions to establish a robust system suited for HTS or compound profiling.

Absorbance and donor quenching activities by compounds are also frequently cited limitations of the technique (Comley, 2006). An additional disadvantage of the TR-FRET readout is the complexity of the biochemical system with several assay components and their multiple equilibria. These interactions may be disturbed by the compounds actually tested for their potency against the protease and thus lead to false positive results.

Limitations of the TR-FRET approach in terms of peptide sequence flexibility must be considered as well. The prerequisites and limitations for successful peptide labeling, for example, for the LANCE Ultra approach include: (1) need for a cysteine in position zero, i.e., the peptide's N terminus, (2) no other free cysteine allowed in the peptide sequence, (3) overall charge different from zero, (4) no disulfide bonds in the peptide, (5) no tryptophan or histidine at the C terminal position, and (6) maximum length of 30 amino acids (personal communication).

The principal limitations of the FRET approach as outlined in the FRET quench section above also hold true for the TR-FRET readout. Feasibility is limited to substrates in which short distances between donor and acceptor dye do not disturb the enzymatic activity. The flexibility of the peptide conformation makes the prediction of the effective distance between the dyes and consequently the prediction of the FRET effect difficult. The distance between donor and acceptor cannot be easily approximated by the mean hydrodynamic radii of the dyes.

2.3.3   FLUORESCENCE POLARIZATION (FP)

FP is an alternative readout principle for endopeptidase activity assays. FP or anisotropy measurements allow the detection of changes in the rotational correlation time of particles. These differences in the rotational correlation (or relaxation) time are related to different masses of particles. The experimental determination of steady-state fluorescence anisotropy requires the linear polarization of the light used for the excitation of the probe as well as linear polarization of the emitted fluorescence. Based on data of an appropriate experiment, the fluorescence anisotropy can be calculated as:

$$r = (I_{vv} - g*I_{vh})/(I_{vv} + 2*g*I_{vh})$$

where I_{vv} and I_{vh} are the vertically and horizontally polarized fluorescence emission intensities measured with respect to vertically polarized excitation light (Lakowicz, 2006). The factor g is used to correct the usually imperfect optical system. The g factor is given by:

$$g = I_{hv}/I_{hh}$$

For horizontally polarized excitation light, I_{hv} and I_{hh} are the intensities of the vertically and horizontally polarized fluorescence emissions, respectively. It is strongly recommended to use the background-corrected intensity values for the calculations. The background correction is determined by subtracting the intensity value obtained for the buffer with a certain polarizer setting from the intensity value for the sample obtained with identical settings.

Practically, the g factor can be determined using a free fluorophore with a known anisotropy value as a standard. Free fluorophores like TAMRA (5,6-carboxytetramethylrhodamine) and fluorescein with r_{true} values of $r_{true} = 0.0202$ ($p_{true} = 30$ mPU, see below) and $r_{true} = 0.0134$ ($p_{true} = 20$ mPU, see below), respectively, can be used (some sources report 27 mPU for fluorescein in solution):

$$g = [I_{vv} *(1 - r_{true})]/[I_{vh} * (2*r_{true} + 1)]$$

The anisotropy measurements yield only the relative amounts of the different species in solutions. The fraction of the product can be calculated as:

$$F_{prod} = (r_{sub} - r)/[(r_{sub} - r) + (Q*(r - r_{prod})]$$

where the factor Q is necessary to correct for differences in the fluorescence intensities (FIs) of the fluorophore in the parent substrate (FI_{sub}) and in the product (FI_{prod}) (Marks et al., 2005):

$$Q = FI_{prod}/FI_{sub}$$

It should be noted that all relations given in the formulas above are restricted to fluorescence anisotropy. However, it is popular in the screening community to use fluorescence polarization over fluorescence anisotropy. Polarization and anisotropy are related by:

$$P = 3*r/(2 + r) \quad \text{or} \quad r = 2*P/(3 - P)$$

Fluorescence polarization values are usually shown in millipolarization units (mPU); mPU = APU = 1000 mPU. A prerequisite for fluorescence anisotropy-based assays is that the enzymatic cleavage event leads to a significant change in the mass of the molecular moiety carrying the fluorophore. When short peptides are used as protease substrates in an FP assay, the difference in size between substrate and the fluorophore containing product is in many cases too small to generate changes in the polarization value needed to develop assays robust enough for high-throughput compound testing.

However, the development of robust FP-based protease assays is feasible as described by Tirat et al. (2005). The authors developed FP assays for UCH-L3 and USP2 based on both undecapeptides and on ubiquitin constructs as substrates, C terminally labeled with TAMRA. Dynamic ranges of ≥100 mPU at 100% substrate conversion sufficient for automated compound testing under initial velocity conditions were observed only with large substrates based on ubiquitin. In general, our experience is that robust FP assays for proteases can be developed if the molecular weight (MW) of the proteolytic product carrying the fluorophore is smaller than 10 kDa and the MW of the intact, uncleaved substrate is at least 10 times larger than the cleavage product when fluorophores with short lifetimes of 1 to 4 ns (such as fluorescein or rhodamine) are used.

The concept of MW enhancers was developed to increase polarization value changes caused by large size differences between substrate and product when small peptide substrates were used. The most frequently used approach is based on the attachment of a biotin molecule to the terminus of the peptide on the opposite site of the scissile bond to the fluorophore. After the incubation of the substrate with the protease, streptavidin is added to the assay. Streptavidin binds to biotin and enhances the masses of the substrate and the unlabeled product, thus causing an increase of the polarization

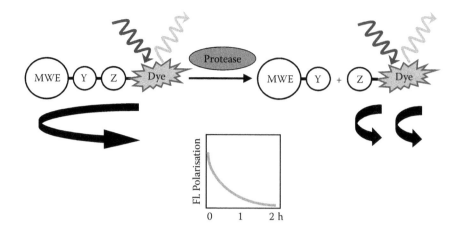

FIGURE 2.6 Fluorescence polarization readout principle. In the intact peptidic substrate (amino acids symbolized by X, Y and Z) labeled with a fluorophore (dye) and a molecular weight enhancer (MWE) at the opposite sites of the scissile bond, a high polarization is observed (long black arrow). After cleavage of the substrate between amino acids X and Y by a protease, the MWE and dye are decoupled and a low polarization of the labeled cleavage product is generated (short black arrows). A decrease of the polarization dependent on the enzymatic velocity is recorded.

value of the substrate (Turek-Etienne et al., 2003; Flotow et al., 2002; Leissring et al., 2003; Levine et al., 1997; see Figure 2.6). Examples FP-1 and FP-2 in Table 2.2 indicate the dynamic ranges that can be achieved with FP assays with MW enhancers.

Other MW enhancers have been described. Peptides with oligonucleotides as enhancers were attached to substrates for a polarization assay with caspase-3 (Lopez-Calle et al., 2002). As streptavidin has four binding sites, fluorescence self-quenching effects between the dye molecules can occur with increasing binding degree resulting in depolarization (Pope et al., 1999). The alternative MW enhancer approaches may overcome this limitation.

In contrast to the other fluorescence-based methods described, FP offers the opportunity to measure the binding affinities (K_D) of inhibitors in addition to the inhibition of the catalytic activity. The application of this approach is straightforward if binding of a peptide with a MW below 10 kDa to the protease is known. The peptide can be labeled with a fluorophore distant from the binding site without influencing the binding properties. Such binding studies can help to explain the conformational changes of proteases (Nomura et al., 2006), but can also be used to determine binding affinities of protease inhibitors by competition experiments. The approach is much more difficult where a small molecule inhibitor must be labeled, because significant efforts in terms of structural analysis and synthetic chemistry are required to avoid or minimize the influence of the fluorophore on the binding characteristics. Moreover, FP is a ratiometric readout and is thus less sensitive to autofluorescence and quenching artifacts by test samples (Levine et al., 1997; Owicki, 2000). Fluorescent compound interference with a readout can be minimized by using red-shifted dyes such as Cy3B (excitation and emission at 530 and 580 nm respectively) and Cy5 (excitation and emission at 640 and 680 nm respectively) (Turek-Etienne et al., 2003). One major disadvantage of the FP method is that the FP value is basically calculated from two separate intensity measurements. Hence, small changes in FP occur as small changes on already high signals. The resulting sensitivity of this readout is usually sufficient for applications like HTS in which a graduation of the readout in 10% steps is adequate. Other applications like the compound profiling require smaller graduations. In these cases, the resolution of an FP measurement conducted with a standard microtiter plate (MTP) reader may be too inaccurate.

2.3.4 FLUORESCENCE LIFETIME

Fluorescence lifetime (FLT) is a recent addition to the portfolio of readouts for biochemical protease assays. Lifetime is an intrinsic property that corresponds to the average time fluorophore electrons remain in the excited state before relaxing to the ground state. The FLT assay principle is described in Figure 2.7. In the case of a single emitting species, the probability of observing a photon at a certain time point after the fluorophore is excited follows an exponential decay (Moger et al., 2006; Eggeling et al., 2003). Changes in the physico-chemical environment of the fluorophore can lead to changes in the fluorescence lifetime. This effect can be utilized for monitoring proteolyses by introducing a specific fluorophore–quencher interaction—incorporation of the fluorophore and quencher on either side of the scissile bond of the substrate. Cleavage of the substrate leads to a separation of the interaction partners, resulting in an increase in fluorescence lifetime of the fluorophore. The use of this parameter as readout in biological assays is more beneficial compared to conventional optical readouts such as absorption, luminescence, and fluorescence intensity and may even replace current readout technologies used in drug discovery (Jäger et al., 2003; Turconi et al., 2001a).

In most applications the fluorescence lifetime is determined in the time domain, i.e., the time-dependent decay of the fluorescence emission after employing repetitive brief excitation pulses, for example, time-correlated single photon counting (TCSPC; O'Connor and Phillips, 1984). Digital electronics correlate the arrival of the fluorescence photons at the detector in relation to the excitation pulses. In a smaller number of applications, the fluorescence lifetime is calculated from the phase domain, i.e., the phase shift and demodulation of sinusoidal modulated light (Clegg and Schneider, 1996). The sensitivity of the fluorescence lifetime readout and its intrinsic nature (independent of the set-up or adjustment of the instrument or overall fluorophore concentration) result in high statistical accuracy and make FLT a valuable tool for HTS applications. The fluorescence lifetimes of the different fluorescent species present in a sample can be determined and quantified taking background and autofluorescence signals into account (Jäger et al., 2003). Similar to fluorescence anisotropy, the FLT determined for a mixture of different species is also the mean value of the

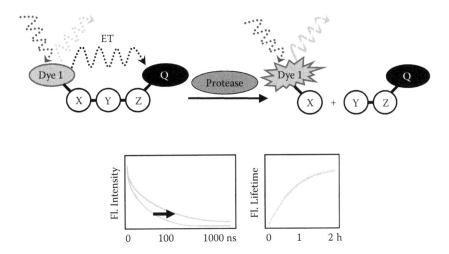

FIGURE 2.7 Fluorescence lifetime readout principle. In the intact peptidic substrate (amino acids symbolized by X, Y and Z) labeled with a fluorophore (dye) and a fluorescence quencher (Q) at the opposite sites of the scissile bond, a short fluorescence lifetime is observed (light gray dotted arrow) after excitation with a pulsed laser (dark gray dotted arrow) due to an energy transfer to the quencher (ET). After cleavage of the substrate between amino acids X and Y by a protease, the energy transfer is disrupted and a longer fluorescence lifetime is observed (light gray continuous arrow). An increase in fluorescence lifetime dependent on the enzymatic velocity is recorded.

lifetimes of the individual species in the mixture. Accordingly, the fraction of the formed product can be calculated as:

$$F_{prod} = (\tau_{sub} - \tau)/[(\tau_{sub} - \tau) + (Q^*(\tau - \tau_{prod})]$$

where τ_{sub} is the fluorescence lifetime of the substrate and τ_{prod} is the fluorescence lifetime of the product. As with fluorescence anisotropy, the factor Q is necessary to correct for differences in FI of the fluorophore in the parent substrate (FI_{sub}) and in the product (FI_{prod}): $Q = FI_{prod}/FI_{sub}$. However, the applicability of the FLT for biological reactions is difficult to predict because FLT changes are not always predictable due to environmental quenching, solvent, or polarity effects. The lifetimes of most fluorophores are typically in the range of 0.1 to 10 ns when determined in the red region of the spectrum. Moger et al. (2006) determined the lifetimes of DY-633 at 0.2 ns, EVOblue at 0.65 ns, and MR-121 at 1.85 ns.

The same publication indicated that fluorescence lifetimes of compounds from the compound collection of Pfizer classified as problematic due to their autofluorescence characteristics resulted in false positive results in many FI-based assays. For most compounds, the fluorescence lifetimes were below 1 ns. Thus, FLT measurements with a reporter fluorophore displaying a lifetime significantly longer than 1 ns are suitable for application in protease assays for compound testing. However, for applications under initial velocity conditions with a substrate turnover below 20%, fluorophores with lifetimes of a few nanoseconds are still critical because the dynamic range of the assay is then too low with lifetimes below 1 ns.

Under confocal settings S/N ratios sufficient for high-throughput applications can be achieved with these fluorophores (Moger et al., 2006). Under standard optical settings, dynamic ranges far above 1 ns are required for robust screening assays.

Especially attractive for protease assays among the fluorophores suitable for fluorescence lifetime measurements is a group of dyes that can be quenched by natural amino acids in the peptide substrate. MR121, ATTO651, 2,3-diazabicyclo[2.2.2]oct-2-ene (DBO), and Puretime™ 14 (PT 14) belong to this class of dyes (Marme et al., 2004; Hennig et al., 2006, 2007). The long lifetime of DBO allows a time-gated intensity measurement, separating the probe signal from any unwanted, short-lived fluorescence intensity arising from compounds or other assay components, thereby increasing assay robustness. The DBO-based FLT assay strategy was applied to both endopeptidases and carboxypeptidases. Hennig et al. (2007) describe an assay for carboxypeptidase A using peptidic substrates, with DBO included into the non-primed-site sequence and a tryptophan as a C terminal amino acid in the S1′ position serving as a quencher.

Smith et al. (2004) described the synthesis and the fluorescence properties of acridones and quinacridones. First pilot studies of protease (and kinase) activity assays demonstrated that the fluorescence lifetimes of certain acridone dyes are significantly reduced by a tyrosine residue in close proximity to the dye (Graves et al., 2005, 2006). Recently, the application of the PT14 acridone dye in a kallikrein 7 assay was described (Doering et al., 2009). The authors showed that the fluorescence lifetime of PT 14 was quenched by a tyrosine and even more efficiently by a tryptophan in close proximity to the dye. This allows the reduction from two labels within the peptide substrate (as in FRET assays) to one label plus one natural amino acid on either side of the scissile bond. As a result, artifacts in the interaction between protease and substrate on the one hand and between substrate and compound on the other hand, created by the introduction of two artificial dyes into the natural recognition sequence of a protease, are reduced. The enzymatic cleavage leads to a spatial separation of the probe and the tryptophan, resulting in an increase in the FLT of PT14. This increase correlates with the progress of substrate turnover.

Protease assays based on the FLT of PT14 are especially attractive due to the long lifetime of the dye of 14 ns that allows the discrimination of the lifetimes of short-lived fluorescent compounds from that of PT14. Thus, the rate of false-positive and -negative results can be reduced. The major

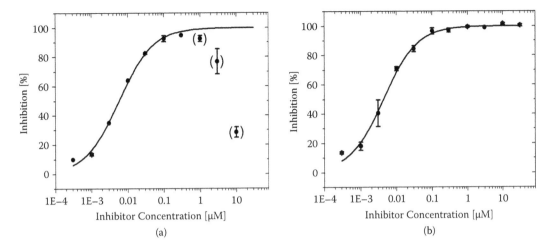

FIGURE 2.8 Dose–response curves for a protease inhibitor with autofluorescence characteristics. (a) Result of standard fluorescence intensity-based assay employing an EDANS/DABCYL-labeled substrate at a concentration of 2 µM. The profiling data (IC$_{50}$ of 6 nM and Hill coefficient of 0.9) could be obtained only after excluding the four data points for compound concentrations above 0.3 µM from the curve fit (excluded data are marked with brackets; the point at 30 µM and –128% inhibition is not shown). (b) Result obtained with the novel fluorescence lifetime-based assay employing a PT14-labeled substrate at a concentration of 1 µM. The profiling data (IC$_{50}$ of 4 nM and Hill coefficient of 0.9) were obtained without manipulation of data.

advantage of this novel assay format over the previously used standard fluorescence intensity assays employing EDANS and DABCYL as a fluorophore and quencher pair can be illustrated by the direct comparison of representative dose–response curves obtained for a proprietary inhibitor with autofluorescence characteristics. Upon excitation at 350 nm, the compound autofluorescence interferes with the intensity readout (Figure 2.8). In the case of the FRET quench assay, the four data points obtained for the compound concentrations above 0.3 µM were excluded manually prior to the curve fitting to yield IC$_{50}$ data. For the FLT-based assay, no interference was observed and the curve fit could be conducted automatically without exclusion of data points. For the example shown here, the data sets from both assay formats yield similar IC$_{50}$ values in the range of 4 to 6 nM. The FLT-1 to FLT-5 examples in Table 2.2 illustrate the dynamic ranges that can be achieved with FLT assays based on PT14.

PT14 is comparatively cost efficient and thus affordable for high-throughput applications. We know of only one disadvantage of FLT as a readout for biochemical protease activity assays: the requirement for a detection device equipped with a pulsed laser diode and TCSPC capabilities. According to our knowledge, such readers are currently offered by only two companies (Tecan and Edinburgh Instruments).

2.4 KINETIC PARAMETERS

The catalytic efficiency of an enzyme is indicated by its k_{cat}/K_M value, the value combining the effectiveness of both the productive substrate binding and the subsequent conversion of substrate molecules into product (Copeland, 2000). This value is the apparent second-order rate constant for enzyme action under conditions in which the binding site of the enzyme is largely unoccupied by substrate. The k_{cat}/K_M value is the index for comparing the relative rates of cleavage of alternative, competing substrates. The K_M is the Michaelis constant, an apparent dissociation constant and hence a measure of substrate affinity. This value equals the concentration of substrate needed to reach half maximum velocity of the enzyme reaction.

The K_M value also affects the substrate concentration employed in the assay. The assay is more sensitive to weak competitive inhibitors when working at low substrate concentrations. In drug discovery, protease assays are usually run with substrate concentrations at or below the K_M value, because the focus is mainly on the identification and characterization of competitive inhibitors (Cheng and Prusoff, 1973; Copeland, 2005). The turnover number k_{cat} is a first-order rate constant that defines the maximum number of substrate molecules converted to product per time unit under saturating conditions (i.e., at high excess of substrate).

For competitive, reversible enzyme inhibition, the lowest measurable IC_{50} value is half of the enzyme concentration used in the assay (Cheng and Prusoff, 1973). From a practical view, k_{cat}/K_M values of 10^4 M^{-1} s^{-1} and above are desirable for inhibitor profiling assays. With an enzyme-substrate pair characterized by a k_{cat}/K_M value of 10^4 M^{-1} s^{-1}, an assay can usually be run with a protease concentration in the single-digit nanomolar range in an automated setting.

2.4.1 DETERMINATION OF KINETIC PARAMETERS K_M AND k_{CAT} UNDER INITIAL VELOCITY CONDITIONS

The Michaelis constant, K_M, can be obtained from measurements conducted at a constant enzyme concentration and different substrate concentrations. Under conditions in which the substrate concentration [S] is significantly higher than the enzyme concentration and the substrate consumption is below 20%, the initial enzymatic velocity v_0 can be approximated by the Henri–Michaelis–Menten equation:

$$v_0 = (v_{max} *[S])/([S] + K_M)$$

where v_{max} is the maximal velocity at saturating substrate concentrations. The initial enzyme velocity is defined by the product formed per time unit. Therefore the signal measured in the experiment ($\Delta FI/\Delta t$, $\Delta FP/\Delta t$, $\Delta FLT/\Delta t$, etc.) must be converted into numbers of product molecules (moles) per time unit. The Michaelis–Menten equation can then be applied to the curve obtained by plotting the measured enzyme velocities against the corresponding substrate concentration leading to the values for K_M and v_{max} by a non-linear regression fit.

The k_{cat} value is defined as $k_{cat} = v_{max}/[E_0]$ where $[E_0]$ is the employed enzyme concentration.

For the determination of the initial rate constants and the kinetic parameters, the experimental data can be fitted using non-linear regression programs such as Origin 7.5SR6 (OriginLab Corporation, Northampton, Massachusetts).

In cases of extremely high K_M values (no significant changes in initial velocity are obtained by changing the substrate concentration) or where an increase in substrate concentration is limited by the experimental set-up (limitations based on a non-linear detection system), the k_{cat}/K_M value can be determined directly. If the initial substrate concentration $[S]_0$ used in an enzymatic reaction is less than 10% of the K_M value ($[S]_0 \ll 0.1 \times K_M$), the progress of the complete substrate turnover follows a single exponential curve (Mayer-Almes and Auer, 2000; Palmier and Van Doren, 2007). Although the individual parameters k_{cat} and K_M cannot be determined under these conditions, the apparent rate constant obtained is sufficient for the ranking of alternative substrates.

2.5 FLOW CHART FOR ASSAY DEVELOPMENT

The starting point of all successful assay development processes is a protein preparation that ideally contains only the enzyme of interest. The amount of enzyme needed for the whole assay development process depends mainly on the catalytic efficiency of the enzyme with the substrate. Only 10 µg of a protease can be sufficient for proper determination of the k_{cat}/K_M value, the determination of the endpoint linearity by an enzyme titration, and the assay validation by IC_{50} determination for a selection of known reference inhibitors. Our laboratory defined as good practice for the development of protease activity assays the following flow chart.

1. The identification of a suitable substrate sequence via literature search or by experimental methods (see Section 2.2 covering substrate finding approaches).
2. The definition of the assay strategy and readout determines the requirements for the labeling of the substrate (see 2.3 covering assay formats).
3. If the substrate is a synthetic peptide, a stock solution can be prepared by dissolving the lyophilized peptide in anhydrous, Ar-saturated dimethylsulfoxide (DMSO). The stock solution with a typical substrate concentration of 5 mM can be kept in aliquots at −20°C.
4. It is recommended to test for substrate solubility under the desired buffer conditions. The substrate stock solution can be diluted in the final buffer for the enzymatic reaction to the desired final assay concentration, e.g., 2 μM. After spinning the solution in a centrifuge at moderate speed for approximately 15 min, it should be checked visually for precipitation. If precipitation occurs, the pre-dilution of the substrate in a different buffer is recommended. If the solubility of the substrate is generally low in all acceptable buffer conditions, a substrate redesign should be initiated.
5. The first enzyme titration, i.e., the recording of progress curves at a constant substrate concentration and different protease concentrations, should yield the increase of the initial rate with increasing protease concentration. The cleavage products should be analyzed (for example, by LC/MS) to identify the primary cleavage site and should be compared with the expected cleavage position.
6. The substrate should then be characterized by kinetic constants of the reaction, ideally determined under initial velocity conditions (see Section 2.4 covering kinetic parameters).
7. If protease inhibitors should be identified and characterized, the assay should be tested for signal stability and endpoint linearity in the next step. The progress curves recorded for substrate concentrations near or below the K_M value should be linear for up to 2 h. Such signal stability is a prerequisite to run the assay with a pre-incubation time of 1 h for enzyme and inhibitor. This long pre-incubation is recommended to ensure also that the IC_{50} values for slowly binding inhibitors are correctly determined. An additional hour for the incubation of enzyme, inhibitor and substrate after this pre-incubation is recommended.
8. Instabilities observed in the test described above must be handled by adjusting the buffer conditions (change of detergent, addition of reducing reagents, and/or modification of dilution pattern and liquid handling protocols).
9. The final step is the validation of the assay. Inhibition constants of known, published inhibitors of the protease of interest should be reproduced.

2.6 SUMMARY AND CONCLUSION

With more than 500 members, the proteases constitute one of the largest families of enzymes in the human genome and are an important class of targets for drug discovery. Biochemical assays based on fluorescence readouts are the most frequently used technologies to elaborate the enzymatic mechanisms of proteases and to test the inhibitory potencies of drug-like chemical compounds. These assay formats exist for all classes of proteases and substrate specificities.

The development of protease assays is very straightforward if (1) the substrate recognition sequence for the protease of interest is known and (2) short peptides can be used as substrates. Extensive amounts of information about substrates are available from electronically searchable databases in the public domain. In addition, several strategies to experimentally identify substrates for proteases of unknown specificity have been described. The most recent method is the so-called PICS technology that covers all the sequences relevant in a human proteome. The MS/MS-based identification of the cleavage products and the subsequent identification of single substrate peptides by database searches in libraries of the respective proteome provide information about the consensus sequence of the substrate. In cases where proteins are required as protease substrates, the substrate design must be hypothesis-driven, based on the recognition/activation mechanism. Often, the substrate used in an assay is derived from a natural protease substrate that is biologically or chemically labeled with one or two fluorophores.

Most biochemical assays used to determine the potency of inhibitors against (1) aminopeptidases and (2) endopeptidases with minor contributions to the substrate binding efficiency by the S′ site, are based on the measurement of FI as readout. AMC and Rh110 are the most common fluorophores in these cases. For studies of enzyme kinetics, the application of AMC is sufficient. If an assay is intended to determine the inhibitory potencies of small chemical compounds, it is recommended to use red-shifted dyes like Rh110 to minimize interference of compound fluorescence with the fluorescence originating from the assay dye.

FRET-based readouts are the most common methods applied for endopeptidase activity assays. In general, the FRET readout is simple, numerous assay examples are published, many fluorescently labeled substrates are commercially available, and labeling procedures are well established. Furthermore, broad range of fluorescence donors and acceptors is available. FRET quench substrates allow fast and cost-efficient operations in an industrialized HTS environment. However, in many cases, the assays suffer from fluorescence artifacts from the tested compounds. This can lead to false-positive and false-negative results because the most frequently used FRET quench dyes are excited in the ultraviolet range.

TR-FRET measurements present interesting alternatives to the FRET quench format to minimize compound interference. Time-gated measurements allow the exclusion of compound autofluorescence from the detection. The technology is very popular due to its sensitivity, ease of use, and simplicity of assay development. The high costs for the reagents are the major disadvantage of TR-FRET-based protease assays.

In general, all FRET assay formats are limited to substrates in which short distances between donor and acceptor dye do not disturb the interaction between protease and substrate. The flexibility of the peptide conformation makes the prediction of the effective distance between the dyes and consequently the prediction of the FRET effect difficult. The distance between donor and acceptor cannot be easily estimated by the mean hydrodynamic radii of the dyes.

FP is an interesting alternative readout for endopeptidase activity assays. A prerequisite for FP-based assays is that the enzymatic cleavage event leads to a significant change in the mass of the molecular moiety carrying the fluorophore. When short peptides are used as protease substrates, the small dynamic signal range originating from a small difference in size between substrate and the fluorophore containing proteolytic product can be overcome with the concept of MW-enhancers, e.g., by attaching a streptavidin to the substrate via a biotin molecule at the opposite site of the scissile bond of the fluorophore. The MW enhancer principle can be applied generically and the development of such FP assays is straightforward. The operation of FP assays in a high-throughput setting is easy and cost-efficient. However, the FP method lacks the precision to determine small changes if performed on standard MTP readers. The FP value is calculated from intensity measurements of two channels, one carrying the parallel and the other the perpendicular polarized components of the fluorescence emission. Changes in the FP are detected as small changes on a signal of high intensity.

FLT has been added to the portfolio of readouts for biochemical protease assays recently. In our opinion, FLT is the most promising readout format among those discussed in this chapter. Fluorophores like the acridone-based PT14 exhibit relatively long fluorescence lifetimes of more than 10 ns. This long FLT can be discriminated from background fluorescence that usually has a much shorter lifetime. The fluorescence of the dye is quenched by the amino acids natural tyrosine or tryptophan. Thus, also the synthesis of substrates for endopeptidase assays requires only a single labeling step with a water-soluble dye. The substrate synthesis is thereby simplified and becomes more cost efficient. To our knowledge, the only disadvantage of FLT as a readout for biochemical protease activity assays is the moderate availability of instrumentation such as microtiter plate (MTP) readers equipped with pulsed laser diode and TCSPC capabilities.

In general the development of biochemical fluorescence-based protease assays is straightforward. Knowledge of the substrate specificity determines how much effort must be used to design a proper assay substrate. The selection of the fluorescence-based readout and a suitable dye (pair) should be driven by (1) the purpose of the assay, (2) the required sensitivity and robustness, (3) the specification of the desirable throughput, and (4) cost calculations. Other boundaries can be the availability of MTP reader hardware or a required fit to other assay formats that are run in parallel to the protease assays on the same screening and reader devices. In principle, robust assay operation under initial velocity conditions with sufficient dynamic range is possible with all readout technologies described in this chapter.

REFERENCES

Aibe, K. et al. 1996. Substrate specificity of recombinant osteoclast-specific cathepsin K from rabbits. *Biol. Pharm. Bull.* 19, 1026–1031.

Andrews, D.L. and Demidov, A.A. 1999. *Resonance Energy Transfer.* Chichester: John Wiley & Sons.

Angliker, H. et al. 1995. Internally quenched fluorogenic substrate for furin. *Anal. Biochem.* 224, 409–412.

Attucci, S. et al. 2002. Measurement of free and membrane-bound cathepsin G in human neutrophils using new sensitive fluorogenic substrates. *Biochem. J.* 366, 965–970.

Banks, P., Gosselin, M., and Prystay, L. 2000. Impact of a red-shifted dye label for high-throughput fluorescence polarization assays of G protein-coupled receptors. *J. Biomol. Screen.* 5, 329–334.

Barret, A.J. 2004. Bioinformatics of proteases in the MEROPS database. *Curr. Opin. Drug Disc. Dev.* 7, 334–341.

Barrett, A.J. and MacDonald, J.K. 1986. Nomenclature: protease, proteinase and peptidase. *Biochem. J.* 237, 935–940.

Berger, A. and Schechter, I. 1976. On the size of the active site in proteases I: papain. *Biochem. Biophys. Res. Comm.* 27, 157–162.

Bratanova, E.K. and Petkov, D.D. 1987. N-antraniloylation converts peptide p-nitroanilides into fluorogenic substrates of proteases without loss of their chromogenic properties. *Anal. Biochem.* 162, 213–218.

Burbaum, J.J. and Sigal, N.H. 1997. New technologies for high-throughput screening. *Curr. Opin. Chem. Biol.* 1. 72–78.

Cai, S.X. et al. 2001: Design and synthesis of rhodamine 110 derivative and caspase-3 substrate for enzyme and cell-based fluorescent assay. *Bioorg. Med. Chem. Lett.* 11, 39–42.

Carmel, A., Kessler, E., and Yaron, A. 1977. Intramolecularly quenched fluorescent peptides as fluorogenic substrates of leucine aminopeptidase and inhibitors of clostridial aminopeptidase. *Eur. J. Biochem.* 73, 617–625.

Chen, R.F. 1968. DANSYL labeled proteins: determination of extinction coefficient and number of bound residues with radioactive dansyl chloride. *Anal. Biochem.* 25, 412–416.

Cheng Y.C. and Prusoff W.H. 1973. Relationship between the inhibition constant (K_I) and the concentration of inhibitor which causes 50% inhibition (IC_{50}) of an enzymatic reaction. *Biochem. Pharmacol.* 22, 3099–3108.

Clegg, R.M. and Schneider, P.C. 1996. Fluorescence lifetime-resolved imaging microscopy: a general description of lifetime-resolved imaging measurements. In *Fluorescence Microscopy and Fluorescent Probes,* Slavik, J., Ed. New York: Plenum, pp. 15–33.

Comley, J. 2006. TR-FRET based assays: getting better with age. *Drug Disc. World.* Spring, 23-37.

Comley, J. 2007: *TR-FRET-Based Assay Trends 2007.* Cambridge, U.K. HTStec.

Copeland, R.A. 2000. Enzymes; a practical introduction to structure, mechanism, and data analysis. New York: Wiley-VCH..

Copeland, R. 2005. Evaluation of enzyme inhibitors in drug discovery. Hoboken: John Wiley & Sons.

Diamond, S.L. 2007. Methods for mapping protease specificity. *Curr. Opin. Chem. Biol.* 11, 46–51.

Doering, K. et al. 2009. A fluorescence lifetime-based assay for protease inhibitor profiling on human kallikrein 7. *J. Biomol. Screen.* 1, 1–9.

Eggeling, C. et al. 2003. Highly sensitive fluorescence detection technology currently available for HTS. *Drug Disc. Today* 8, 632–641.

Ferrer, M. et al. 2003: Miniaturizable homogeneous time-resolved fluorescence assay for carboxypeptidase B activity. *Anal. Biochem.* 317. 94–98.

Florentin, D., Sassi, A., and Roques, B.P. 1984: A highly sensitive fluorometric assay for enkephalinase, a neutral metalloendopeptidase that releases tryrosine-glycine-glycine from enkephalins. *Anal. Biochem.* 141, 62–69.

Flotow, H. et al. 2002. Development of a plasmepsin II fluorescence polarization assay suitable for high-throughput antimalarial drug discovery. *J. Biomol. Screen.* 7, 367–371.

Förster T. 1948. Zwischenmolekulare Energiewanderung und Fluoreszenz. *Ann. Phys.* 2, 55–75.

George, J. et al. 2003. Evaluation of an imaging platform during the development of a FRET protease assay. *J. Biomol. Screen.* 8, 72–80.

Gerhartz, B., Nistroj, A.J., and Demuth, H.U. 2002. Enzyme classes and mechanisms. In *Proteinase and Peptidase Inhibition*, Smith, H.J. and Simons, C., Eds. London: Taylor & Francis, pp. 1–20.

Gershkovich, A.A. and Kholodovych, V.V. 1996. Fluorogenic substrates for proteases based on intramolecular fluorescence energy transfer (IFETS). *J. Biochem. Biophys. Meth.* 33, 135–162.

Grant, S.K., Sklar, J.G., and Cummings, R.T. 2002. Development of novel assays for proteolytic enzymes using rhodamine-based fluorogenic substrates. *J. Biomol. Screen.* 7, 531–540.

Graves, P. et al. 2005. Whatever happened to fluorescence lifetime? Poster presented at 11th Annual SBS Conference, Geneva.

Graves, P. et al. 2006. Discriminating fluorescence lifetime assays for screening. Poster presented at 12th Annual SBS Conference, Seattle.

Graziano, V. et al. 2006: Enzymatic activity of the SARS coronavirus main proteinase dimer. *FEBS Lett.* 580, 2577–2583.

Gurtu, V., Kain, S.R., and Zhang, G. 1997. Fluorometric and colorimetric detection of caspase activity associated with apoptosis. *Anal. Biochem.* 251, 98–102.

Harris, J.L. et al. 2000. Rapid and general profiling of protease specificity by using combinatorial fluorogenic substrate libaries. *Proc. Natl. Acad. Sci. USA*, 7754–7759.

Hassiepen, U. et al. 2007. A sensitive fluorescence intensity assay for deubiqitinating proteases using ubiquitin-rhodamine110-glycine as substrate. *Anal. Biochem.* 371, 201–207.

Hennig, A. et al. 2007. Design of peptide subsrates for nanosecond time-resolved fluorescence assays of proteases: 2,3-diazabicycle[2.2.2]oct-2-ene as a non-invasive fluorophore. *Anal. Biochem.* 360, 255–265.

Hennig, A. et al. 2006. Nanosecond time-resolved fluorescence protease assays. *Chem. BioChem.* 7, 733–737.

Horton, R.A. et al. 2007. A substrate for deubiquitinating enzymes based on time-resolved fluorescence energy transfer between terbium and yellow fluorescent protein. *Anal. Biochem.* 360, 138–143.

Jäger, S., Brand, L., and Eggeling, C. 2003. New fluorescence techniques for high-throughput drug discovery. *Curr. Pharm. Biotechnol.* 4, 463–467.

Karvinen, J. et al. 2002. Homogeneous time-resolved fluorescence quenching assay (LANCE) for caspase 3. *J. Biomol. Screen.* 7, 223–231.

Knight, C.G., Willenbrock, F., and Murphy, G. 1992. A novel coumarin-labelled peptide for sensitive continuous assays of the metalloproteinases. *FEBS Lett.* 296, 263–266.

Konstantinidis, A.K. et al. 2007. Longer wavelength fluorescence energy transfer depsipeptide substrates for hepatitis C virus NS3 protease. *Anal. Biochem.* 368, 156–167.

Lakowicz, J.R. 2006. *Principles of Fluorescence Spectroscopy*. New York: Springer.

Lauer-Fields, J.L. et al. 2007. Screening of potential a disintegrin and metalloproteinase with thrombospondin motifs-4 inhibitors using a collagen model fluorescence resonance energy transfer substrate. *Anal Biochem.* 373, 43–51.

Lecaille, F. et al. 2003. Probing cathepsin K activity with a selective substrate spanning its active site. *Biochem. J.* 375, 307–312.

Leissring, M.A. et al. 2003. Kinetics of amyloid β-protein degradation determined by novel fluorescence polarization-based assays. *J. Biol. Chem.* 278, 37314–37320.

Lestienne, P. and Bieth, J.G. 1980: Activation of human leukocyte elastase activity by excess substrate, hydrophobic solvents and ionic strengths. *J. Biol. Chem.* 255, 9289–9294.

Levine, L.M. et al. 1997. Measurement of specific protease activity utilizing fluorescence polarization. *Anal. Biochem.* 247, 83–88.

Li, H.X. et al. 2007. Substrate specificity of human kallikreins 1 and 6 determined by phage display. *Prot. Sci.* 17, 664–672.

Liu, J. et al. 1999: Fluorescent molecular probes V: a sensitive caspase-3 substrate for fluorometric assays. *Bioorg. Med. Chem. Lett.* 9, 3231–3236.

Lopez-Calle, E., Fries, J., and Jungmann, J. 2002. Methods and means for detecting enzymatic cleavage and linkage reactions. Patent WO/2002/059352.

Malfroy, B. and Burnier, J. 1987. New substrates for enkephalinase (neutral endopeptidase) based on fluorescence energy transfer. *Biochem. Biophys. Res. Comm.* 143, 58–66.

Mao, S.S. et al. 2008. A time-resolved, internally quenched fluorescence assay to characterize inhibition of hepatitis C virus nonstructural protein 3-4A protease at low enzyme concentrations. *Anal Biochem.* 373, 1–8.

Marks, B.D. et al. 2005. Multiparameter analysis of a screen for progesterone receptor ligands: comparing fluorescence lifetime and fluorescence polarization measurements. *Assay Drug Devel. Technol.* 3, 613–622.

Marme, N. et al. 2004. Highly sensitive protease assay using fluorescence quenching of peptide probes based on photoinduced electron transfer. *Angew. Chem. Int. Ed.* 43, 3798–3801.

Mathis, G. 1999. HTRF technology. *J. Biomol. Screen.* 4, 309–314.

Mayer-Almes, F.J. and Auer, M. 2000. Enzyme inhibition assays using fluorescence correlation spectroscopy: a new algorithm for the derivation of k_{cat}/K_M and K_i values at substrate concentrations much lower than the Michaelis constant. *Biochemistry*, 39, 13261–13268.

Moger, J. et al. 2006. The application of fluorescence readouts in high-throughput screening. *J. Biomol. Screen.* 11, 765–772.

Nishino, N., Makinose, Y., and Fujimoto, T. 1992. 2-Aminobenzoyl-peptide-2,4-dinitroanilinoethylamides: facile fluorescent detection system for sequence specific proteases. *Chem. Lett.* 1, 77–80.

Nomura, A.M. et al. 2006. One functional switch mediates reversible and irreversible inactivation of a herpesvirus protease. *Biochemistry* 45, 3572–3579.

O'Connor, D.V. and Phillips, D. 1984. *Time-Correlated Single Photon Counting.* New York: Academic Press.

Okun, I. et al. 2006. Screening for caspase-3 inhibitors: a new class of potent small-molecule inhibitors of caspase 3. *J. Biomol. Screen.* 11, 277–285.

Owicki, J.C. 2000. Fluorescence polarization and anisotropy in high-throughput screening: perspectives and primer. *J. Biomol. Screen.* 5, 297–306.

Palmier M.O. and Van Doren S.R. 2007. Rapid determination of enzyme kinetics from fluorescence: overcoming the inner filter effect. *Anal Biochem.* 371, 43–51.

Pham, W. et al. 2004. Developing a peptide-based near-infrared molecular probe for protease sensing. *Bioconj. Chem.* 15, 1403–1407.

Pope, A.J., Haupts, U.M., and Moore, K.J. 1999. Homogeneous fluorescence readouts from miniaturized high-throughput screening: theory and practice. *Drug Disc. Today* 4, 350–362.

Preaudat, M. et al. 2002: A homogeneous caspase-3 activity assay using HTRF technology. *J. Biomol. Screen.* 7, 267–274.

Rawlings, N.D., Morton, F.R., and Barrett, A.J. 2006. MEROPS: the peptidase database. *Nucl. Acid Res.* 34, D270–D272.

Réhault, S. et al. 1999. New sensitive fluorogenic substrates for human cathepsin G based on the squence of serpin-reactive site loops. *J. Biol. Chem.* 274, 13810–13817.

Richardson, P.L. 2002. The determination and use of optimized protease substrates in drug discovery and development. *Curr. Pharm. Dis.* 8, 2559–2281.

Schilling, O. and Overall, C.M. 2008. Proteome-derived, database-searchable peptide libraries for identifying protease cleavage sites. *Nat. Biotechnol.* 26, 685–694.

Silverman, L., Campbell, R., and Broach, J.R. 1998. New assay technologies for high-throughput screening. *Curr Opin. Chem. Biol.* 2, 397–403.

Smith, J.A., West, R.M., and Allen, M. 2004. Acridones and quinacridones: novel fluorophores for fluorescence lifetime studies. *J. Fluoresc.* 14, 151–171.

St. Hilaire, P.M. et al. 1999. Fluorescence-quenched solid phase combinatorial libraries in the characterization of cysteine protease substrate specificity. *J. Comb. Chem.* 1, 509–524.

Stambolieva, N.A., Ivanov, I.P., and Yomtova, V.M. 1992. N-antraniloyl-Ala-Ala-Phe-4-nitroanilide, a highly sensitive substrate for subtilisins. *Arch. Biochem. Biophys.* 294, 703–706.

Stein, R.L. and Strimpler, A.M. 1987. Catalysis by human leukocyte elastase: aminolysis of acyl enzymes by amino acid amides and peptides. *Biochemistry*, 26, 2238–2242.

Tanaka, T. et al. 1985. Human leukocyte cathepsin G: substrate mapping with 4-nitroanilides, chemical modification and effect of possible cofactors. *Biochemistry*, 24, 2040–2047.

Tirat, A. et al. 2005. Synthesis and characterization of fluorescent ubiquitin derivatives as highly sensitive substrates for deubiquitinating enzymes UCH-L3 and USP-2. *Anal. Biochem.* 343, 244–255.

Toth, M.V. and Marshall, G.R. 1990. A simple, continuous fluorometric assay for HIV protease. *Int. J. Pept. Protein Res.* 36, 544–550.

Turconi S. et al. 2001a. Developments in fluorescence lifetime-based analysis for ultra-HTS. *Drug Disc. Today* 6, 27–39.

Turconi, S. et al. 2001b. Real experience of uHTS: a prototypic 1536-well fluorescence anisotropy-based uHTS screen and application of well-level quality control procedures. *J. Biomol. Screen.* 6, 275–290.

Turek-Etienne, T.C. et al. 2003. Evaluation of fluorescent compound interference in 4 fluorescence polarization assays: 2 kinases, 1 protease and 1 phosphatase. *J. Biomol. Screen.* 8, 176–184.

Turk, B. 2006. Targeting proteases: success, failures and future prospects. *Nat. Rev. Drug Disc.* 5, 785–799.

Turro, N.J. 1965. *Molecular Photochemistry.* New York: W.A. Benjamin.

Van der Meer, B.W., Coker, G.I., and Chen, S.Y. 1994. *Resonance Energy Transfer: Theory and Data.* New York: VCH.

Von Geldren, W.T., Holleman, W.H., and Opgenorth, T.J. 1991. A fluorescence assay for endothelin-converting enzyme. *Pep. Res.* 4, 32–35.

Wang, G.T. et al. 1990. Design and synthesis of new fluorogenic HIV protease substrates based on resonance energy transfer. *Tetrahedron Lett.* 31, 6493–6496.

Xia, L. et al. 1999. Localization of rat cathepsin K in osteoclasts and resorption pits: inhibition of bone resorption cathepsin K-activity by peptidyl vinyl sulfones. *Biol. Chem.* 380, 679–687.

Zou, J. et al. 2005. ADAM33 enzyme properties and substrate specificity. *Biochemistry* 44, 4247–4256.

3 Assay Development for Nuclear Hormone Receptors

Chang Bai and Eric N. Johnson

CONTENTS

3.1 INTRODUCTION

The world of drug discovery has focused on pharmaceutical intervention in many pathways by altering the signaling cascades of a diverse array of targets. Arguably one of the most prevalent targets for drug discovery is the GPCR family (~1000 members) followed by the kinases (~500 members). The nuclear receptor (NR) family is comprised of a much smaller subset of drug targets (48 members), but the diverse biological functions augmented by NRs and the success of several NR agonists and antagonists in the clinic makes them very appealing targets for drug discovery (Moore et al., 2006; Bai et al., 2007).

The nuclear receptor family consists of a unique class of transcription factors (Mangelsdorf et al., 1995). NRs generally contain two main structural domains: a highly conserved N terminal DNA-binding domain that targets the receptor to DNA sequences and a C terminal ligand-binding domain. In addition, many members of the NR family contain a less conserved N terminal extension that varies significantly in length. While the transactivation function domain normally resides within the ligand-binding domain, an additional activation function domain (AF-1) is typically found in the less conserved N terminal extension. For some nuclear hormone receptors such as the androgen and glucocorticoid receptors that contain long N termini, the AF-1 function is prevalent. Such structural differences of the receptor subfamily should be considered during assay design.

Nearly half of the nuclear receptors are either known drug targets or are being pursued as targets for drug development. There are many examples of nuclear receptor ligands used in the treatment of a diverse array of diseases. The replacement of steroid receptor agonists such as thyroid and estrogen were among the most commonly used pharmaceutical agents in the history of drug discovery, demonstrating the importance and utility of this target class. Several NR antagonists were also developed for the treatment of hormone-sensitive diseases, that is, cyproterone acetate, flutamide, and bicalutamide for prostate cancer (targeting androgen receptor), spironolactone for blood pressure (targeting mineralocorticoid receptor), and tamoxifen for breast cancer (targeting estrogen receptor) (Chen. 2008; Gillatt, 2006; Tobias et al., 2004).

Although ligands for many nuclear hormone receptors have been reported, the cognate ligands for some nuclear hormone receptors still remain to be discovered. They are generally referred to as orphan receptors. The pharmaceutical industry's interest in this target class is exemplified by the fact that after a ligand was identified for a nuclear receptor, the receptor almost immediately became a candidate for drug development. It is expected that the continuous de-orphaning and elucidation of receptor function will reveal new horizons for drug development, for example, the de-orphaning of peroxisome proliferator-activated receptors (PPARs). Ligands for PPARα and PPARγ (rosiglitazone and pioglitazone) were developed for type 2 diabetes and hyperlipidemia (Greene, 1999; Lehmann et al., 1995).

Because of the ubiquitous expression patterns of many NRs and the cross reactivities often seen with NR agonists and antagonists, treatment with such agents often resulted in undesirable effects along with physiological benefits. Significant progress has been made in refining the pharmacological and physical properties of these agents. One example is the use of ointments or inhalers for glucocorticoid agonists to extend the retention time at the disease tissues and minimize drug exposure into the blood to reduce side effects. Additionally, the selectivity of the spironolactone NR ligand, which also binds to AR, was significantly improved with the development of eplerenone, resulting in reduced breast and gynecomastia pain in comparison with spironolatone.

Perhaps the most exciting area for drug development targeting nuclear hormone receptors lies among the so-called selective modulators typified by the successful development and launch of tamoxifen and raloxifene, two drugs that target estrogen receptors, are different from the traditional pharmacological agonists or antagonists that mimic or block the functions of cognate endogenous ligands, respectively. The modulators exhibit function and tissue selectivity; they have the capacity to display profound antagonistic activities in some tissues while remaining silent or even agonistic in other tissues. This concept opened the door to improve upon current therapies that target nuclear hormone receptors and allowed the use of NR ligands for therapeutic indications that were previously limited due to safety concerns. Currently, much of the drug discovery effort on nuclear hormone receptors focuses on strategies to reduce the adverse effect profiles of known agents by improving the tissue and functional selectivity of ligands.

When designing assays to find agonists and antagonists of NRs, several parameters must be considered. As mentioned, the structure of a receptor may lend itself to one type of assay versus another. An assay that utilizes the AF-1 function may be applied to NRs that contain extended N termini, while an assay that focuses on the AF-2 function may be broadly applicable to NRs. In addition to NR structure, the mechanism of activation must also be evaluated when designing an assay to identify agonists or antagonists. The binding of a ligand to the NRs often triggers the translocation of the nuclear hormone receptor from the cytoplasm to the nucleus. Thus nuclear translocation can be the basis to identify agonists or antagonists.

However, some receptors are constitutively expressed in the nucleus and this type of receptor would not be amenable to a nuclear translocation assay. The activities of nuclear receptors may be dependent upon complex interactions with a number of coregulatory proteins, commonly known as coactivators or corepressors, and modifications by post-translational means. Cell type-specific expression levels of receptors and coregulators may contribute to some, but not all, of the molecular bases for gene and functional selectivity of receptor activity. Therefore selecting a cell line that expresses both the target receptor and the necessary cofactors may be required to design an appropriate assay.

Because the function of many NRs is to modify gene expression, much of the assay design regarding the identification of agonists or antagonists of NRs focuses on monitoring transcriptional activity using carefully chosen reporter-based assays or assays that measure the expression of native messenger proteins. Recently, monitoring the expression of endogenous genes has been emphasized because it is believed to result in an assay with a greater relevance to biology. Several technologies have emerged that are flexible and yet inexpensive for the purposes of high-throughput screening as well as the support of medicinal chemistry programs, while still powerful enough to capture the expression of a diverse spectrum of target genes.

Biochemical ligand displacement assays are frequently used as initial screening tools to "weed out" low affinity ligands and select for compounds that have significant affinities for the intended targets. It is also one of the favored formats for a receptor selectivity screen because it is straightforward and generates results that are easy to compare with closely related receptors. Other assays measure ligand and receptor signaling cascades, that is, ligand-dependent translocation of receptor, utilization of coactivator or corepressor, or assays that directly quantify levels of target protein rather than mRNA. Recent progress has been made in developing multiplexed technologies that allow measurement of multiple target mRNAs or protein expression in a single assay reaction.

This chapter will describe some of the most common assay technologies that one can use to establish a nuclear receptor program. While some nuclear receptor assay technologies are powerful tools in mechanistic studies and functional elucidation such as location analysis using chromatin immunoprecipitation or whole genome microarray, the throughput and/or the cost factors have prevented their routine use as primary screening assays for medicinal chemistry programs. They are generally applied to the assay paradigm when a limited number of leads has been selected via other methods. These assays are outside the scope of this article; readers are encouraged to read other papers.

3.2 ASSAY DEVELOPMENT

3.2.1 DISPLACEMENT OF RADIO LIGANDS

The displacement of radio-labeled ligands has been a common method of drug discovery for many years (Smith and Sestili, 1980). The principle of the assay is fairly simple. A known ligand of a NR is labeled with radioactivity. The radio-labeled ligand is incubated with cell extracts containing the NHR of interest in the presence or absence of putative ligands. The reaction is filtered and washed, removing the unbound radioactivity and retaining, in theory, only the ligand bound to the NHR. Quantification of the radioactivity bound to the membrane reveals the fraction of radioactive ligand bound to the receptor. A compound that competes with the NHR ligand for binding to the receptor will result in decreased radioactive counts.

Later iterations of this assay incorporated the use of scintillation proximity assay (SPA) beads. Rather than using filtration to remove the unbound radio ligand, the proximity of the bead to the radioactivity would result in a signal that could be quantified. We have utilized radio ligand binding assays to detect ligands of receptors by imaging SPA beads that produce light in close proximity to radioactivity.

Development of these assays can be difficult. A radio-labeled ligand that binds to the target receptor with high affinity is required, along with sufficient receptor and sufficient radioactivity to cause the SPA bead to signal. These aspects may limit the ability of a specific assay to be miniaturized into a 1536-well plate format. However, for some receptors and ligands, it has proven an effective way to screen.

While this assay has been used in many screening campaigns and is capable of identifying ligands of NHRs, it lacks a functional component; that is, using this technology, one cannot conclude that a compound that displaces the radio-labeled ligand is an agonist, a partial agonist, or an antagonist. An additional functional assay is required to determine the efficacy (as an agonist, partial agonist, or antagonist) of the compound. Furthermore, the possible presence of allosteric ligands that use a different interaction surface may escape such a screening paradigm. Such ligand displacement assays can identify only compounds that share the same binding pockets with the labeled ligand.

3.2.2 TRANSCRIPTION ASSAYS

Perhaps the most widely used assays to study NHR biology are those that measure transcription. Because we learned early on that stimulation of NHRs will result in the transcription of various genes, biologists developed functional assays that exploit this attribute to detect NHR agonists or antagonists. Several variations of assays measure transcription. Many scientists use reporter genes to demonstrate the efficacy of NHR agonists. Constructs have been produced in which the ligand binding domain of an NHR (SF-1), the Gal4 DNA binding domain, and a Gal4-luciferase reporter system have been created to detect stimulators of SF-1 (Madoux et al., 2008). One of the benefits of using a luciferase reporter system is that the amount of signal is often very high and thus the signal to basal window is often quite large. In addition, because these systems detect luminescence and not fluorescence, there is no interference with fluorescent compounds.

Scientists have used an analogous system with the ligand-binding domains of NHRs driving the expression of a β-lactamase reporter (Wilkinson et al., 2008). The use of β-lactamase as a reporter system offers the advantage of a ratiometric assay. Dyes, upon excitation at 405 nm, evoke an emission at 520 nm. However, in the presence of β-lactamase, the dye is cleaved such that upon excitation at 405 nm, an emission occurs at 450 nm. By comparing the emissions at 520 nm and 450 nm, the amount of β-lactamase can be determined with an added indication of cell number. Cells that fail to load with dye (suggesting issues with cell health) exhibit decreases in emission at both wavelengths. Furthermore, the β-lactamase system facilitates clonal selection of cells using FACS analysis of cells before and after treatment.

Finally, GFP has often been employed as a reporter gene. Similar to the above examples, GFP expression has been engineered under the control of the glucocorticoid response element (GRE). When glucocorticoids activate the glucocorticoid receptor (GR), the GR translocates to the nucleus, binds the GRE, and stimulates transcription of GFP (Zhang et al., 2008). Similar to β-lactamase reporters, GFP reporters offer the advantage of using FACS to select clonal populations of cells. The assay of GR agonist activity utilizes the detection of fluorescence at 505 nm after stimulation at 470 nm.

Reporter gene assays have added tremendously to the landscape of screening for NHR agonists and antagonists. Unlike binding assays, reporter gene assays offer the ability to study receptor function. As in all assays, reporter gene assays exhibit negative attributes. One major drawback is that the agonistic or antagonistic activities of a ligand can be altered when a receptor is overexpressed in such a system. Thus, during creation of a cell line expressing a receptor, care must be taken to achieve physiological levels of receptor expression. In addition, a compound that interferes with the biology of the reporter, such as a compound that blocks luciferase or β-lactamase activity, will appear as blocking the NHR target. Similarly, a compound that inhibits the transcriptional or translational machinery of the cell independent of the NHR will also appear identical to a compound that blocks the NHR target. Therefore, proper counter screens and controls are required to identify the true positives from the hits of the screen.

The first and most important parameter in developing a transcription-based assay to identify agonists or antagonists of a NHR is the generation of an appropriate cell line that obviously must contain the NHR of interest. However, there are costs and benefits based on whether the expression of the target should be engineered or endogenous. In a cell line in which the expression of the target is engineered, the scientist can control the amount of receptor expressed in the cell, the genotype of a particular receptor (perhaps a mutation is of interest for a particular disease), and the cellular background. In addition, with an engineered cell line, a relatively obvious counter screen can be performed in those cells lacking the receptor of interest or expressing another NHR.

Conversely, there are arguments for studying receptors that are endogenously expressed in cells. For example, it is likely that the NHR may bind to cofactors. Using an endogenously expressed receptor in a cell line that is relevant to the biology of interest may allow that receptor to interact in the appropriate configuration and concentration to other proteins expressed in the cell. This type of cell line may be extremely beneficial for linking the hits identified in the assay to a biological phenotype, particularly when all the cofactors may not be known. In addition to the expression of the receptor, it must be decided whether other cofactors should be engineered into the cell line. Again, engineering the expression of these proteins can allow control of the concentration of these proteins and choices as to which proteins should or should not be expressed. However, endogenous expression of these potential cofactors presumes that the concentration of cofactor is equivalent to a biologically relevant system.

Regardless of the expression of the receptor and cofactors, the reporter gene must be introduced to the cells. There are several options for the introduction of the reporter gene construct into the cells including chemical-based transfection ($CaCl_2$), lipid-based transfection, and electroporation.

The clonal selection of the cell line is a very important aspect of assay development. Using either fluorescently activated cell sorting (FACS) or limiting dilution a clonal population of cells should be identified. Each population should be evaluated using the functional reporter assay and selected based on the fact that it has a low level of background signal and a high level of signal upon receptor stimulation.

With the appropriate cell line available for the screen, optimization of the high-throughput screening assay can begin. A matrix experiment is generally performed in which a range of cell densities is plated for assay (1,000 to 10,000 cells per well in a 1536-well plate). This cell number determination is performed in conjunction with a dose titration of receptor agonist. The end result is a combination of cell density and appropriate agonist concentration for the assay. If the assay is to select an antagonist, then optimization should occur with an approximate EC_{70-80} concentration of agonist such that the window between stimulated and unstimulated is great, but the assay remains sensitive enough to identify antagonists.

After a suitable cell number has been identified, additional experiments should be completed to assess the optimal time required for the cells to be stimulated with agonist and "loaded" with dye. In general a range of stimulation times between 2 and 6 hr followed by a range of times incubated with dye ranging from 6 to 12 hr can be tested. It might seem that the longer the incubation time, the more signal would be generated and the greater the assay window, and thus this test would not be necessary. However, as the cells are stimulated for longer periods or are incubated with dye for longer periods, it is possible that the basal activity of the system will increase, resulting in an increased background and a net decrease in signal-to-basal ratio. In addition, it is also possible that the cells will not tolerate the prolonged incubation times and may exhibit some toxicity on their own or be more susceptible to compound-induced toxic effects.

Finally the detection of the signal must be optimized on the instrument of choice. Appropriate filters can work to decrease background and enhance signal-to-basal ratios. In addition, the type of reader can affect both the sensitivity of the assay and throughput. A reader that evaluates one well at a time may convey a greater sensitivity to the assay but drastically reduces throughput. Conversely, a reader that evaluates an entire plate very rapidly may increase throughput, but be less sensitive. The conditions necessary to obtain a sufficient assay window in an acceptable time frame must be determined for each assay.

A sample protocol for the detection of a β-lactamase reporter assay is listed below:

- Dispense 1500 cells per well in a 1536-well plate
- Dispense 10 nL of experimental compounds
- Incubate for 10 min at room temperature
- Dispense 10 nL of agonist
- Incubate the plate for 22 hrs at 37°C, 5% CO_2
- Dispense 1 µL of 6X dye in loading buffer
- Incubate for 2 hr and read (excitation, 405 nm; emission, 450 nm and 520 nm)

3.2.3 COREGULATOR–NR INTERACTION ASSAYS

The utilization of coregulatory proteins is a well known mechanism by which nuclear receptors activate or repress expression of downstream genes. Ligand binding to a nuclear receptor results in the recruitment of a number of proteins that help assemble components of the basic transcriptional machinery at the chromatin site to initiate and/or maintain transcription (O'Malley and McKenna, 2008). It is generally thought that the expression and utilization of specific coregulatory proteins are important determinants for the functional selectivity of a given ligand. An assay that tracks the formation of such specific protein complexes is viewed as a means to monitor the tissue and functional selectivities of a given ligand.

Since the discovery of this diverse family of coregulator proteins, a number of technologies has been utilized to address this need, such as FRET-based assays (Jeyakumar and Katzenellenbogen et al., 2009), combinatorial peptide phage display, yeast- or mammalian-based two-hybrid systems (Bai and Elledge, 1996; Luo et al., 1997), immunoprecipitation, surface plasmon resonance (Lavery, 2009), and GST pull-down assays. Each exhibits certain advantages. The relatively high-throughput of short peptide-based assays, including an array-based format, is attractive for quick analysis of a large number of peptides against compounds. However, the limited lengths of the cofactor peptides used in such assays often misrepresent the physiological nature of the receptor and cofactor interaction. Therefore, one must be careful not to draw conclusions based only on interactions detected in such an assay. On the other hand, immunoprecipitation is a very useful tool to dissect the interactions of receptors and coregulators although its low throughput has limited its implementation as an assay for primary screening.

The mammalian two-hybrid system exhibits a greater throughput than the immunoprecipitation assay. The mammalian two-hybrid system is a natural extension of the original yeast two-hybrid system initially designed by Stan Fields. Vectors suitable for such applications are commercially available. In general, the receptor of interest and coregulatory proteins are cloned into a transactivation domain vector such as the pVP16 vector (Clontech) and a DNA binding domain vector such as the pM. For each receptor and coregulatory protein, careful design is needed to ensure that the basal transcription activity driven by the DNA binding domain plasmid is sufficiently low such that the interaction of a coregulatory protein and the receptor can be clearly detected as an increase in the transcriptional signal.

A panel of coregulatory proteins can be cloned into DNA binding vectors. Coregulatory proteins play an important role in transcription regulation; the p160 family and p300 may also contain intrinsic histone acetyltransferase activities that can directly impact chromatin structure by promoting a transcriptionally active chromatin configuration. It is not surprising that many of the coregulatory proteins tend to display high levels of transactivation activity in the mammalian two-hybrid assays. The transactivation masks the transcription activity generated from ligand-induced interactions between coregulators and NRs in the assay.

To ensure a meaningful quantification, selected deletions or point mutations can be introduced into the coregulatory protein sequence, to reduce transactivation activities. Such mutations should be designed to exert minimal effects on the global structures of the coregulators to keep their surfaces for interaction with NRs intact. A cell line that is amenable to easy transfection is preferred. The suitability of cell lines varies, depending on the receptor, the coregulatory protein, and the physiological state of the expected interaction. One may argue that a cell line that matches a particular tissue environment is better suited for assay development as it expresses other factors that could facilitate a given interaction. Ideally, a stable line that carries the expression of a reporter and the NR fused with VP16 may be generated. A transient transfection experiment can be carried out with a large panel of coregulators. After overnight incubation, ligands are added. The reporter signals can be measured after a further incubation of 24 to 48 hr.

The mammalian two-hybrid system, although relatively low throughput, can be a valuable tool for compound profiling for medicinal chemistry programs. Since it can tolerate a relatively large size of protein and the proteins are expressed in a mammalian cell background, it provides a significant advantage over many other methods that rely on short peptides.

3.2.4 NUCLEAR TRANSLOCATION ASSAYS

The translocation of a nuclear hormone receptor to the nucleus is an early stage of receptor activation. A number of assay technologies can detect this nuclear translocation event. For example, using

GFP-tagged NHRs, one can detect the translocation of the receptor from the cytosol to the nucleus. Using imaging analysis, one can evaluate whether a compound of interest caused a translocation of GR from the cytosol to the nucleus. A modification of this screening strategy employs enzyme fragment complementation to identify this translocation event. That is, a nuclear hormone receptor, such as GR, is expressed as a fusion protein with a portion of β-galactosidase. A complementary portion of β-galactosidase is constitutively expressed in the nucleus. Upon ligand binding to GR, the receptor translocates to the nucleus and the two portions of β-galactosidase form a functional enzyme capable of metabolizing a substrate and producing light. By quantifying the light produced in the reaction, one can extrapolate the amount of GR in the nucleus.

The development of this type of nuclear translocation assay can be relatively simple because the manufacturers have identified conditions to maximize the response. While a cell titration coupled with agonist and/or antagonist titrations is still required, the optimal conditions for the chemistry involved in the detection are generally provided by the assay reagents. In addition, time and temperature of incubations should be optimized for each receptor and cell line. A sample protocol of a 1536-well plate assay in which the detection of the nuclear translocation of GR is measured is listed below:

- Dispense 6000 cells per well
- Dispense 30 nL of compounds
- Incubate the plate for 3 hrs at 37°C, 5% CO_2
- Dispense 3 µL of detection reagent
- Incubate plate for 1 hr at room temperature
- Read plate (90-s exposure, high gain, 1 × binning)

3.2.5 DETECTION OF MESSAGE

Because NHRs are generally intimately linked with transcription, the quantification of message is an important way in which the activity of an NHR can be measured. A wide spectrum of assays has been designed to detect changes in message. Some of those techniques such as gene chip microarray studies allow the evaluation of a wide array of transcripts (even genome-wide) on a single sample at a time. Other technologies, such as qPCR, allow the study of one or a few messages at a time, but are amenable to studying many different compounds in a single plate (commonly 384-well plates are used and 1536-well PCR instruments are in development).

A number of technologies appear to occupy the "middle ground" such as a quantitative nuclease protection assay that can profile 4 to 16 messages per sample in a 384- or 96-well plate, respectively. Bead-based technologies allow 1 to 100 samples to be profiled in a given well of a 96- or 384-well plate. Taken together, these options provide the flexibility to run high-throughput screening campaigns or follow-up assays to screening campaigns using the quantification of message as a primary readout. Because the expression of endogenous genes is measured, these assays are considered more closely related to physiological conditions. With the continuous improvement of the technology, it is expected that we will see much more use of such assays in the future.

3.2.6 FOLLOW-UP ASSAYS FOR HTS CAMPAIGNS

Many of the assay technologies used for high-throughput screening utilize engineered cells. For example, in reporter gene assays, the cells must be transfected with the reporter gene construct. In nuclear translocation assays, the cells must express a tagged NHR and perhaps also a tag within the nucleus of the cell. While these assays offer the benefit of relatively large signal-to-basal windows, robotic-friendly protocols, and an easy-to-interpret counter screen by using a parental cell or a cell expressing a different NHR, they may exhibit a decreased relevance to the biology of interest. For that reason, the implementation of an appropriate follow-up assay in a physiologically relevant system may prove very useful.

We have run a screening campaign in which the activation of GR was measured by its translocation to the nucleus using the DiscoveRx PathHunter system. To complement the results of that assay and evaluate the function of GR as a result of compound incubation, we ran a cytokine secretion assay in which putative GR agonists were incubated with primary human monocytes prior to challenge with lipopolysaccharide (LPS) (Patel et al., 2009). Treatment with a GR agonist such as dexamethasone will prevent the LPS-evoked secretion of inflammatory cytokines such as TNFα.

The selection of an ultra high-throughput screening assay such as the PathHunter nuclear translocation assay completed in 3456-well plates allows rapid screening of large libraries of compounds. The further profiling of those hits in a primary cell assay such as the LPS-induced cytokine secretion assay in human monocytes provides functional data regarding compound performance that should better predict success in animal models and humans. Because the primary cells lack engineering, there may be less chance for off-target or assay-related artifacts. For example, the assay presumes that the target (GR in this case) is expressed at the appropriate concentration and location to mimic the patient population. Furthermore, the presence of necessary cofactors endogenously expressed in the primary cells should have a better chance to replicate the biological system of interest than an over-expression system.

A sample protocol for the 384-well plate cytokine secretion assay is listed below:

- Dispense 30,000 primary human monocytes (45 µL) per well of a 384-well plate
- Dispense 2.5 µL of compounds
- Incubate for 1 hr at 37°C, 5% CO_2
- Dispense 2.5 µL of LPS to a final concentration of 200 ng/mL
- Incubate overnight (approximately 18 hr)
- Centrifuge the plates to pellet cells and remove supernatant
- Freeze supernatant at –20°C until assay
- Detect amount of cytokine in supernatant

For the MesoScaleDiscovery multispot human cytokine assay:

- Dispense 10 µL of cell supernatant per well of a 384-well plate
- Seal and incubate plate at room temperature for 2 hr
- Dispense 10 µL of detection antibody solution into each well of plate
- Seal and incubate plate at room temperature for 2 hr
- Wash 3 times with PBS containing 0.05% Tween-20
- Aspirate wash solution and replace with 35 µL MSD Read buffer
- Analyze plate in MSD Sector Imager 6000

3.3   SUMMARY AND CONCLUSIONS

Over the past decade, nuclear receptors have become a significant target class in the world of drug discovery. Compounds that target different NRs are used in a wide array of therapeutic indications, including diabetes, osteoporosis, breast and prostate cancer, arthritis, asthma, hypertension, and hyperlipidemia. While some of the early assays designed to find agonists and antagonists of NRs were limited in throughput and disease relevance, scientists now have many choices at their disposal for screening large collections of small molecules to find appropriate drug candidates.

The choice of assay type in a drug discovery campaign is complicated. Are you seeking an agonist, an antagonist, or a selective modulator? What tools (cell lines, recombinant proteins, small molecule probes, etc.) are at your disposal? How many compounds do you want to profile? What are your cost and time constraints? Each of these issues can modify your determination of the optimal assay for your particular project.

The concept of selective modulation of nuclear receptors has expanded the spectrum of activity that one can target for NR drug development. Since many of the NRs display a broad expression and regulate the expression of a large number of genes, the focus of NR drug development in recent years was to eliminate undesirable side effects while maintaining the intended therapeutic efficacy. NRs constitute one of the most important families for drug development. With the successful launch of several selective modulators of NRs and the demonstration of selectivity of several more in preclinical settings, it is hoped that many more drugs targeting nuclear receptors with better selectivity will reach the clinic in the coming years. However, the mechanism by which the current selective modulators result in tissue selectivity remains unclear. Additionally, one assay paradigm successfully used for one given NR may not be suitable for another NR. A better understanding of unique regulation and associated proteins is an important step in attempting to develop therapies for a given NR. Furthermore, other technology platforms such as genetic methods to dissect NR functions, proteolysis assays, and new crystal structure methods attempting to detect ligand-induced structural changes continue to be improved. These tools, although outside the scope of this chapter, will likely play an increasing role in NR drug discovery in the future.

REFERENCES

Bai, C. and Elledge, S.J. 1996. Gene identification using the yeast two-hybrid system. *Meth. Enzymol.* 273, 331–347.

Bai, C., Flores, O., and Schmidt, A. 2007. Opportunities for development of novel therapies targeting steroid hormone receptors. *Expert Opin. Drug Discov.* 2, 725–737.

Chen, T. 2008. Nuclear receptor drug discovery. *Curr. Opin. Chem. Biol.* 12, 418–426.

Gaddam, K.K., Pratt-Ubunana, M.N. and Calhoun, D.A. 2006. Aldosterone antagonists: effective add-on therapy for the treatment of resistant hypertension. *Expert Rev. Cardiovasc. Ther.* 4, 353–359.

Gillatt, D. 2006. Anti-androgen treatments in locally advanced prostate cancer: are they all the same? *J. Cancer Res. Clin. Oncol.* 132, S17–S26.

Greene, DA. 1999. Rosiglitazone: a new therapy for type 2 diabetes. *Expert Opin. Invest. Drugs* 8, 1709–1719.

Jeyakumar, M. and Katzenellenbogen, J.A. 2009. A dual-acceptor time-resolved Förster resonance energy transfer assay for simultaneous determination of thyroid hormone regulation of corepressor and coactivator binding to the thyroid hormone receptor. *Anal. Biochem.* 386, 73–78.

Lavery, D.N. 2009. Binding affinity and kinetic analysis of nuclear receptor/co-regulator interactions using surface plasmon resonance. *Meth. Mol. Biol.* 505, 171–186.

Lehmann, J.M. et al. 1995. An antidiabetic thiazolidinedione is a high affinity ligand for peroxisome proliferator-activated receptor gamma (PPARγ). *J. Biol. Chem.* 270, 12953–12956.

Luo, Y. et al. 1997. Mammalian two-hybrid system: a complementary approach to the yeast two-hybrid system. *Biotechniques* 22, 350–352.

Madoux, F. et al. 2008. Potent, selective and cell penetrant inhibitors of SF-1 by functional ultra-high-throughput screening. *Mol. Pharm.* 73, 1776–1784.

Mangelsdorf, D.J. et al. 1995. The nuclear receptor family: the second decade. *Cell* 83, 835–839.

Moore, J.T., Collins, J.L., and Pearce, K.H. 2006. The nuclear receptor superfamily and drug discovery. *Chem. Med. Chem.* 1, 504–523.

O'Malley, B.W. and McKenna, N.J. 2008. Coactivators and corepressors: what's in a name? *Mol. Endocrinol.* 22, 2213–2214.

Patel, A. et al. 2009. A combination of ultra high-throughput PathHunter and cytokine secretion assays to identify glucocorticoid receptor agonists. *Anal. Biochem.* 385, 286–302.

Smith, R.G. and Sestili, J. 1980. Methods for ligand-receptor assays in clinical chemistry. *Clin. Chem.* 26, 543–550.

Tobias, J.S. 2004. Recent advances in endocrine therapy for postmenopausal women with early breast cancer: implications for treatment and prevention. *Ann. Oncol.* 15, 1738–1747.

Wilkinson, J.M. et al. 2008. Compound profiling using a panel of steroid hormone receptor cell-based assays. *J. Biomol. Screen.* 13, 755–765.

Zhang, Y., Guo, F., and Jiang, C. 2008. An efficient and economic high-throughput cell screening model targeting the glucocorticoid receptor. *J. Drug Targeting* 16, 58–64.

4 Assay Development for G Protein-Coupled Receptors and Ion Channels

Joseph G. McGivern, Yi-xin Qian, and Paul H. Lee

CONTENTS

4.1 INTRODUCTION

G protein-coupled receptors (GPCRs) and ion channels are expressed in the plasma membranes of virtually all cells. They usually serve as transmembrane signal transducers and/or transport proteins and are key regulators of cellular function. Following their activation, a cascade of downstream signaling events is triggered and this can lead to immediate and delayed effects on cell physiology and gene expression. The importance of the GPCR and ion channel families in the practice of medicine is reflected in the myriad clinical drugs that target these proteins. GPCR and ion channel modulator drugs are used to treat major illnesses, including cardiovascular, metabolic, neurologic, and infectious diseases as well as some forms of cancer.

Altogether, GPCRs and ion channels constitute more than 40% of all marketed drugs (Overington, Al-Lazikani, and Hopkins 2006) and, by our estimates, generate in excess of $65 billion in annual sales worldwide (Lundstrom 2006; McGivern and Worley 2007). However, these target classes remain considerably underexploited. Existing drugs act on only a subset of the known members of each family and are associated with side effects that may stem from their suboptimal selectivity profiles. Therefore, significant opportunity exists to discover and develop new and improved therapeutics that target members of these families. To deliver on this promise, the pharmaceutical industry has invested heavily in understanding better the structure–function relationships of membrane proteins and in new assay technologies that will aid the discovery of novel drugs that modulate the functions of these cell-surface proteins.

Modern drug discovery for GPCRs and ion channels favors random high-throughput screening (HTS) methods to identify compounds that not only interact with these proteins but that also modulate their function. Ligand binding assays are the best ways to characterize the physical interaction between a drug and a target protein, and are used most commonly to identify compounds that interact with specific binding pockets on ligand-activated receptors and ion channels. Often, the specific binding pocket is the orthosteric site where the endogenous ligand binds and it is not difficult to imagine that compounds that can somehow obstruct access of this ligand to its site might also exert agonist or antagonist effects in a properly configured functional assay. Nevertheless, ligand binding assays are not useful for determining the effect of a drug on protein function and in the case of voltage-gated channels, there is no physical ligand and so their relevance is even less clear. In addition, the value of ligand binding assays is limited because receptors and channels are large proteins that contain many possible drug binding pockets. Therefore, as ligand binding assays query the pharmacology of only a single site, functionally active compounds may actually appear to be inactive in a binding assay. Consequently, cell-based assays have emerged as preferred approaches to GPCR and ion channel drug discovery.

Functional assays allow drug effects on protein function to be measured and offer several advantages, including (1) the ability to differentiate compounds that behave as antagonists, agonists, and inverse agonists, (2) the ability to identify allosteric modulators of receptor and ion channel function, and (3) the feasibility to implement a screen for novel or surrogate ligands of orphan receptors. Some assay formats can be applied to the study of both GPCRs and channels, but many of today's cutting-edge methods are unique to each class. Therefore, we have divided this chapter into two main sections covering assay development for (1) G protein-coupled receptors and (2) ion channels.

4.2   ASSAY DEVELOPMENT FOR G PROTEIN-COUPLED RECEPTORS

4.2.1   BIOLOGY OF G PROTEIN-COUPLED RECEPTORS

GPCRs are characterized by seven transmembrane domains oriented with an extracellular N terminus and an intracellular C terminus. They are activated by a diverse array of extracellular substances including biogenic amines, neuropeptides, hormones, chemokines, odorants, amino acids, free fatty acids, photons, and metabolic intermediates (Pierce, Premont, and Lefkowitz 2002; He et al. 2004; Wise, Jupe, and Rees 2004). The process of GPCR activation is initiated when an agonist binds to its recognition site on the receptor. This binding event induces a conformational change in the receptor and the formation of an active agonist–receptor complex that interacts with heterotrimeric $G_{\alpha\beta\gamma}$ proteins.

Activation of the receptor promotes the exchange of guanosine diphosphate (GDP) for guanosine triphosphate (GTP) on the G_α subunit. The GTP-bound G_α subunit then dissociates from both the agonist–receptor complex and the $G_{\beta\gamma}$ dimer and both the G_α subunit and the $G_{\beta\gamma}$ dimer are then free to activate specific effector proteins such as adenylate cyclase, phospholipases, phosphodiesterases, and ion channels, leading to the initiation of downstream signaling processes. The G_α subunit is inactivated when GTP is hydrolyzed to GDP. The resulting GDP-bound G_α subunit can reassociate with the $G_{\beta\gamma}$ dimer and then the heterotrimeric $G_{\alpha\beta\gamma}$ complex is available again for subsequent rounds of receptor activation. GPCR-mediated signaling is ultimately terminated by mechanisms that involve protein–protein interaction-dependent receptor desensitization and internalization (Kostenis 2006).

The G_α proteins can be classified on the basis of similarities in their amino acid sequences and coupling to effector proteins. The four major categories are $G_{\alpha s}$, $G_{\alpha i}$, $G_{\alpha q}$, and $G_{\alpha 12}$ and these are responsible for activating different signaling pathways in cells. $G_{\alpha s}$ and $G_{\alpha i}$ stimulate and inhibit adenylate cyclase, respectively. Adenylate cyclase is an enzyme that catalyzes the formation of cyclic adenosine monophosphate (cAMP), an intracellular second messenger

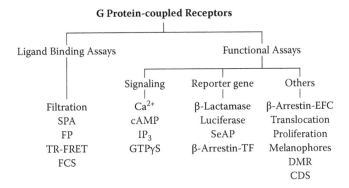

FIGURE 4.1 Assays commonly used in GPCR research. SPA = scintillation proximity assay; FP = fluorescence polarization; TR-FRET = time-resolved fluorescence resonance energy transfer; FCS = fluorescence correlation spectroscopy; SeAP = secreted alkaline phosphate; TF = transcription factor; EFC = enzyme fragment complementation; DMR = dynamic mass redistribution; CDS = cellular dielectric spectroscopy.

that activates protein kinase A (PKA). $G_{\alpha q}$ activates phospholipase C (PLC), which catalyzes hydrolysis of membrane phosphatidylinositol (PI), producing the intracellular second messengers inositol 1,4,5-triphosphate (IP_3) and diacylglycerol (DAG). In turn, IP_3 evokes Ca^{2+} release from the endoplasmic reticulum and DAG, in combination with Ca^{2+}, activates protein kinase C (PKC). The $G_{\alpha 12}$ family consists of $G_{\alpha 12}$ and $G_{\alpha 13}$. These regulate a Na^+-H^+ exchange protein (NHE-1) and RhoA (Cabrera-Vera et al. 2003). A further subclass of promiscuous G proteins ($G_{\alpha 15/16}$) has been identified and found to link a variety of receptors to the PLC-IP_3-Ca^{2+} signaling pathway (Offermanns and Simon 1995).

Figure 4.1 illustrates the commonly used assay formats in GPCR research. These assays can be divided into two major groups: non-functional/biochemical methods, represented here by the various forms of ligand binding assays, and functional/cell-based methods. Cell-based assays can be subdivided based on whether they involve measuring the levels of intracellular second messengers, the expression of reporter genes, or protein–protein interactions.

In the remainder of this section, we will provide a brief overview of non-functional/biochemical and functional/cell-based assay formats that are commonly used in GPCR research. We will also describe in detail the practical aspects of developing a non-functional/biochemical (ligand binding) assay for GPR23 and a functional/cell-based (reporter gene) assay for the calcitonin receptor (CTR). The reader is directed to several recent reviews for discussion of other GPCR assay formats, including their pros and cons (Thomsen, Frazer, and Unett 2005; Williams and Sewing 2005; Eglen, Bosse, and Reisine 2007; Xiao et al. 2008).

4.2.2 Non-Functional and Biochemical Assays for GPCRs

For many years, ligand binding assays have been among the most popular ways to study the interactions of compounds with GPCRs and ion channels (Noel, Mendonca-Silva, and Quintas 2001). They can be run in saturation, displacement, or kinetic modes; displacement is the most commonly used mode in HTS. Displacement assays query the ability of a test compound to displace a traceable ligand from a large unlabeled protein target, usually found in a membrane homogenate prepared from tissues or cells expressing the receptor or channel of interest. A high affinity ligand that recognizes the receptor is an absolute requirement and luckily many such ligands are available commercially, including agonist and antagonist molecules.

Many ligand binding assays employ traceable ligands that incorporate a radioactive isotope of hydrogen (^{3}H), carbon (^{14}C), or iodine (^{125}I); a benefit of using ^{125}I is the very high specific activity (2000 Ci/mmol) that can be achieved, which greatly increases the sensitivity of the assay and enables membranes that have lower levels of receptor expression to be used. It is necessary to distinguish the bound fraction of traceable ligand from its free fraction. In radioactive ligand binding assays, it is usually straightforward to calculate the fraction of traceable ligand bound to the receptor by scintillation counting following separation of bound from free by filtration or by scintillation proximity methods that do not require a separation step.

Scintillation proximity assays (SPAs) represent a particularly powerful way to perform ligand binding experiments and are used widely in HTS and later stages of drug discovery due to their ease of automation (Carpenter et al. 2002; Glickman, Schmid, and Ferrand 2008). In receptor binding SPA, membranes prepared from cells expressing the target of interest are coupled to wheat germ agglutinin (WGA)-coated microscopic beads that contain scintillant material that emits light upon excitation by β-particles that are released when a radioactive isotope decays in close proximity after binding to the target. SPAs are homogeneous and rely on the ability of water to dissipate the energy from radiolabeled ligand that is distant (>1.5 µm) from the bead surface. In contrast, β-particles emitted by radioligand bound to the immobilized receptor are close enough to excite the scintillant material. The emitted light can be captured by a detection instrument equipped either with photomultiplier tubes such as TopCount® or MicroBeta® TriLux (PerkinElmer Life and Analytical Sciences, Inc.) or with a sensitive charge-coupled device (CCD) camera such as LEADseeker™ (GE Healthcare Life Sciences) or ViewLux™ (PerkinElmer Life and Analytical Sciences, Inc.).

SPA offers the advantage of not requiring the separation steps that can promote ligand dissociation from the receptor in traditional filtration-based assays. Consequently, SPA permits lower affinity ligands to be used and the binding equilibrium to be maintained throughout the whole experiment. The use of radioactive material is associated with significant challenges, hazards, and costs and so it is preferable to use alternative methods, where possible. For instance, ligands can be labeled with fluorophores to enable non-radiometric binding assays that rely on detection methods such as fluorescence polarization (FP), time-resolved fluorescence resonance energy transfer (TR-FRET), and fluorescence correlation spectroscopy (FCS) (Inglese et al. 1998; Lee and Bevis 2000; Rudiger et al. 2001). However, most fluorophores are bulky and not all ligands can tolerate their addition without a negative impact on the pharmacology. This problem can preclude the use of fluorometric binding assays unless the fluorescent ligand can be demonstrated to retain sufficiently high binding affinity and functional activity.

4.2.3 Development of Scintillation Proximity Assay for GPR23

GPR23 (P_2Y_9) is a novel receptor that binds lysophosphatidic acid (LPA), although its physiological importance is not fully understood (Noguchi, Ishii, and Shimizu 2003). In order to develop a GPR23 binding assay, we prepared membranes from a Chinese hamster ovary (CHO) cell line that expressed human GPR23 under the control of a tetracycline-inducible expression system. Under normal conditions, these cells expressed GPR23 at a very low level but could be induced by tetracycline (or a derivative) to express the receptor at a high level.

Expression of GPR23 was induced by treating the cells overnight with doxycycline (1 µg/mL). The next day, the cells were harvested and centrifuged at 1900 g for 10 min at 4°C to produce a cell pellet, which was washed with ice-cold phosphate-buffered saline (PBS) and suspended in a homogenization buffer containing 25 mM HEPES (pH 7.4), 10% sucrose, and EDTA-free Complete™ protease inhibitor (Roche Diagnostics). The cells were left on ice for 30 min before being homogenized using a glass homogenizer. The cell lysate was centrifuged at 1300 g for 10 min at 4°C. The supernatant was collected and then centrifuged at 142,000 g for 1 hr at 4°C to produce a crude membrane particulate fraction. This fraction was suspended in a storage buffer containing 25 mM HEPES (pH 7.4), 10 mM MgCl$_2$, and EDTA-free Complete™ protease inhibitor and was then homogenized. The protein content of the membrane preparation was assessed using the DC protein assay (Bio-Rad Laboratories) and subsequently the membranes were divided into aliquots containing >1 mg/mL of protein.

The GPR23 binding SPA was developed in a 384-well format with luminescence measured on a TopCount system. Reagents were added into white flat-bottomed polystyrene plates (Corning, #3652) in the following order: 10 μL unlabeled compound in assay buffer, 20 μL premixed beads and cell membranes, and 10 μL [^{3}H]LPA (Figure 4.2A). The composition of the assay buffer was similar to that of the storage buffer, but also included 0.2% bovine serum albumin (BSA). Several different

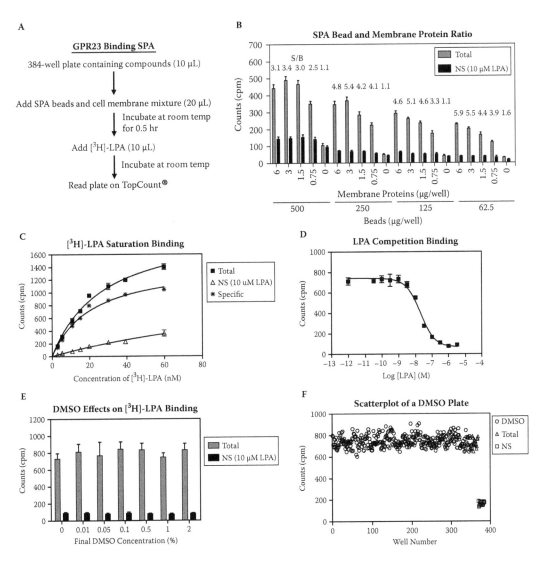

FIGURE 4.2 Development of SPA-based ligand binding for GPR23. (A) Protocol for GPR23 binding assay. (B) Titration of beads and GPR23 membrane protein to determine optimal mixture ratio. (C) Saturation binding of [^{3}H]LPA to GPR23. (D) Competitive displacement of [^{3}H]LPA by unlabeled LPA. (E) Determination of DMSO tolerance of GPR23 assay. (F) Scatter plot of signals obtained in each well of 384-well plate. All wells in columns 1 through 23 contained 1.25% DMSO and the 16 wells in column 24 contained DMSO and 10 μM LPA before addition of mixed beads and membrane.

types of SPA imaging beads are available including YSi-WGA, polyvinyltoluene-WGA (PVT-WGA) and PVT-WGA treated with polyethyleneimine (PEI) that can help to reduce non-specific binding. Due to their lower density, PVT-WGA beads settle more slowly in solution than YSi-WGA beads and are preferred for automated assays. When we compared the PVT-WGA and PVT-WGA-PEI beads, we observed no difference in the total or non-specific binding and so PVT-WGA beads were used in subsequent work.

To optimize the signal-to-background (S/B) ratio of our assay, we varied systematically the concentrations of membrane proteins and beads (Figure 4.2B). We found that with 20 nM [^{3}H]LPA, total [^{3}H]LPA binding generally increased directly with increasing amounts of both membrane proteins and beads. However, the non-specific binding (in the presence of 10 μM unlabeled LPA) also increased and so the S/B ratio was reduced. Based on these results, we determined that the optimal condition for our GPR23 binding SPA was 20 μL of buffer containing 250 μg of PVT-WGA beads premixed with 2 μg of GPR23 membranes.

We used the GPR23 SPA to determine the equilibrium binding affinity of [^{3}H]LPA for GPR23 receptors. Using the conditions described above, various concentrations of [^{3}H]LPA were added to each well of the plate and incubated at room temperature for 2 hr in the dark prior to reading. Figure 4.2C shows the saturation binding curve for [^{3}H]LPA and GPR23. The specific [^{3}H]LPA binding was calculated by subtracting the non-specific signal in the presence of 10 μM unlabeled LPA from the total signal. Under these assay conditions, the specific binding was concentration-dependent and reached a plateau, consistent with a receptor-mediated mechanism.

We calculated the dissociation constant (K_d) of [^{3}H]LPA to be 21 nM and the receptor density (B_{max}) to be 18 pmol/mg protein, indicating a high level of GPR23 expression. Importantly, membranes prepared from uninduced cells displayed no specific [^{3}H]LPA binding, suggesting that the signal we observed with membranes prepared from induced cells was GPR23-dependent. In order to establish an optimized displacement binding assay, it is generally recommended to use the labeled ligand at a concentration close to its K_d value. If the concentration used is much less than K_d, then the ligand may be depleted during the experiment, leading to a rightward shift of the displacement curve and a resulting error in the IC_{50} value. On the other hand, if the concentration used is much greater than K_d, then not only would the non-specific background signal increase, but there would also be an increase in the amount of radioisotope used and the amount of radioactive waste generated.

One advantage of SPA is that the signal can be measured repeatedly after adding the reagents, allowing the kinetics and stability of the binding reaction to be determined. We found that the specific [^{3}H]LPA binding increased with time, reaching a maximum value 2 hr after radioligand addition. Thereafter, the signal remained stable for another 3 hr before gradually declining (data not shown). Knowledge of receptor binding kinetics and signal stability will facilitate efficient scheduling of a primary screening campaign, during which it is preferable to determine the signal at equilibrium. When the signal remains stable for a long time after reaching equilibrium, larger batches of test plates can be processed via automation, whereas when the signal is short-lived, it is possible to process only smaller batches of plates.

Having established that our assay protocol permitted measurement of the specific binding of [^{3}H]LPA to GPR23, we examined next the ability of unlabeled LPA to displace [^{3}H]LPA from the receptor. Different concentrations of unlabeled LPA (1 pM to 3 μM) were added to each well, followed by 20 μL of premixed beads and membrane. After incubation at room temperature for 30 min, 10 μL of [^{3}H]LPA (final concentration 20 nM) were added and all reagents incubated at room temperature for 2 hr in the dark prior to reading. We found that unlabeled LPA displaced [^{3}H]LPA in a concentration-dependent manner with a calculated inhibitory constant (K_i) of 14 nM (Figure 4.2D).

During pharmacology and HTS experiments, small molecules are typically dissolved in 100% dimethylsulfoxide (DMSO) and so we determined the effect of this substance on assay performance. We found that total binding and non-specific binding were not significantly affected by DMSO up to a final concentration of 2% (Figure 4.2E). This is advantageous because the DMSO tolerance of a GPR23 cell-based assay would likely be much lower than 2% and so small molecules could be studied in the GPR23 binding SPA with less dilution of the original DMSO-based stock solution i.e., at higher test concentrations.

Figure 4.2F shows a scatter plot of the individual signals in the 384 wells of a representative plate in a DMSO trial run (final concentration is 1.25%) that was used to simulate assay performance during a primary screen. The wells in column 23 contained DMSO only, and these represented our high control wells. The total binding signal in these 16 wells was 730 ± 41 counts per minute [c.p.m., mean $\pm$ standard deviation (S.D.) of mean]. The wells in column 24 contained 10 µM unlabeled LPA in 1.25% DMSO, and these represented our low control wells. We estimated the non-specific binding signal in these 16 wells to be 163 ± 16 c.p.m. (mean $\pm$ S.D.). The wells in columns 1 through 22 contained DMSO only and these can be considered to simulate the behavior of the test wells of a primary screening plate containing inactive compounds. Not surprisingly, the signal in these 352 wells was of similar magnitude to the total signal in the high control column. We calculated that our assay had a S/B ratio of 4.5 and a Z′ factor of 0.7, indicating that this GPR23 SPA would be acceptable for use in primary screening (Zhang, Chung, and Oldenburg 1999).

4.2.4 FUNCTIONAL CELL-BASED ASSAYS FOR GPCRs

GPCR activation initiates multistep signaling processes. A variety of novel cell-based assay technologies have emerged in the last decade or so and it is now possible to measure both proximal and distal events that result from GPCR activation, including G protein activation, accumulation or depletion of intracellular second messengers, gene transcription, and protein–protein interaction-dependent receptor desensitization and internalization.

The GTPγS binding assay provides a relatively simple method for measuring a functional consequence of agonist binding to a GPCR. This assay involves measuring the binding of the radioactive non-hydrolyzable GTP analogue, [^{35}S]GTPγS, as a way to determine the extent of G protein activation (Harrison and Traynor 2003). This assay provides a readout that is proximal to the initial receptor activation event and offers the advantage that the measured event is independent of the G protein's coupling preference. Following G protein activation, the dissociated G_α subunit and $G_{\beta\gamma}$ dimers independently modulate the activity of specific effector molecules, which in turn control the levels of intracellular second messengers such as IP_3, cAMP, and Ca^{2+}.

The concentrations of cellular IP_3 and cAMP can be measured using methods that employ inositol phosphate or cAMP sequestering antibodies, respectively (Williams 2004; Trinquet et al. 2006). These methods rely on competitive binding between the cell-derived messengers and exogenous labeled versions. Intracellular Ca^{2+} concentration can be monitored using Ca^{2+}-sensitive fluorescent dyes such as Fluo-3, Fluo-4, and Fura-2 (Kassack et al. 2002) or Ca^{2+}-sensitive photoproteins such as aequorin (Dupriez et al. 2002). Further downstream, GPCR modulation of gene transcription can be monitored using reporter gene assays in which expression of enzymes such as β-lactamase, luciferase, or secreted alkaline phosphatase has been placed under the control of appropriate transcription factors such as nuclear factor of activated T cells (NFAT) or cAMP response element binding protein (CREBP) (George, Bungay, and Naylor 1997; Zlokarnik et al. 1998; Kunapuli et al. 2003; Abdel-Razaq, Bates, and Kendall 2007). (NFAT represents a family of ubiquitous transcription factors activated by the phosphatase calcineurin that in turn is activated by the Ca^{2+}-sensing protein, calmodulin.)

Of particular relevance to the current section, β-lactamase is a bacterial enzyme that normally cleaves β-lactam antibiotics such as the penicillins and cephalosporins. CCF4 is a FRET-based fluorescent substrate of β-lactamase that contains two fluorophores, coumarin and fluorescein, that are linked by a cephalosporin backbone. When CCF4 is intact, the coumarin and fluorescein are in close proximity and excitation of coumarin by light at wavelength 409 nm results in efficient transfer of energy to fluorescein, leading to emission of green fluorescence (peak at 520 nm). When the expression of β-lactamase increases, CCF4 is proportionally cleaved and this disrupts the transfer of energy such that excitation at 409 nm results in emission of blue fluorescence (peak at 447 nm). The level of β-lactamase expression that results from GPCR activation can thus be quantified from the ratio of blue to green fluorescence using an emission filter set of 460 nm and 530 nm on a fluorescent plate reader, such as the Analyst-GT® (MDS Analytical Technologies).

Compared to other assay formats, the β-lactamase reporter gene assay presents several advantages. It is non-invasive and stable, and so the signal from CCF4 can be read multiple times in live cells in order to track expression of β-lactamase, which aids in determining the optimal timing of the various steps in the assay protocol. Due to signal amplification, the β-lactamase reporter gene assay is often more sensitive to receptor activation than assays that measure the concentrations of second messengers. Consequently, this assay format can enable identification of compounds with relatively poor efficacy and/or potency (George, Bungay, and Naylor 1997; Kunapuli et al. 2003). The ratiometric readout permits detection and elimination of cytotoxic compounds without the need for an additional counter-screening assay and serves as an in-well control to minimize experimental noise due to well-to-well variation in cell density. Finally, the β-lactamase reporter gene assays can be configured independently of a receptor's G protein coupling preference. For instance, the promiscuous G protein $G_{\alpha15/16}$ couples non-$G_{\alpha q}$-coupled receptors to the PLCβ-IP$_3$-Ca^{2+}-NFAT signaling pathway (Offermanns and Simon 1995; Coward et al. 1999), which enables development of functional assays for endogenous or surrogate ligands of orphan GPCRs (Bresnick et al. 2003).

A number of drawbacks must also be considered when using β-lactamase reporter gene assays. These assays rely on fluorescence signals generated downstream of the receptor and so there is an increased probability that false positives might be identified during a primary screen. This would include compounds that modulate β-lactamase expression by interfering with the signaling pathway invoked by receptor activation. Consequently, additional control experiments are required to determine the receptor specificities of the hits identified in a primary screen. This may involve testing the hits in counterscreening assays that measure functional responses in the parental β-lactamase cells and/or in a clonal cell line that expresses a different GPCR coupled to the same signaling pathway. Alternatively, the hits can be tested in receptor binding or second messenger assays to confirm on-target activities.

Functional protein–protein interaction assays can also be used to study receptor activation. Following GPCR activation, the protein β-arrestin is recruited to interact with an intracellular domain of the receptor. This protein–protein interaction leads to receptor desensitization and internalization, which can be studied by monitoring the subcellular localization of the β-arrestin–GPCR complex. This is often accomplished by using a microscope-based imaging system to determine the subcellular localization of a fusion protein consisting of a fluorescent protein linked to β-arrestin or to the GPCR in fixed cells (Heding 2004).

Alternative non-imaging-based, live cell assays have been developed to permit investigation of the interaction of β-arrestin and an activated GPCR. One such assay technology relies on enzyme fragment complementation (EFC). The C terminus of the GPCR is tagged with an enzyme donor peptide fragment of β-galactosidase (β-gal) whereas the β-arrestin protein is tagged with an enzyme-acceptor peptide fragment of β-gal. Following receptor activation, the two proteins associate, which leads to the formation of active β-gal and substrate hydrolysis (von Degenfeld et al. 2007). A second such assay utilizes a non-mammalian protease and a transcription factor (TF) to induce expression of a reporter gene. The TF is tethered to the C terminus of the GPCR by a linker that contains a specific cleavage site for the protease. The protease is fused to β-arrestin and upon GPCR activation, the recruitment of β-arrestin leads to cleavage of the linker by the protease and release of the TF. The TF can then translocate to the nucleus of the cell and activate the reporter gene (Barnea et al. 2008). These novel protein–protein interaction assay technologies are ideal for identifying surrogate ligands of orphan GPCRs because the signals are generated specifically by the target receptor.

4.2.5 DEVELOPMENT OF REPORTER GENE ASSAY FOR CALCITONIN RECEPTOR

The calcitonin receptor (CTR) is a GPCR that binds the peptide calcitonin (CT) and protects the skeleton by inhibition of bone resorption by osteoclasts (Sexton, Findlay, and Martin 1999). In order to develop a stable cell line expressing human CTR, the Freestyle 293F cell line (Invitrogen) was chosen because it tested negative for expression of receptor activity-modifying proteins (RAMPs) 1 through 3 in a bDNA analysis experiment. RAMPs are a family of integral membrane proteins that associate with CTR to generate high affinity amylin receptors, AMY_1 and AMY_2, that display different ligand affinity and specificity compared to CTR (Hay, Poyner, and Sexton 2006). The 293F cells were transfected with a plasmid containing the CTR gene and a CMV promoter. As CTR is coupled to $G_{\alpha s}$ and its activation increases the intracellular level of cAMP, the cells were also transfected with a plasmid containing the β-lactamase reporter gene under control of a CRE in its promoter region. Thus, activation of CTR by human CT (hCT) is expected to increase the level of cAMP in the CTR-CRE-bla cells and this will stimulate expression of the β-lactamase enzyme.

Following transfection, the CTR-CRE-bla cells were grown in the presence of appropriate antibiotics for 2 weeks and sorted using a flow cytometer. In these preliminary experiments, the pool of cells was divided into two groups, only one of which was stimulated with 1 nM hCT. After 4 hr, both groups of cells were incubated in the presence of CCF4 (1 µM) for 2 hr. When excited by light at 409 nm on a flow cytometer, cells that expressed β-lactamase emitted a blue fluorescent signal. A pool of responders was collected along with four 96-well plates of single cell clones.

We found that 25% of cells in the hCT-stimulated group expressed β-lactamase, compared to <1% of cells in the unstimulated group. The single cell clones were grown in the 96-well plates until there were sufficient cells for continued maintenance in culture and functional testing in a concentration–response experiment using hCT. We were most interested in the clones that displayed both the highest blue to green fluorescence ratio as well as EC_{50} values for hCT that were in close agreement with literature values. These clones were subjected to further analysis using additional CTR agonists such as amylin to identify the best clone, which we considered to be the one displaying the correct rank order of ligand pharmacology and the highest S/B ratio.

Having selected the best CTR-CRE-bla clone, we determined its performance at different cell densities using the protocol described in Figure 4.3A. Cells were seeded at 2,500, 5,000, 10,000 or 20,000 per well in poly-D-lysine coated 384-well black clear-bottomed plates (BD Biosciences, BioCoat™, #354663) in assay medium [Dulbecco's modified Eagle minimal essential medium (D-MEM) with 10% dialyzed fetal bovine serum (FBS), 25 mM HEPES (pH 7.3), 0.1 mM non-essential amino acids, penicillin 100 units/mL, and streptomycin 100 µg/mL]. The plates were kept in an incubator at 37°C with 5% CO_2 in air for 16 to 20 hr and subsequently various concentrations of hCT (10 µL) were added to each well and the cells incubated for a further 4 hr. Subsequently, 10 µL of CCF4 (1 µM) were added to each well and then the plate was kept in the dark at room temperature for 1.5 hr prior to reading on an Analyst-GT. Figure 4.3B shows the hCT-induced β-lactamase activity in the best CTR-CRE-bla clone; response ratios were calculated according to Equation 4.1:

$$\text{Response ratio} = \frac{\text{signal 460 nm}_{stim} / \text{signal 530 nm}_{stim}}{\text{signal 460 nm}_{unstim} / \text{signal 530 nm}_{unstim}} \qquad (4.1)$$

We found no significant difference in the response ratio between cells plated at 2,500, 5,000 or 10,000 cells per well. However, when the cells were plated at 20,000 cells per well, we observed a lower response ratio due to a higher unstimulated background. We chose a cell density of 10,000 cells for all subsequent assay optimization steps because it revealed the least well-to-well variation compared to the other cell densities.

To determine the optimal CCF4 loading time, the cells were stimulated with various concentrations of hCT for 4 hr. Subsequently, the signal from each well on the plate was read 1, 1.5, and 2 hr

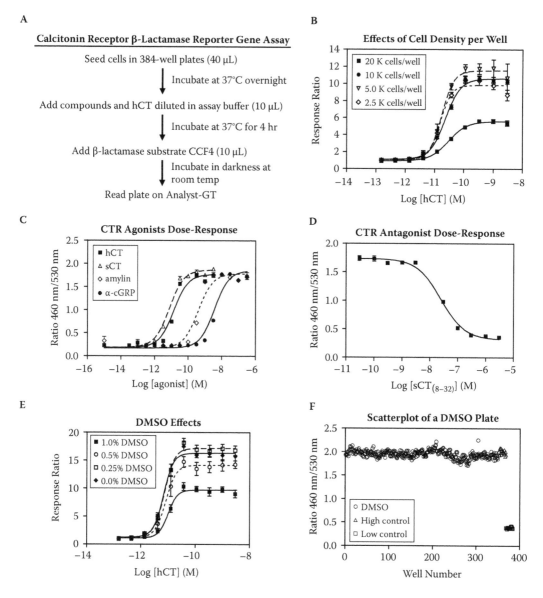

FIGURE 4.3 Development of β-lactamase reporter gene assay for CTR. (A) Protocol for CTR β-lactamase reporter gene assay. (B) Concentration–response curves for hCT in the presence of different numbers of CTR-CRE-bla cells per well. (C) Concentration–response curves for four CTR agonists in the β-lactamase assay; rank order of potency was sCT > hCT > amylin > α-CGRP. (D) Concentration–response curve of CTR antagonist $sCT_{(8-32)}$ on hCT-induced β-lactamase activity. (E) Determination of DMSO tolerance of CTR assay. (F) Scatter plot of signals obtained in each well of 384-well plate. All wells in columns 1 through 23 contained 100 pM hCT and 0.4% DMSO; the 16 wells in column 24 contained DMSO only.

after the addition of the substrate. The EC_{50} values of hCT were similar at all three time points but a loading time of 2 hr yielded the largest response ratio, and so we chose a 2-hr loading time for all subsequent assay optimization steps.

Having established the optimal conditions for the β-lactamase assay, we determined the rank order of the potencies of selected peptide agonists of CTR. All peptides were dissolved in water before dilution in assay buffer. The cells were stimulated with various concentrations of hCT, salmon

CT (sCT), amylin, and α-calcitonin gene-related peptide (α-CGRP) for 4 hr at 37°C (Figure 4.3C). All agonists displayed similar maximum blue to green fluorescence ratios of ~1.7 and assay windows of ~8. All agonists produced complete concentration–response curves and we calculated the EC_{50} values of sCT, hCT, amylin, and α-CGRP to be 7, 14, 400, and 3700 pM, respectively, all of which are similar to literature values (Hilton et al. 2000).

The effect of the CTR antagonist peptide, $sCT_{(8-32)}$, was tested in the presence of 100 pM hCT to determine both the receptor specificity of the β-lactamase-dependent signal and the assay performance in the antagonist mode. The antagonist peptide was added along with hCT in the first addition step of the protocol, and they were incubated together for 4 hr at 37°C. We found that $sCT_{(8-32)}$ inhibited hCT-induced β-lactamase activity in a concentration-dependent manner with an IC_{50} of 25 nM (Figure 4.3D). Overall, our data demonstrate that the CTR-CRE-bla cells respond to known CTR modulators with expected potencies.

Cell-based assays are generally less DMSO-tolerant than biochemical assays. Due to the long compound incubation time of the β-lactamase assay, it is particularly important to pay attention to the effect of this agent on assay performance. Therefore, the EC_{50} value of hCT was determined in the presence of 0%, 0.25%, 0.5%, and 1.0% DMSO. The response ratio for β-lactamase activity in the presence of different DMSO concentrations is shown in Figure 4.3E. We found that DMSO reduced the response ratio in a concentration-dependent manner. For instance, the maximal response ratio was reduced by approximately 20% and 50% in the presence of 0.5% and 1% DMSO, respectively. The EC_{50} value of hCT was similar in 0% and 0.5% DMSO (8 and 7 pM, respectively). Based on these results, we determined the maximum permissible concentration of DMSO to be 0.5% during pharmacology and HTS experiments.

Figure 4.3F shows a scatter plot of the individual signals in 384 wells of a representative plate in a DMSO trial run (final concentration 0.4%); as in the GPR23 assay, this run was used to simulate assay performance during a screening run. All wells in columns 1 through 23 received 10 µL of hCT (500 pM) and DMSO (2%) in assay buffer during the first addition step of the assay protocol. The wells in columns 1 through 22 were considered to simulate the wells of a screening plate containing inactive test compounds. The wells in column 23 represented the high control wells of our plate. The wells in column 24 received 10 µL of assay buffer containing DMSO (2%) and represented the low control wells of our plate.

The plate was incubated for 4 hr at 37°C and then CCF4 (10 µL of 1 µM) was added to each well and left for 2 hr prior to reading. We found that hCT stimulated a robust β-lactamase signal with very little variation among wells. The 16 high and 16 low control wells revealed blue to green fluorescence ratios of 2.0 ± 0.1 and 0.4 ± 0.02 (mean ± S.D.), respectively. The blue to green fluorescence ratio for the 352 wells in columns 1 through 22 was 2.0 ± 0.1 (mean ± S.D.), which is similar to the high control ratio. We calculated the Z′ value to be 0.9, which is indicative of an excellent assay (Zhang, Chung, and Oldenburg 1999). This β-lactamase assay could also be configured to identify agonists of CTR. In the agonist screening format, the wells in columns 1 through 22 would receive test compounds in the absence of hCT. Compounds that activate CTR and stimulate β-lactamase activity would be identified by higher blue to green fluorescence ratios compared to the low control wells.

4.3 ASSAY DEVELOPMENT FOR ION CHANNELS

4.3.1 BIOLOGY OF ION CHANNELS

Normal cellular physiology depends on the ability of ions to move across cell membranes. Ion channels are highly specialized transport proteins that facilitate this movement by forming regulated ion-permeable routes through the membranes of nearly all excitable and non-excitable cells. They are found in plasma membranes and in the membranes of intracellular organelles. Plasma membrane ion channels display significant structural and functional diversity (Ashcroft 2006) and can be classified broadly as voltage-gated or ligand-gated (Briggs and Gopalakrishnan 2007; McGivern and Worley 2007), although some are subject to modulation by both types of stimuli (Adelman et al. 1992). The primary basis of ion channel diversity lies in almost 400 individual genes that encode a variety of pore-forming (α) and auxiliary (β, γ, δ, etc.) subunits (e.g., see http://www.iuphar-db.org/ for a voltage-gated ion channel database and http://www.ebi.ac.uk/compneur-srv/LGICdb/LGICdb.php for a ligand-gated ion channel database), many of which are subject to splice variation.

The α subunit is the most essential component of an ion channel. It is an integral membrane protein that requires a phospholipid environment to maintain a functional three-dimensional structure. Most α subunits are capable of forming functional channels when expressed alone in an artificial system, but they are often associated in native systems with transmembrane or cytosolic auxiliary subunits that modulate and fine-tune the properties of the channel. It is important to note that ion channel subunits often co-assemble in a tissue-specific manner and sometimes the expression patterns of individual subunits may be altered in disease. Therefore, when one is developing an assay using a heterologously expressed ion channel target, it is always preferable to employ a combination of subunits appropriate to the tissue and/or disease of interest.

The crystal structures of several mammalian and bacterial ion channels have revealed that a functional channel is formed by up to five α subunits arranged in a circular manner with the pore located in the center (Jiang et al. 2002, 2003; Long, Campbell, and Mackinnon 2005; Jasti et al. 2007; Hilf and Dutzler 2008). The pore of the channel is usually lined by several α helical segments contributed by each of the α subunits. The α helices contain charged amino acids that determine the ionic charge preference of the channel. Some channels also contain a so-called pore (P) loop that confers on them a higher degree of specificity for individual ionic species.

Voltage-gated ion channels are formed by α subunits that contain voltage-sensing regions that can move in response to changes in cell membrane potential. This voltage sensor is usually found in the fourth transmembrane α helix (S4) and is coupled directly to the pore-forming α helices so that when it moves it can induce a conformational change in the pore that causes the channel to open. Extracellular ligand-gated ion channels (ionotropic receptors) open and close in response to the binding and unbinding of a first messenger, such as the neurotransmitter glutamate, by virtue of an extracellular receptor region functionally coupled to the pore. Not all channels have an inherent ability to sense their ligands. For instance, some Ca^{2+}-activated K^+ channels co-assemble with the Ca^{2+}-binding protein, calmodulin, to sense and respond to changes in the intracellular concentration of Ca^{2+} (Maylie et al. 2004).

Ion channels are attractive drug targets. Currently more than 30 ion channel targeting drugs are marketed worldwide for use as local anesthetics during minor surgery (blockers of voltage-gated Na^+ channels) and to treat a variety of conditions, including epilepsy, chronic pain, cardiac arrhythmia, hypertension, angina, and type II diabetes. Although there are exceptions (Gee et al. 1996), most drugs modulate ion channel function by interfering with either the gating and/or permeation processes after binding to receptor sites on the pore-forming α subunits (Hockerman et al. 1997).

Despite their attractiveness as drug targets, ion channels remain underexploited. Historically they have not been pursued aggressively due to lack of suitable screening assay formats and incomplete understanding of the relationships between their structural and functional properties. As with GPCRs, the assay formats used in ion channel research can be non-functional/biochemical or functional/cell-based (Figure 4.4). An example of a non-functional assay commonly used for safety pharmacology testing involves measuring the displacement of radio- or fluorescently labeled ligands from their binding sites on the cardiac voltage-gated K^+ channel, hERG (Finlayson et al. 2001; Chiu et al. 2004; Singleton et al. 2007). For the most part though, functional/cell-based assays are preferred and various formats exist, including electrophysiology (see below), ion flux (Ford et al. 2002), and membrane potential- or ion-sensitive fluorescent dye-based methods (Molokanova and Savtchenko 2008). A comparison of multiple assay platforms that can be applied to the voltage-gated Na^+ channel, $Na_V 1.7$, has been published recently (Trivedi et al. 2008).

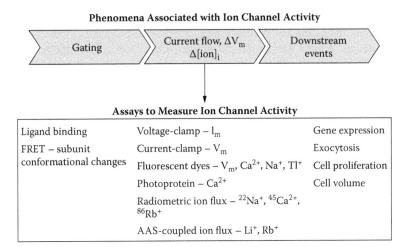

FIGURE 4.4 Assays commonly used in ion channel research. FRET = fluorescence resonance energy transfer; I_m = transmembrane current; V_m = transmembrane potential; ΔV_m = change in transmembrane potential; $\Delta[ion]_i$ = change in concentration of intracellular ion; AAS = atomic absorption spectrometry.

Several requirements must be satisfied before a functional/cell-based assay can be deemed acceptable for use in high-throughput screening. A mammalian host cell line engineered to reconstitute as closely as possible the pore-forming and auxiliary subunit composition of the native human ion channel target of interest is required. The ion channel must be expressed in the cell membrane at a level sufficient to produce ionic current of measurable amplitude during electrophysiology assays such as the manual patch-clamp method (Hamill et al. 1981). Functional/cell-based ion channel assays require a mechanism to activate the channel and ideally the activation mechanism will be physiologically relevant. In the remainder of this section, we will provide an overview of voltage-clamp assays that are critical components of ion channel research. We will also describe in detail the practical aspects of developing an automated planar voltage-clamp assay for the voltage-gated K^+ channel, $K_V1.3$. The reader is directed to published reviews for details of published common ion channel assay formats (Worley and Main 2002; Treherne 2006).

N.B.: Ion channel α subunits are dynamic proteins that undergo activity-dependent conformational changes in structure. They usually exist in a number of inter-convertible states (closed/resting, open/activated, and inactivated/desensitized) and transitions between these states usually occur during gating e.g., in response to ligand binding or unbinding or to changes in cell membrane potential. This phenomenon has implications for ion channel drug discovery in general because of the need to consider whether a particular ion channel state may be relevant in a specific disease setting. This would necessitate appropriate configuration of the screening assay so that it can be used to identify compounds that interact with the relevant state. This phenomenon also has specific implications for pharmacological studies because many ion channel drugs exhibit different affinities for the various states of ion channels, as explained by the modulated receptor hypothesis (Hille 1977). For an ion channel drug that displays state-dependent channel interactions, its apparent potency in a cell-based assay will increase significantly under conditions in which the preferred state predominates.

4.3.2 VOLTAGE-CLAMP-BASED ASSAYS FOR ION CHANNELS

Unlike enzymes or G protein-coupled receptors, the consequence of ion channel activation does not involve the depletion of a substrate or the generation of a product. Rather, when an ion channel opens (or activates), it permits the passive movement of Na^+, Ca^{2+}, K^+, or Cl^- ions down their electrochemical gradients. By regulating the flow of ionic charges across membranes, activated ion channels contribute tiny amounts of electrical current to the theoretical circuits inside cells. Due to the complex nature of the gating and rectification properties of many ion channels, it is preferable to measure ionic currents at fixed membrane potentials using voltage-clamp methods such as whole-cell patch-clamp. This method can be used to study small mammalian cells and has a high detection sensitivity by virtue of the high resistance (GΩ) seals that form between the tip of the recording electrode and the cell membrane.

Whole-cell patch-clamp provides control over the cell membrane potential and simultaneously allows the current passing through voltage-gated and ligand-gated channels to be quantified. Using the whole-cell patch-clamp, voltage-gated ion channels can be opened by applying a voltage step to the cell via the recording electrode. Depending on the polarity of the voltage step, this will either polarize or depolarize the cell membrane. By definition, the electrical nature of the voltage step suggests that the stimulus is physiologically relevant, although the clamped nature of the stimulus is certainly not natural. Ligand-gated ion channels can also be studied using this method by applying a chemical activator, preferably an endogenous ligand, while maintaining the membrane potential at a level different from the reversal potential of the current.

Although the whole-cell patch-clamp method delivers high content kinetic information and is the gold standard against which other ion channel technologies are compared, it remains very low throughput in nature and requires a highly skilled operator to design and perform the experiments and interpret the data. Therefore, in response to the demands of drug discovery groups in the pharmaceutical industry, a number of technology companies have recently developed automated, voltage-clamp systems such as the PatchXpress® (MDS Analytical Technologies; QPatch® (Sophion Bioscience), Patchliner® (Nanion Technologies), and IonWorks® (MDS Analytical Technologies).

Each of these hardware systems couples banks of amplifiers with custom-designed planar chips for current recording purposes. The design of these chips involves an array of wells (8, 16, 48, or 384) with the bottom formed by a planar strip of glass or plastic substrate. A series of apertures (1 to 2 µm in diameter), each of which is functionally equivalent to the tip of a patch-clamp electrode, is engineered in the planar substrate to match the pattern of the wells. Currently, most planar chips are constructed with a single aperture at the bottom of each well, with one notable exception (see population patch-clamp PatchPlate below). During an experiment, cells in suspension are deposited into the wells and suction is applied to attract a random cell to the aperture(s), where it will hopefully form a seal with sufficiently high resistance to permit resolution of whole-cell ionic currents. Most of these systems have been applied to the study of recombinant ion channels expressed in mammalian host cells, but progress is now being made with native ion channels in certain immune system cells (Estes et al. 2008).

One unique advantage of automated voltage-clamp systems is that they enable the voltage- and frequency-dependent effects of ion channel modulators to be investigated during the primary screening phase of drug discovery. During primary screen follow-up activities, customized voltage-clamp protocols can be developed for each individual primary and selectivity target to enable valid comparisons of IC$_{50}$ values for each target under conditions that promote equivalent proportions of the relevant states of the channels (Tao et al. 2006).

The planar chips used by the PatchXpress, QPatch, and Patchliner systems incorporate a glass-based substrate that permits the formation of $G\Omega$-level seal resistances between the cells and the plate. Consequently, the noise level on these systems tends to be low and they are thus capable of distinguishing small ionic currents (<100 pA) reliably. Their planar chips contain 8, 16, or 48 wells with a single aperture in the center bottom position of each well. These systems permit parallel recordings from several individual cells on the chip. However, due to variation in the performance of randomly selected cells, only a proportion are expected to complete any given experimental protocol and so the capacity to acquire pharmacology data may not match initial expectations (Xu et al. 2003).

Reasons for failure to make a successful recording include poor seal formation, low channel expression, and unacceptable seal and/or current stability during the experiment. These systems utilize a ruptured membrane-based whole-cell recording configuration and permit some exchange of solutes between the cell cytoplasm and the intracellular recording solution. They also feature extracellular fluid exchange capability, which makes them useful for conducting complete concentration–response experiments on individual cells.

By way of comparison, and of particular relevance to the current chapter, several significant differences surround the aforementioned $G\Omega$ systems and IonWorks. The planar chip (PatchPlate) used in IonWorks is plastic-based and does not usually permit formation of $G\Omega$-level seals. As a result, its detection sensitivity tends to be lower than that of the $G\Omega$ and cells with robust ion channel expression are required for an acceptable assay. The two types of PatchPlate are the standard and population patch-clamp (PPC™), both of which have 384 wells (8 rows, 48 columns). The standard PatchPlate has a single aperture in the center bottom position of each well. This type of PatchPlate enables IonWorks to record whole-cell currents from many single cells distributed throughout different wells on the plate. However, when IonWorks is used with standard PatchPlates, it suffers from a similar failure rate as the $G\Omega$ systems, which necessitates replicate tests in pharmacology experiments to maximize the probability of obtaining usable data. In contrast, PatchPlate PPC features an array of 64 apertures at the bottom of each well. This modified design enables an ensemble current to be recorded from the population of cells that covers the array of apertures in each well, with the result that the success rate can approach 100% (Finkel et al. 2006).

Each well on a PatchPlate can be thought of as having two fluid-containing chambers separated by the planar substrate that forms the bottom of each well. The upper chamber contains extracellular recording solution whereas the lower chamber contains intracellular recording solution. IonWorks makes current recordings using a perforated patch-based whole-cell configuration. Solute exchange between the cytoplasm of the cell and the intracellular solution is limited to what can pass through the pores formed by the amphotericin-B permeabilizing agent. This antibiotic is included in the intracellular solution during one of the automated steps of the experimental protocol (Figure 4.5A) and delivers estimated access resistances of 5 to 10 MΩ in each well.

IonWorks relies on an integrated 48-channel electrode head to control the membrane potential and record ionic currents. As each PatchPlate has 384 wells, it is necessary for the electronics head to step sequentially through the PatchPlate to obtain eight sets of recordings, with the interval between successive sets being protocol-dependent. During this sequence, voltage control is discontinuous and the cell membrane potential is free to change according to the predominant ionic permeability at any given moment. Liquid and cell handling in IonWorks is accomplished by a 48-channel pipettor (fluidics head). Due to the design of the system, it is not possible for both the fluidics and electrode heads to occupy the space above the PatchPlate simultaneously, and so IonWorks cannot maintain control over the cell membrane potential during compound application. Therefore, IonWorks is most suited to the study of some voltage-gated, Ca^{2+}-activated and slowly desensitizing ligand-gated ion channels (John et al. 2007; Wittel et al. 2007). It is generally not so useful for the study of rapidly desensitizing ionotropic receptors, although the use of an allosteric modulator that would decrease the rate of desensitization could make it feasible to study such a target on this system. Another limitation of IonWorks is that it may not always deliver pharmacology results that agree with those from whole-cell patch-clamp experiments (Sorota et al. 2005; Bridgland-Taylor et al. 2006). The unique advantage of IonWorks is that it offers a much greater recording capacity than the $G\Omega$ systems and for this reason it has filled a niche in ion channel drug screening.

4.3.3 Development of Automated Voltage-Clamp Assay for $K_V1.3$

Ca^{2+} is an important second messenger in cells of the immune system, including T lymphocytes (Feske 2007). Normally, its cytoplasmic concentration is maintained within a narrow range (e.g., <0.1 μM) by a variety of pumps that serve to sequester Ca^{2+} in intracellular stores such as the

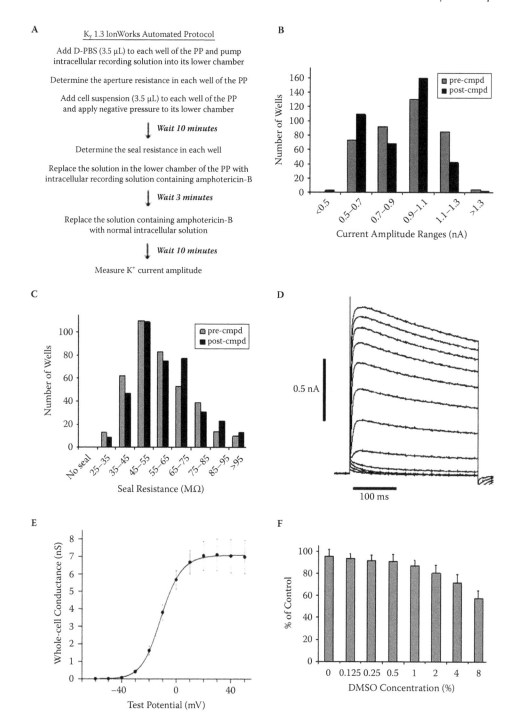

FIGURE 4.5 Development of automated electrophysiology assay for $K_V1.3$. (A) Description of IonWorks process for $K_V1.3$ assay (PP = PatchPlate). (B) Distribution of pre-DMSO and post-DMSO K+ current amplitudes in typical PatchPlate PPC. (C) Distribution of pre- and post-compound seal resistances in typical PatchPlate PPC. (D) Family of K+ currents recorded from a representative single well of a PatchPlate PPC. Currents were evoked by a series of 300-ms test pulses, stepping in 10-mV increments between –60 and +50 mV. (E) Average conductance–voltage relationship of K+ currents in 372 wells of a PatchPlate PPC. (F) Determination of DMSO tolerance of $K_V1.3$ assay.

endoplasmic reticulum or expel it to the extracellular compartment. However, the cytoplasmic concentration of Ca^{2+} may rise during T lymphocyte activation due to IP_3 receptor-dependent Ca^{2+} release from stores as well as Ca^{2+} influx into cells via Ca^{2+} release-activated Ca^{2+} (CRAC) channels. This increase in cytoplasmic Ca^{2+} activates calcineurin, which in turn activates transcription factors, including NFAT, leading to increased cytokine production (e.g., interleukin-2 [IL-2]) and cell proliferation. Calcineurin inhibitors such as cyclosporine block the calcineurin signaling pathway and inhibit the activation of T lymphocytes (T_{EM}) (Liu 1993). Such drugs are broadly immunosuppressive and are used to prevent rejection of transplanted organs and tissues. However, they are associated with serious side effects and risks.

Modulation of Ca^{2+} signaling processes upstream of calcineurin represents an alternative approach to the control of cytokine production and cell proliferation. The voltage-gated K^+ channel, $K_V1.3$, is expressed at moderate levels in quiescent T lymphocytes but is upregulated in activated T_{EM} cells, where it helps to set the cell membrane potential, thereby indirectly influencing Ca^{2+} signaling (Chandy et al. 2004). Blockers of $K_V1.3$ can inhibit cytokine release and T cell proliferation, probably as a consequence of membrane depolarization and a reduced driving force on Ca^{2+} entry through the CRAC channel. Thus K^+ channel-mediated control of Ca^{2+} signaling in immune cells has emerged as a novel area for the discovery of pharmacological agents that might have clinical utility in the treatment of T_{EM} cell-dependent autoimmune diseases such as multiple sclerosis and non-autoimmune diseases such as asthma. In particular, the $K_V1.3$ channel is an attractive target for modulating the function of T lymphocytes. For this reason, we developed an IonWorks-based assay to screen for blockers in an ion channel-focused library of small molecules.

The $K_V1.3$ cell line that we used was a legacy cell line generated by transfecting the gene encoding the human $K_V1.3$ α subunit (KCNA3) into a HEK host cell line. The pool of transfected clones was sorted into single cells using a flow cytometer and then a non-radiometric Rb^+ efflux method was used to identify the clone with an acceptable S/B ratio and appropriate pharmacology, based on the inhibitory potencies of several reference compounds. Fortunately, the clone selected by Rb^+ efflux also performed very well on the IonWorks system and we did not have to generate a new clone. Although it was not required in this case, IonWorks has the capability to assist with clonal selection.

Cells were maintained in tissue culture flasks in a humidified environment containing 5% CO_2 in air at 37°C. The culture medium used for cell maintenance was F-12 Nutrient Mixture containing GlutaMAX™ (Invitrogen) supplemented with 10% FBS, 1% non-essential amino acids, and 750 μg/mL of geneticin (G418). To maximize performance of the $K_V1.3$ cells in the electrophysiology experiments, we did not allow them to exceed 80% confluence in culture. Immediately prior to the electrophysiology experiments, cells were harvested using a mixture of Versene and trypsin (Invitrogen). Once detached, the cell suspension was collected and centrifuged to produce a cell pellet that was re-suspended in Dulbecco's PBS (D-PBS) containing 0.9 mM Ca^{2+} and 0.5 mM Mg^{2+}. Gentle trituration produced a uniform cell suspension with approximately 2×10^6 cells/mL. This cell density is within the range typically used in IonWorks experiments; we have found that lowering the cell density can compromise the ability to obtain an acceptable success rate, whereas increasing it can have negative effects on the potencies of some small molecules.

Electrophysiology experiments on IonWorks involve a number of manual and software-programmable automated steps (Figure 4.5A). For K^+ current recordings, the extracellular solution was D-PBS containing 110 mM K^+, whereas the custom-made intracellular solution (pH adjusted to 7.35) contained (in mM) K-gluconate (90), KF (20), NaCl (2), $MgCl_2$ (1), EGTA (10), and HEPES (10). We evaluated and compared the performance of the $K_V1.3$ cells on both standard and PPC PatchPlate types. For all experiments, the holding potential was set at −70 mV because previous manual patch-clamp experiments demonstrated that this will shift the $K_V1.3$ channels to a resting state. Outward K^+ currents were evoked by depolarizing the cells using test pulses of 300 ms in duration. The pseudo-steady-state K^+ current was measured by averaging the current amplitude toward the end of the test pulse (at 280 to 290 ms). In a typical standard PatchPlate experiment, slowly inactivating outward K⁺ currents could be evoked by depolarizing the cells to a test potential of +40 mV (data not shown). The average K⁺ current amplitude in successful wells was almost 2 nA, but the overall success rate was considered unsatisfactory for use in primary screening: 41% of the wells failed. Therefore, we did not characterize further the $K_V1.3$ assay using the standard PatchPlate.

In contrast, the performance of the $K_V1.3$ assay on IonWorks was much better with PatchPlate PPC. As with the standard PatchPlate, preliminary experiments revealed that slowly inactivating currents could be evoked by depolarizing the cells to +40 mV. On a typical plate, the average K⁺ current amplitude at this test potential was 0.9 ± 0.2 nA ($n = 375$ wells, one plate). The success rate for recording a K⁺ current of amplitude >0.5 nA at this test potential was >95% ($n = 1152$ determinations, three plates). Also, the activation and inactivation kinetics of the K⁺ current were consistent among wells and importantly, the K⁺ current amplitudes and seal resistances were stable during the course of the experiment (Figure 4.5B and C).

In the next series of experiments, we characterized the voltage dependence of $K_V1.3$ activation by applying a series of test potentials between –60 and +50 mV (Figure 4.5D). The activation of $K_V1.3$ was strongly voltage-dependent with a threshold potential of –40 mV. We were particularly interested in selecting a test potential that would fully activate the $K_V1.3$ channels for use in subsequent pharmacology and screening experiments. The K⁺ currents (I_K) evoked at each test potential were converted to K⁺ conductances (G_K) by applying a modified form of Ohm's law (Equation 4.2):

$$G_K = \frac{I_K}{V_{test} - V_{rev}}$$
(4.2)

In Equation 4.2, V_{test} is the test potential and V_{rev} is the reversal potential. In our system, V_{rev} for K⁺ was calculated using the Nernst equation as –78 mV at +20°C and this is the value we used in our calculations. The K⁺ conductances were plotted as a function of test potential (Figure 4.5E) and the data were described by the following Boltzmann function (Equation 4.3),

$$G_K = \frac{G_{K(max)}}{1 + \exp \frac{V_{test} - V_{\frac{1}{2}}}{k}}$$
(4.3)

In Equation 4.3, $G_{K(max)}$ is the maximum K⁺ conductance, $V_{\frac{1}{2}}$ is the test potential that activates 50% of the $K_V1.3$ channels, and k is the slope factor of the curve. In a single PatchPlate PPC experiment we estimated the average $G_{K(max)}$ to be 7 ± 1 nS, indicating robust channel expression, the average $V_{\frac{1}{2}}$ to be –11 mV, and the average k to be 9. The activation $V_{\frac{1}{2}}$ for human $K_V1.3$ determined using IonWorks was approximately 10 to 15 mV more depolarized than the $V_{\frac{1}{2}}$ for rodent $K_V1.3$ channels obtained using manual patch-clamp methods (Grissmer et al. 1994; Vicente et al. 2006). We found that channel activation was maximal at test potentials more positive than +20 mV and thus chose a test potential of +40 mV to fully activate the $K_V1.3$ channels in all subsequent experiments.

As our intent was to use the automated electrophysiology assay to screen a library of small molecules against $K_V1.3$, we next determined the tolerance of the assay to DMSO (range of final concentrations ~0.1% to 8%). At high concentrations of DMSO (>1%) we noted a significant inhibitory effect on K⁺ current, whereas at low concentrations (<0.5%) we found <10% effect on current amplitude (Figure 4.5F). Based on these results, we concluded that the maximum permissible concentration of DMSO during pharmacology and screening experiments was 0.5%.

Next we examined the pharmacological sensitivity of the $K_V1.3$ assay. We added various concentrations of test compounds to each well of the PatchPlate and allowed equilibration for 5 to 10 min. To estimate the blocking effect on the $K_V1.3$ channel, the amplitude of the post-compound K⁺ current was expressed as a percentage of the pre-compound current amplitude. When these percent-of-control values were plotted as a function of concentration, the IC_{50} value for each compound could be estimated using the following form of the logistic equation:

$$\% \text{ of control} = y_{min} + \left(\frac{y_{max} - y_{min}}{1 + \left(\frac{\text{conc.}}{IC_{50}} \right)^n} \right) \tag{4.4}$$

In Equation 4.4, y_{min} is the minimum y value of the curve, y_{max} is the maximum y value of the curve, conc. is the test concentration, and n is the Hill slope of the curve. We found that the $K_V1.3$ current was sensitive to inhibition by the *Stichodactyla helianthus* peptide, ShK, as well as ShK-Dap[22], charybdotoxin (all pre-dissolved in water, with 0.1% BSA present during the experiment), and 4-aminopyridine (4-AP), which was dissolved directly in D-PBS (Figure 4.6A).

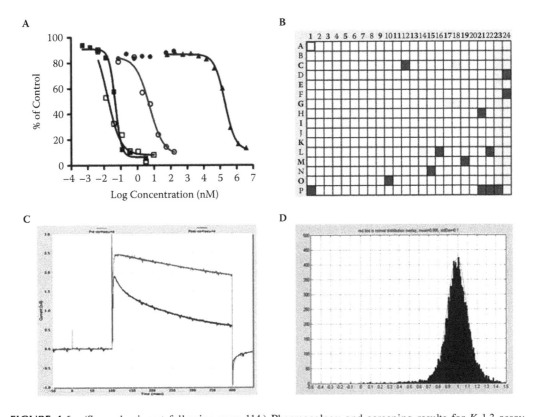

FIGURE 4.6 (See color insert following page 114.) Pharmacology and screening results for $K_V1.3$ assay. (A) Concentration–response curves for inhibition of K⁺ currents by 4-AP (▲) and ShK-Dap[22] (□), ShK (■), charybdotoxin (○), and dendrotoxin (●, data not fitted with curve) peptides. The estimated IC_{50} values are listed in Table 4.1. Dendrotoxin is a selective $K_V1.1$ blocker and did not inhibit the $K_V1.3$ current in our assay. (B) View of typical compound plate from screening campaign. Thirteen wells failed to form seals (blue squares). The yellow squares identify wells in which K⁺ current was inhibited by >30%. All successful wells in columns 23 and 24 are yellow because they contain 3.33 mM of 4-AP, a standard blocking agent for $K_V1.3$; randomly located yellow wells represent unconfirmed hits. (C) Representative pre-compound (red) and post-compound (blue) K⁺ current traces from well H6 of the plate in (A). The test compound appears to be an open channel blocker, as evidenced by the increasing current inhibition at the later times during the pulse. (D) Frequency histogram of normalized K⁺ current amplitudes (post- and pre-compound) for the ion channel-focused library. The histogram shows a normal distribution curve with an average value of 0.995 and a standard deviation of 0.1. Therefore, a hit was defined in POC terms as a compound with POC <70 i.e., an inhibitory effect of >30%.

TABLE 4.1
Summary of Pharmacology of $K_V1.3$

Substance	IonWorks	Patch-clamp
ShK	133 pM	11 pM [a]
ShK-Dap[22]	17 pM	23 pM [a]
Charybdotoxin	5 nM	3 nM [b]
Dendrotoxin	Inactive up to 167 nM	250 nM [b]
4-Aminopyridine	196 µM	195 µM [b]

Potencies of several peptides and 4-AP at human $K_V1.3$ (IonWorks) are compared with published values for murine $K_V1.3$, expressed in L929 cells. Patch-clamp references are: (a) Kalman et al. 1998 and (b) Grissmer et al. 1994.

The inhibitory potencies of three of these molecules agreed with published values for murine $K_V1.3$ (Table 4.1). The one exception was ShK that appeared to be 10-fold less potent than both published values and our own in-house values obtained by manual patch-clamp methods. It has been reported that ShK has a slow on-rate ($\tau = 20$ min) for block of $K_V1.3$ (Middleton et al. 2003). This phenomenon may contribute to the reduced potency of ShK in our automated electrophysiology assay, because our protocol included a compound incubation time of only 5 to 10 min. Longer compound incubation times may improve the potency of ShK but would be associated with greater run-down in the K^+ current amplitude. We also tested the $K_V1.1$-selective blocker, dendrotoxin, and not surprisingly it did not inhibit the $K_V1.3$ current at concentrations up to 167 nM, which is well above its IC_{50} value for $K_V1.1$ in our hands (17 pM; data not shown).

We concluded that the performance of the $K_V1.3$ assay using PatchPlate PPC was acceptable for pharmacology experiments and proceeded to screen an ion channel focused library of 12,160 small molecules at $K_V1.3$. These compounds were provided in round-bottomed, 384-well polypropylene plates (Corning, #3657). There was a single compound in each well of columns 1 to 10 and 13 to 22 (750 nL of 1 mM in DMSO per well), and these were diluted further with D-PBS to yield a concentration of 15 µM (50 µL, 1.5% DMSO) in the compound plate. When the compounds were added to the PatchPlate, they were further diluted three fold from the compound plate to achieve a test concentration of 5 µM. The wells in columns 11 and 12 of the compound plate contained DMSO only (750 nL) and served as negative control wells in which we estimated stability of the K^+ current during the experiment. For the positive control wells of the compound plate (columns 23 and 24), we added 50 µL of 10 mM 4-AP; this enabled us to verify that the $K_V1.3$ cells in each PatchPlate were sensitive to inhibition by 4-AP (3.3 mM).

Using a single IonWorks instrument to screen four plates per day, we completed the entire campaign in fewer than 10 days. As expected, most compounds had little effect on the amplitude of the K^+ current. However, active compounds were identified on most screening plates (Figure 4.6B). Some of the active compounds appeared to display state-dependent interactions with the channel, as evidenced by acceleration of the current inactivation kinetics, suggestive of open-state channel block (Figure 4.6C).

When the compound activities were binned in 1% intervals, a frequency histogram could be plotted (Figure 4.6D). This histogram was fitted by a normal distribution curve, with a mean value of 99.5% and S.D. of 10%. We defined a hit as any compound that reduced the K^+ current amplitude by a percentage that was more than three standard deviations i.e., >30%. Using this method, 258 initial hits were identified, corresponding to a 2.1% hit rate. These hit compounds are likely to have rapid on-rates due to the bias that would have been introduced by the relatively short incubation time. A total of 858 wells failed during the screen i.e., we experienced a 5.9% failure rate; the compounds in all failed wells were retested in duplicate along with the initial hits for confirmation purposes. Only compounds that inhibited the K^+ current by >30% in both retest wells were accepted as confirmed. Using this approach, we confirmed the activity of 172 initial hits (67%) and these compounds progressed to IC_{50} determinations. Of the 172 confirmed hits, 35 compounds had IC_{50} values <1 µM and 60 had IC_{50} values of 1 to 10 µM.

When additional IonWorks instruments are available, larger primary screening campaigns can be accommodated. For instance, with two instruments and an efficient process established (up to seven plates per day per unit), we have carried out a primary screen of >30,000 compounds at a Ca^{2+}-activated K^+ channel in a similar time frame to the current study (Wittel et al. 2007). Of course, an automated electrophysiology system with capacity to handle more than seven plates per day would be very beneficial to the drug screening process. In conclusion, due to its evolving potential to deliver the high-throughput required for large-scale electrophysiology studies, automated electrophysiology has become the current platform of choice in our laboratory for primary screening at voltage-gated and some Ca^{2+}-activated ion channels.

ACKNOWLEDGMENTS

We would like to thank Soo Hang Wong for her support in the GPR23 and CTR assay development. We also thank Invitrogen for providing the CTR-CRE-bla cell line and BioFocus for providing the $K_V1.3$ cell line.

REFERENCES

Abdel-Razaq, W., T.E. Bates, and D.A. Kendall. 2007. The effects of antidepressants on cyclic AMP-response element-driven gene transcription in a model cell system. *Biochem. Pharmacol. 73*, 1995–2003.

Adelman, J.P. et al. 1992. Calcium-activated potassium channels expressed from cloned complementary DNAs. *Neuron 9*, 209–216.

Ashcroft, F. M. 2006. From molecule to malady. *Nature 440*, 440–447.

Barnea, G. et al. 2008. From the cover: genetic design of signaling cascades to record receptor activation. *Proc. Natl. Acad. Sci.* USA *105*, 64–69.

Bresnick, J.N. et al. 2003. Identification of signal transduction pathways used by orphan G protein-coupled receptors. *Assay Drug Dev. Technol. 1*, 239–249.

Bridgland-Taylor, M.H. et al. 2006. Optimization and validation of a medium-throughput electrophysiology-based hERG assay using IonWorks HT. *J. Pharmacol. Toxicol. Meth. 54*, 189–199.

Briggs, C.A. and M. Gopalakrishnan. 2007. Ion channels: Ligand gated. In *Comprehensive Medicinal Chemistry*, vol. 2, ed., Triggle, D.J. and Taylor, J.B., Eds. Elsevier: Oxford.

Cabrera-Vera, T.M. et al. 2003. Insights into G protein structure, function, and regulation. *Endocrinol. Rev. 24*, 765–781.

Carpenter, J.W. et al. 2002. Configuring radioligand receptor binding assays for HTS using scintillation proximity assay technology. *Meth. Mol. Biol. 190*, 31–49.

Chandy, K.G. et al. 2004. K^+ channels as targets for specific immunomodulation. *Trends Pharmacol. Sci. 25*, 280–289.

Chiu, P.J. et al. 2004. Validation of a [^{3}H]astemizole binding assay in HEK293 cells expressing HERG K^+ channels. *J. Pharmacol. Sci. 95*, 311–319.

Coward, P. et al. 1999. Chimeric G proteins allow a high-throughput signaling assay of G_i-coupled receptors. *Anal. Biochem. 270*, 242–248.

Dupriez, V.J. et al. 2002. Aequorin-based functional assays for G protein-coupled receptors, ion channels, and tyrosine kinase receptors. *Recept. Chann. 8*, 319–330.

Eglen, R.M., R. Bosse, and T. Reisine. 2007. Emerging concepts of guanine nucleotide-binding protein-coupled receptor (GPCR) function and implications for high-throughput screening. *Assay Drug Dev. Technol. 5,* 425–451.

Estes, D.J. et al. 2008. High-throughput profiling of ion channel activity in primary human lymphocytes. *Anal. Chem. 80,* 3728–3735.

Feske, S. 2007. Calcium signaling in lymphocyte activation and disease. *Nat. Rev. Immunol. 7,* 690–702.

Finkel, A. et al. 2006. Population patch-clamp improves data consistency and success rates in the measurement of ionic currents. *J. Biomol. Screen. 11,* 488–496.

Finlayson, K. et al. 2001. [^{3}H]Dofetilide binding to hERG transfected membranes: a potential high-throughput preclinical screen. *Eur. J. Pharmacol. 430,* 147–148.

Ford, J.W. et al. 2002. Potassium channels: gene family, therapeutic relevance, high-throughput screening technologies and drug discovery. *Progr. Drug Res. 58,* 133–168.

Gee, N.S. et al. 1996. The novel anticonvulsant drug, gabapentin (Neurontin), binds to the $\alpha_2\delta$ subunit of a calcium channel. *J. Biol. Chem. 271,* 5768–5776.

George, S.E., P.J. Bungay, and L.H. Naylor. 1997. Evaluation of a CRE-directed luciferase reporter gene assay as an alternative to measuring cAMP accumulation. *J. Biomol. Screen. 2,* 235–240.

Glickman, J.F., A. Schmid, and S. Ferrand. 2008. Scintillation proximity assays in high-throughput screening. *Assay Drug Dev. Technol. 6,* 433–455.

Grissmer, S. et al. 1994. Pharmacological characterization of five cloned voltage-gated K$^+$ channels, types K$_V$1.1, 1.2, 1.3, 1.5, and 3.1, stably expressed in mammalian cell lines. *Mol. Pharmacol. 45,* 1227–1234.

Hamill, O.P. et al. 1981. Improved patch-clamp techniques for high-resolution current recording from cells and cell-free membrane patches. *Pflugers Arch. 391,* 85–100.

Harrison, C. and J.R. Traynor. 2003. The [^{35}S]GTPγS binding assay: approaches and applications in pharmacology. *Life Sci. 74,* 489–508.

Hay, D.L., D.R. Poyner, and P.M. Sexton. 2006. GPCR modulation by RAMP's. *Pharmacol. Ther. 109,* 173–197.

He, W. et al. 2004. Intermediates as ligands for orphan G protein-coupled receptors. *Nature 429,* 188–193.

Heding, A. 2004. Use of the BRET 7TM receptor/β-arrestin assay in drug discovery and screening. *Expert Rev. Mol. Diagn. 4,* 403–411.

Hilf, R.J. and R. Dutzler. 2008. X-ray structure of a prokaryotic pentameric ligand-gated ion channel. *Nature 452,* 375–379.

Hille, B. 1977. Local anesthetics: hydrophilic and hydrophobic pathways for the drug-receptor reaction. *J. Gen. Physiol. 69,* 497–515.

Hilton, J.M. et al. 2000. Identification of key components in the irreversibility of salmon calcitonin binding to calcitonin receptors. *J. Endocrinol. 166,* 213–226.

Hockerman, G.H. et al. 1997. Molecular determinants of drug binding and action on L-type calcium channels. *Annu. Rev. Pharmacol. Toxicol. 37,* 361–396.

Inglese, J. et al. 1998. Chemokine receptor-ligand interactions measured using time-resolved fluorescence. *Biochemistry 37,* 2372–2377.

Jasti, J. et al. 2007. Structure of acid-sensing ion channel 1 at 1.9 A resolution and low pH. *Nature 449,* 316–323.

Jiang, Y. et al. 2002. Crystal structure and mechanism of a calcium-gated potassium channel. *Nature 417,* 515–522.

Jiang, Y. et al. 2003. X-ray structure of a voltage-dependent K$^+$ channel. *Nature 423,* 33–41.

John, V.H. et al. 2007. Novel 384-well population patch-clamp electrophysiology assays for Ca^{2+}-activated K$^+$ channels. *J. Biomol. Screen. 12,* 50–60.

Kalman, K. et al. 1998. ShK-Dap^{22}, a potent K$_V$1.3-specific immunosuppressive polypeptide. *J. Biol. Chem. 273,* 32697–32707.

Kassack, M.U. et al. 2002. Functional screening of G protein-coupled receptors by measuring intracellular calcium with a fluorescence microplate reader. *J. Biomol. Screen. 7,* 233–246.

Kostenis, E. 2006. G proteins in drug screening: from analysis of receptor-G protein specificity to manipulation of GPCR-mediated signaling pathways. *Curr. Pharm. Des. 12,* 1703–1715.

Kunapuli, P. et al. 2003. Development of an intact cell reporter gene β-lactamase assay for G protein-coupled receptors for high-throughput screening. *Anal. Biochem. 314,* 16–29.

Lee, P.H. and D.J. Bevis. 2000. Development of a homogeneous high-throughput fluorescence polarization assay for G protein-coupled receptor binding. *J. Biomol. Screen. 5,* 415–419.

Liu, J. 1993. FK506 and cyclosporin: molecular probes for studying intracellular signal transduction. *Trends Pharmacol. Sci. 14,* 182–188.

Long, S.B., E.B. Campbell, and R. Mackinnon. 2005. Crystal structure of a mammalian voltage-dependent Shaker family K^+ channel. *Science 309,* 897–903.

Lundstrom, K. 2006. Latest development in drug discovery on G protein-coupled receptors. *Curr. Protein Pept. Sci. 7,* 465–470.

Maylie, J. et al. 2004. Small conductance Ca^{2+}-activated K^+ channels and calmodulin. *J. Physiol. 554.* 255–261.

McGivern, J.G. and J.F. Worley. 2007. Ion channels: Voltage gated. In *Comprehensive Medicinal Chemistry*, Triggle, D.J. and Taylor, J.B., Eds. Elsevier: Oxford.

Middleton, R.E. et al. 2003. Substitution of a single residue in *Stichodactyla helianthus* peptide, ShK-Dap[22], reveals a novel pharmacological profile. *Biochemistry 42,* 13698–13707.

Molokanova, E. and A. Savchenko. 2008. Bright future of optical assays for ion channel drug discovery. *Drug Disc. Today 13,* 14–22.

Noel, F., D.L. Mendonca-Silva, and L.E. Quintas. 2001. Radioligand binding assays in the drug discovery process: potential pitfalls of high-throughput screenings. *Arzneimittelforschung 51,* 169–173.

Noguchi, K., S. Ishii, and T. Shimizu. 2003. Identification of P_2Y_9/GPR23 as a novel G protein-coupled receptor for lysophosphatidic acid, structurally distant from the Edg family. *J. Biol. Chem. 278,* 25600–25606.

Offermanns, S. and M.I. Simon. 1995. $G_{\alpha15}$ and $G_{\alpha16}$ couple a wide variety of receptors to phospholipase C. *J. Biol. Chem. 270,* 15175–15180.

Overington, J.P., B. Al-Lazikani, and A.L. Hopkins. 2006. How many drug targets are there? *Nat. Rev. Drug Disc. 5,* 993–996.

Pierce, K.L., R.T. Premont, and R.J. Lefkowitz. 2002. Seven-transmembrane receptors. *Nat. Rev. Mol. Cell Biol. 3,* 639–650.

Rudiger, M. et al. 2001. Single-molecule detection technologies in miniaturized high-throughput screening: binding assays for G protein-coupled receptors using fluorescence intensity distribution analysis and fluorescence anisotropy. *J. Biomol. Screen. 6,* 29–37.

Sexton, P.M., D.M. Findlay, and T.J. Martin. 1999. Calcitonin. *Curr. Med. Chem. 6,* 1067–1093.

Singleton, D.H. et al. 2007. Fluorescently labeled analogues of dofetilide as high-affinity fluorescence polarization ligands for the human ether-a-go-go-related gene (hERG) channel. *J. Med. Chem. 50,* 2931–2941.

Sorota, S. et al. 2005. Characterization of a hERG screen using the IonWorks HT: comparison to a hERG rubidium efflux screen. *Assay Drug Dev. Technol. 3,* 47–57.

Tao, H. et al. 2006. Efficient characterization of use-dependent ion channel blockers by real-time monitoring of channel state. *Assay Drug Dev. Technol. 4,* 57–64.

Thomsen, W., J. Frazer, and D. Unett. 2005. Functional assays for screening GPCR targets. *Curr. Opin. Biotechnol. 16,* 655–665.

Treherne, J.M. 2006. Exploiting high-throughput ion channel screening technologies in integrated drug discovery. *Curr. Pharm. Des. 12,* 397–406.

Trinquet, E. et al. 2006. D-myo-inositol 1-phosphate as a surrogate of D-myo-inositol 1,4,5-tris phosphate to monitor G protein-coupled receptor activation. *Anal. Biochem. 358,* 126–135.

Trivedi, S. et al. 2008. Cellular HTS assays for pharmacological characterization of $Na_V1.7$ modulators. *Assay Drug Dev. Technol. 6,* 167–179.

Vicente, R. et al. 2006. Association of $K_V1.5$ and $K_V1.3$ contributes to the major voltage-dependent K^+ channel in macrophages. *J. Biol. Chem. 281,* 37675–37685.

von Degenfeld, G. et al. 2007. A universal technology for monitoring G protein-coupled receptor activation *in vitro* and noninvasively in live animals. *FASEB J. 21,* 3819–3826.

Williams, C. 2004. cAMP detection methods in HTS: selecting the best from the rest. *Nat. Rev. Drug Discov. 3,* 125–135.

Williams, C. and A. Sewing. 2005. G protein-coupled receptor assays: to measure affinity or efficacy that is the question. *Comb. Chem. High Through. Screen. 8,* 285–292.

Wise, A., S.C. Jupe, and S. Rees. 2004. The identification of ligands at orphan G protein-coupled receptors. *Annu. Rev. Pharmacol. Toxicol. 44,* 43–66.

Wittel, A. et al. 2007. Assay development and primary screening of non-voltage-gated ion channels using IonWorks Quattro. Automated Electrophysiology Users Meeting, Baltimore, MD.

Worley, J.F. and M.J. Main. 2002. An industrial perspective on utilizing functional ion channel assays for high-throughput screening. *Recept. Chann. 8,* 269–282.

Xiao, S.H, et al. 2008. High-throughput screening for orphan and liganded GPCRs. *Comb. Chem. High-throughput Screen. 11*, 195–215.

Xu, J. et al. 2003. A benchmark study with SealChip planar patch-clamp technology. *Assay Drug Dev. Technol. 1*, 675–684.

Zhang, J.H., T.D. Chung, and K.R. Oldenburg. 1999. A simple statistical parameter for use in evaluation and validation of high-throughput screening assays. *J. Biomol. Screen. 4*, 67–73.

Zlokarnik, G. et al. 1998. Quantitation of transcription and clonal selection of single living cells with α-lactamase as reporter. *Science 279*, 84–88.

5 Assay Development for Heat Shock Proteins

Zhenhai Gao and Susan Fong

CONTENTS

5.1 INTRODUCTION

Heat shock and stress dramatically increase cellular production of several classes of highly conserved chaperone proteins, collectively known as heat shock proteins. These chaperones, including the members of the Hsp60, Hsp70, Hsp90, and Hsp104 families, are adenosine triphosphate (ATP)-dependent molecules that ensure proper folding, conformational maturation, and function of client proteins. The serendipitous discovery by Luke Whitesell et al. (1994) that the anti-tumor agent geldanamycin (GA), a benzoquinone ansamycin antibiotic from *Streptomyces hygroscopicus*, is a bona fide ATP-competitive Hsp90 inhibitor prompted tremendous interest in targeting Hsp90 for treatment of human cancers.

Indeed, 17AAG, a derivative of GA, was the first-in-class Hsp90 inhibitor subjected to clinical trials in patients with hematological and solid tumors (Sausville, Tomaszewski, and Ivy 2003).

Four members of the Hsp90 chaperone family have been found in different cellular compartments. These include the inducible Hsp90α and constitutively expressed Hsp90β in the cytosol, Grp94 in the endoplasmic reticulum, and TRAP1/Hsp75 in the mitochondrial matrix (Sreedhar et al. 2004). GA and 17AAG have proven to be valuable tools for explaining the chaperone functions of Hsp90 and significantly accelerated the pace of identifying Hsp90 client proteins. To date, over a hundred Hsp90 client proteins have been discovered and the list continues to grow rapidly (Pratt and Toft 2003).

Interestingly, many of these client proteins are kinases or transcription factors that are frequently mutated or overexpressed in cancer cells, including Bcr-Abl, c-Met, AKT, Raf-1, HIF-1α, ER, and AR (Goetz et al. 2003). Impairment of Hsp90 functions by ATP-competitive inhibitors has been shown to induce destabilization and eventual degradation of a magnitude of oncogenic proteins in tumor cells. Hsp90 has now emerged as one of the most unique anti-tumor targets in that inhibition of this single target simultaneously blocks multiple signal pathways that are essential in maintaining the uncontrolled growth and the malignant phenotypes of tumors (Workman et al. 2007). Hsp90-specific inhibitors are therefore anticipated to produce a broad spectrum of anti-tumor activities.

The initial proposal of Hsp90 as a therapeutic target, however, was met with concern and skepticism. It was suspected that the pleiotropic Hsp90 inhibitors would produce unacceptable toxicity in human patients by destabilizing a wide array of client proteins that might be essential for normal cell function and survival. Remarkably, 17AAG has exhibited promising anti-tumor activities with manageable side effects in Phase I and Phase II clinical trials, suggesting the potential existence of a therapeutic window for cancer treatment with Hsp90 inhibitors (Pacey et al. 2006). The underlying molecular basis of tumor selectivity of Hsp90 inhibitors is not completely understood and remains the subject of intense research in both academic laboratories and pharmaceutical companies. Nevertheless, the data are consistent with the notion that stressed tumor cells (to cope with transforming pressures including lack of nutrition, accumulation of mutated proteins, etc.) may depend more heavily on heat shock proteins, including but not limited to Hsp90, and hence become more sensitive to Hsp90 inhibition than normal cells (Workman 2004).

Despite the encouraging clinical activities, 17AAG is broadly viewed only as a valuable compound for proof of concept and may be difficult to develop as a marketed drug due to several pronounced undesirable pharmaceutical properties (Chiosis et al. 2003). It is believed that new agents superior to 17AAG are needed to fully realize the therapeutic potential of Hsp90 inhibition. The initial attempts to screen compound libraries for better Hsp90 inhibitors were limited by the lack of sensitive assays suitable for high-throughput screening (HTS). Over the past few years, a series of biochemical and cell-based assays for Hsp90 inhibitors have been developed using state-of-the-art technologies (Du et al. 2007; Howes et al. 2006; Kim et al. 2004; Zhou et al. 2004; Rowlands et al. 2004). Highlighted here are the key assays established in our laboratories that have allowed rapid identification and confirmation of hits from compound library screening, generation of structure–activity relationships (SAR), and determination of *in vitro* binding and cellular potency of Hsp90 inhibitors.

5.2 PRIMARY HOMOGENEOUS TIME-RESOLVED FLUORESCENCE (HTRF) HTS ASSAY

5.2.1 OBJECTIVE AND RATIONALE

A homogeneous and sensitive HTRF binding assay was developed to allow prosecution of an HTS campaign for novel small molecule Hsp90 inhibitors. The HTRF assay was based on a non-radioactive resonance energy transfer between a donor label (europium chelate) and an acceptor label (allophycocyanin [APC]) brought into close proximity by a specific binding interaction.

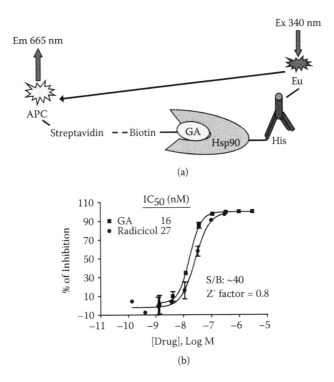

FIGURE 5.1 HTRF biochemical binding assay for Hsp90 inhibitors. (a) HTRF assay using biotin–GA and N-Hsp90α-His. (b) Determination of binding potency (IC$_{50}$) of GA and radicicol for N-Hsp90α-His by HTRF assay set up as described in Section 5.2. N-Hsp90α-His and biotin–GA were at final 100 nM and 30 nM concentrations, respectively.

Fluorescent lanthanide (e.g., europium [Eu]) chelates with long excited state lifetimes were used to avoid interference caused by short-lived emission from acceptor molecules that were directly excited rather than excited by energy transfer. In this assay, the APC–streptavidin and Eu–anti-His antibody were brought together by binding of biotin–GA to the His-tagged N terminal ATP domain (amino acids 9 to 236) of Hsp90α (N-Hsp90α-His) as shown in Figure 5.1A. Excitation of Eu chelate at 340 nm caused APC to produce a unique signal at 665 nm. This assay measured the ability of compounds to disrupt the binding interaction between biotin–GA and Hsp90α, which caused the loss of energy transfer from the donor Eu chelate to the acceptor APC.

5.2.2 PROTOCOL

Biotin–GA was diluted to two times the final concentration in the assay buffer (50 mM Tris, pH 7.5, 6 mM MgCl$_2$, 20 mM KCl, 0.1% bovine serum albumin [BSA]) supplemented with 1 mM DTT, and 25 µL was dispensed into compound plates. This was followed by adding 25 µL N-Hsp90α-His plus Eu-anti-His diluted to two times the final concentration in assay buffer with 1 mM DTT. The mixture was incubated at room temperature for 2 hr, then 25 µL of APC–streptavidin diluted to three times in assay buffer without DTT was added to the plates, followed by an additional 2 hr incubation at room temperature. Plates were read using a Perkin Elmer VICTOR^2 instrument.

5.2.3 VALIDATION AND OPTIMIZATION

For any HTRF binding assay, it is desirable to achieve saturating concentrations with detection reagents. Under those conditions, each binding event would produce a transfer of energy and generation of signal. Ideally, a matrix of all four reagents involved (biotin–GA, APC–streptavidin, Eu-anti-His, and N-Hsp90α-His) would produce the optimal concentrations for each of the components. However, this was not practical, as a matrix of such complexity and magnitude might be unwieldy except to the most highly skilled practitioner. A more workable approach would be to fix the concentration of one or two parameters and vary the others.

To make optimization of the Hsp90 HTRF assay easier and practical, the concentrations of Hsp90 and its binding partner were initially held constant and the detection reagents were varied. Certain concepts were known or had to be assumed. For example, we knew that each streptavidin molecule could bind more than one biotin molecule and the affinity was very strong in the picomolar range. Therefore, it was assumed that adding equal molarities of streptavidin and biotin would result in essentially complete binding of the biotin molecules to streptavidin so that testing a large range of APC–streptavidin combinations was not necessary.

The Eu-anti-His antibody affinity was not provided by the vendor. It was assumed that the antibody affinity was in the lower nanomolar range and was tested in a range starting from low nanomolar to as high as practically could be achieved, taking into consideration the starting concentration of the stock and the overall cost. However, keep in mind that the background signal for an HTRF assay comes from the APC and Eu. Even if saturating levels of detection reagents could be achieved, a balance between the specific binding and the non-specific background signal generation is required.

The reported literature affinity values of GA for Hsp90 were in the 0.1 to 0.4 μM range. As a practical matter, the biotin–GA was initially set at 100 nM with 30 nM Hsp90 in a matrix of APC–SA and Eu-anti-His antibody. This resulted in the best signal-to-background ratio of 40 to 50. The unexpected key to the success of Hsp90 HTRF assays with biotin–GA turned out to be the DTT included as a matter of course in the assay buffer. Removal of the DTT resulted in almost a complete loss of specific signal. We subsequently discovered that DTT actually converted GA to its reduced dihydro-GA form that was recently reported to exhibit substantially higher (~25-fold) binding affinity for Hsp90 (Maroney et al. 2006).

Refinement of the assay consisted of lowering the Hsp90 in matrixes of varying detection reagents to find the lowest protein concentration (for the most sensitive assay) that still produced a good signal-to-background ratio. Ultimately, the optimal concentrations of each component were determined to be 30 nM for Hsp90α, 100 nM for biotin–GA, 5 nM for Eu-anti-His, and 40 nM for APC–streptavidin. Next, we confirmed that at optimal concentrations of biotin–GA, Eu-anti-His, and APC–streptavidin, the assay response was linear to Hsp90 concentration (up to 30 nM). Finally, under optimal conditions, we determined the potencies of the reference compounds: GA and radicicol, another known natural product inhibitor of Hsp90 (Figure 5.1B). It was notable that the observed IC_{50} value of 16 nM for GA in the presence of DTT already reached the lower detection limit of the HTRF assay (~15 nM). The actual binding affinity of the reduced form of GA for Hsp90 could be higher (i.e., IC_{50} < 16 nM), as reported by Maroney et al. in 2006 (5 nM, measured by ThermoFluor microcalorimetry) and determined by a more sensitive AlphaScreen™ assay described in Section 5.4.

5.2.4 OTHER CONSIDERATIONS

With new batches of detection reagents, it is important to ensure that the assay response to Hsp90 concentration remains linear. We found it necessary to make some adjustments of concentrations of each component to keep assay performance optimal. Practically, this was achieved by running an optimization experiment with a small scale matrix of varying concentrations of each assay component.

For biotin–GA, the GA and biotin were connected by a linker. The length of the linker was important. The linker had to be of sufficient length to permit the simultaneous binding of APC–streptavidin plus biotin–GA to N-Hsp90α–His without interference with the formation of a stable tertiary complex. Theoretically, the HTRF detection reagents could have been added simultaneously with the biotin–GA and N-Hsp90α–His, but it was found that pre-binding of the APC–streptavidin to the biotin–GA interfered somewhat (a twofold loss in signal) with biotin–GA binding to N-Hsp90α-His. For this reason, the current assay configuration required the APC–streptavidin to be added in a separate step. We also determined that reading the plates approximately 2 hr after the APC–streptavidin addition seemed to give the best signal. Longer incubation with APC–streptavidin (e.g., overnight) caused a reduction in signal. Speculation for the cause of the reduced signal was that when the biotin–GA was in its off state during steady state equilibrium, the APC–streptavidin was able to bind to it and interfere with its rebinding to the Hsp.

We found that the current HTRF format, albeit robust and sensitive, tended to give many false positive results. Many compounds identified in the HTRF HTS with confirmed inhibition in the same HTRF assays, were nevertheless inactive in the functional Hsp90 ATPase or cell-based assays. It was speculated that the false positive compounds bound non-specifically to the hydrophobic biotin–GA and prevented it from interacting with N-Hsp90α-His. To filter out false positives, a secondary heterogeneous TRF binding assay was developed (described in Section 5.3) using biotin and radicicol, another known natural product inhibitor of Hsp90. Attempts were also made to set up an HTRF assay with the biotin–radicicol in a high-throughput format, but a robust signal could not be achieved.

5.3 SECONDARY DELFIA TRF BINDING ASSAY

5.3.1 OBJECTIVE AND RATIONALE

To confirm the hits from HTS and generate SARs for compounds from medicinal chemistry, a secondary heterogeneous binding assay was developed. This assay used the dissociation-enhanced lanthanide fluorescent immunoassay (DELFIA) technology based on time-resolved fluorescence (TRF) lanthanide chemistry. The europium label was detected by excitation at 340 nm and emission at 615 nm. The resulting long-lived fluorescence signal allowed for delayed detection of emission signals, which reduced auto-fluorescence backgrounds and thereby increased overall assay sensitivity. The current assay method (Figure 5.2A) employed the europium-labeled Eu-anti-His antibody. The assay measured the ability of Hsp90 inhibitors to complete the binding of the biotin–radicicol to the ATP-binding pocket, and was used to determine IC_{50} values and SARs of small molecule Hsp90 compounds.

5.3.2 PROTOCOL

N-Hsp90α-His was diluted to two times the final concentration in the assay buffer consisting of 50 mM HEPES, pH 7, 6 mM $MgCl_2$, 20 mM KCl, and 0.1% BSA. The Hsp90 was dispensed into 96-well polypropylene plates at 50 μL per well. In a separate polypropylene plate, test compounds were diluted to 40 times their final concentration in 100% DMSO. Serial dilutions in DMSO were made in threefold increments, then 2.5 μL of diluted compound were transferred to the 50 μL of Hsp90 and mixed. Background wells received 25 μM (final concentration) radicicol. Biotin–radicicol was diluted into assay buffer at two times the final concentration and 50 μL were added to the Hsp90 compound plate. DMSO was at a final concentration of 2.5%. Samples were incubated at room temperature for 2 hr before 50 μL were transferred to NeutrAvidin-coated plates. Plates were incubated 1 hr, washed three times with DELFIA wash buffer (5 mM Tris, pH 7.5, 0.1% Tween 20, 0.1% sodium azide, 0.9% NaCl), and then 50 μL per well of 3 nM Eu-anti-His diluted into DELFIA assay buffer were added. The plates were next incubated for 2 hr at room temperature, washed four times, and then 50 μL enhancement solution were added. Plates were gently shaken for 7 to 10 minutes before reading in a VICTOR^2 instrument.

5.3.3 OPTIMIZATION AND VALIDATION

To obtain meaningful IC_{50} values, it was important to show that a decrease in protein concentration produces a proportional drop in assay signal. Assays were set up as described above. N-Hsp90α-His

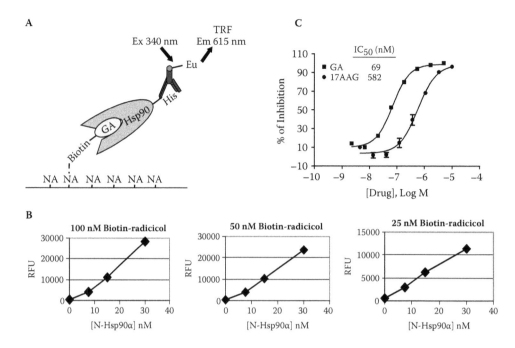

FIGURE 5.2 TRF binding assay for Hsp90 inhibitors. (A) TRF binding assay using biotin–radicicol and N-Hsp90α-His. NA = neutravidin. Eu = europium. (B) Titration of N-Hsp90α-His. TRF assays were set up as described in Section 5.3. N-Hsp90α-His was diluted to 60, 30, and 15 nM and combined 1:1 with 200, 100, or 50 nM biotin–radicicol. (C) Determination of binding potency (IC_{50}) of GA and 17AAG for N-Hsp90α-His by TRF assays. N-Hsp90α-His and biotin–radicicol were at final 30 nM and 100 nM concentrations, respectively. S/B = signal to background.

was diluted to two times the final concentration—60, 30, and 15 nM. The background in this case was no protein rather than 25 μM radicicol. Biotin–radicicol was diluted to 200, 100, and 50 nM and combined at equal volume with the N-Hsp90α to give final biotin–radicicol concentrations of 100, 50, and 25 nM combined with 30, 15, or 7.5 nM N-Hsp90α. Figure 5.2B shows the range of protein concentrations for each concentration of biotin–radicicol that produced a corresponding proportional decrease in signal with each dilution in protein. In all cases this range was from 10 to 30 nM N-Hsp90α. To preserve a high signal-to-background ratio, 30 nM N-Hsp90α and 100 nM biotin–radicicol were chosen for the final assay condition. We then determined the potency of the GA and 17AAG reference compounds (Figure 5.2C). The calculated IC_{50} values for GA and 17AAG were 69 and 582 nM, respectively—comparable to those reported in the literature in the absence of the DTT reducing agent.

5.3.4 Other Considerations

Testing compounds of interest from the HTRF assay in the TRF binding assay helped eliminate compounds that exhibit false positive signals in the HTRF assay. All compounds that were active in the TRF assay also worked in the functional Hsp90 ATPase and cell-based target modulation assays.

Consistency of ambient temperature was important for reproducible results. The signal of the TRF assay seemed to be very sensitive to temperature. At 25°C, the assay signal was only half the signal at 20°C, possibly due to the affinity of the Hsp90 for biotin–radicicol and/or the affinity of the Eu-anti-His antibody. For any binding assay that requires washing and subsequent incubations, the affinity of the binding partners must be tight enough to withstand these steps to some degree. The observation that the TRF signals were higher at a cooler ambient temperature was most likely due to a slower off rate (translating to higher affinity) of the binding partners at the lower temperature.

Setting up the assay at 4°C and employing cold wash buffers produced even higher signals, but running the assay at 4°C was not convenient or practical and could have also caused solubility issues with test compounds. However, if the binding partners are suspected to be of low affinity, running a test assay at 4°C may serve as a proof-of-concept trial. Buffer composition should be reviewed for compatibility with the detection method and assay components. Both the HTRF and TRF assays used the same Hsp90α protein, but the TRF assay suffered a twofold loss of signal in the presence of DTT. It was known that radicicol forms a 1,6 adduct with DTT (Agatsuma et al. 2002) that exhibited reduced biological activities (Maroney et al. 2006; Kwon et al. 1992).

5.4 ALPHASCREEN™ HSP90 ISOFORM SELECTIVITY ASSAY

5.4.1 OBJECTIVE AND RATIONALE

As mentioned in the introduction, the Hsp90 family consists of four members: Hsp90α, Hsp90β, Grp94, and TRAP1/Hsp75. To date, Hsp90α is one of the best characterized isoforms in terms of cellular regulation and functions, and has been the focus of drug development efforts. The understanding of the biology and regulation of the other three Hsp90 members is rather limited and significantly lags behind our knowledge of Hsp90α.

Accumulating evidence suggests that Hsp90 isoforms may be differentially regulated and play distinct physiological roles (Sreedhar et al. 2004). For example, Hsp90β-deficient homozygous mice with normal expression of Hsp90α were not able to differentiate to form placental labyrinths, despite high sequence homology and similar cellular location (cytosol) between the two isoforms. This suggests that the Hsp90α and Hsp90β may play non-overlapping roles in cell differentiation and development. In addition, unlike Hsp90α and Hsp90β, co-chaperones for Grp94 and TRAP1 have not been definitively identified. Currently, the therapeutic advantages and disadvantages of inhibiting all or only a subset of Hsp90 isoforms remain to be established.

Undoubtedly, potent and isoform-selective Hsp90 inhibitors will help elucidate the biology of each isoform and determine whether targeting one particular isoform may provide a better therapeutic window.

Using AlphaScreen technology, homogeneous binding assays for Hsp90α, Hsp90β, and Grp94 were developed. Attempts to devise a TRAP1 binding assay were unsuccessful despite repeated efforts. However, as described in the next section, an assay for measurement of TRAP1 ATPase activity was successful. Combined use of binding and functional ATPase assays offered us the opportunity to determine the selectivity profiles of currently available Hsp90 inhibitors.

AlphaScreen is an energy transfer method that employs donor and acceptor beads; the donor bead absorbs light and energy in the form of a singlet oxygen molecule and is transferred to an acceptor bead to generate a signal. In these assays, the AlphaScreen histidine (nickel chelate) detection kit was used. The kit contained streptavidin-coated beads for binding to biotin–GA and nickel-coated beads for chelation to the His tag on the full length Hsp90 isoforms. Binding of the biotin–GA to the ATP-binding pocket of the His-tagged isoforms (Hsp90α, Hsp90β, and Grp94) brought the streptavidin and nickel-coated beads into close enough proximity to produce a signal (Figure 5.3A).

5.4.2 PROTOCOL

Binding assays were set up in 96-well polypropylene plates. All compounds were first diluted to 40 times final concentration in 100% DMSO and then threefold serial dilutions were made in 100% DMSO. Full-length Hsp90 isoforms were diluted to two times their final concentration in PBS

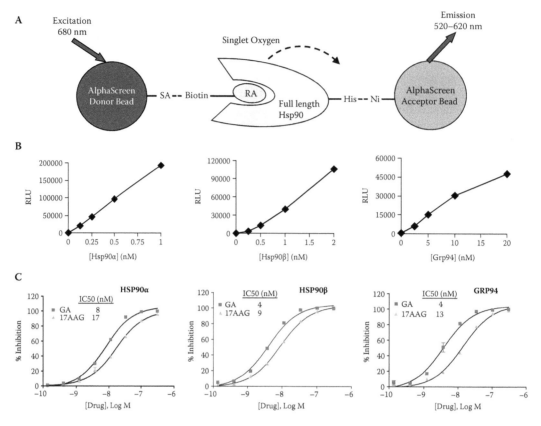

FIGURE 5.3 (A) AlphaScreen™ Hsp90 isoform (full length Hsp90α, Hsp90β, and Grp94) binding assay. (B) Linearity of Hsp90 isoform binding assay response. Hsp90 isoforms were diluted in two-fold increments and binding assays set up as described in Section 5.4. Hsp90α and Hsp90β received 10 nM of b-GA and Grp94 received 30 nM of b-GA. All points represent an average of N = 2. (C) Determination of binding potency of GA and 17AAG for Hsp90α, Hsp90β, and Grp94 by AlphaScreen binding assays. Final assay conditions were 0.25 nM Hsp90α, 1 nM Hsp90β, and 3 nM Grp94, combined with 10 nM biotin–GA for Hsp90α and Hsp90β, and 30 nM biotin–GA for Grp94. Assays were read 4.5 hr post bead additions.

containing 0.1% BSA and 1 mM DTT, and 50 μL per well dispensed into plates, followed by 2.5 μL of DMSO or compounds. Biotin–GA was diluted to 2 times the final concentration and 50 μL per well added to the Hsp90 isoforms. The mixture was incubated for 2 to 2.5 hr at room temperature before 30 μL were transferred to 96-well, half-area white plates. AlphaScreen beads from the histidine detection kit were diluted to 40 μg/mL in PBS plus 0.1% BSA, and 30 μL were then added to the 30 μL in the half-area white plates. Plates were sealed and incubated at room temperature for 1 to 20 hr before reading in the Perkin Elmer Fusion-α instrument. AlphaScreen beads and plates containing beads were continuously kept under darkened conditions. A background signal was determined using a high concentration (10 μM final) of radicicol.

5.4.3 Optimization and Validation

The optimal final concentration of biotin–GA was determined to be 10 nM for the Hsp90α and Hsp90β assays and 30 nM for Grp94, because the signal was less robust for Grp94. At the concentrations chosen, we noted excellent signal-to-background ratios (50 to 100). To test the stability of the AlphaScreen signal, the plates were read, incubated another 1.5 hr at room temperature and then read again. We found that the signals were significantly reduced after a second reading. Some of the signals were reduced by as much as 90%. All subsequent experiments were read only once.

To obtain IC_{50} values, it was important to show that a drop in protein concentration produced a proportional drop in assay signal. The optimal range for Hsp90α was 0.125 to 1 nM, 0.5 to 2 nM for Hsp90β, and 2.5 to 10 nM for Grp94 (Figure 5.3B). The final assay concentrations chosen for the Hsp90 isoforms were 0.25 to 0.5 nM for Hsp90α, 1 nM for Hsp90β, and 3 nM for Grp94. It was notable that the concentration of Hsp90α used in this assay was significantly lower than that used in the HTRF or TRF assays, which represented a dramatic improvement of the detection limits of the biochemical binding assays. Thus, the AlphaScreen assay allowed for the distinction of compounds with potent activities ($IC_{50} < 15$ nM) for Hsp90α. Under optimal conditions, the binding potencies of 17AAG and GA for Hsp90α, Hsp90β, and Grp94 were determined. Figure 5.3C demonstrates that 17AAG and GA bind to all three Hsp90 isoforms equipotently ($\leq$ twofold difference). The presence of DTT along with increased assay sensitivity may explain the lower IC_{50} values of 17AAG and GA for Hsp90α as determined by AlphaScreen versus TRF assays.

5.4.4 OTHER CONSIDERATIONS

False positives may be observed with the AlphaScreen histidine detection kit; certain inhibitors of Ni chelation to histidine are known. Compounds such as imidazoles can displace histidine from nickel chelates, resulting in a loss of signal. Compounds of interest should be tested for AlphaScreen interference by testing against a biotinylated $(His)_6$ peptide provided in the kit.

The timing of when to read the plates after the AlphaScreen reagents have been added should also be evaluated. In most cases, overnight incubation produced good, robust signals. In a few cases, such as with the GRP94 assay, the signal started to decline approximately 2 hr after the AlphaScreen reagents were added.

5.5 ATPASE ASSAY

5.5.1 OBJECTIVES AND RATIONALE

Pure human Hsp90α possesses rather weak intrinsic ATPase activity and hydrolyzes ATP with a k_{cat} of 0.02 min^{-1} and K_m greater than 300 μM (Owen et al. 2002). This hampered initial attempts to screen Hsp90 inhibitors using an ATPase assay. The ATPase activity of Hsp90α can be significantly stimulated by the presence of a co-chaperone (e.g., the Aha1 activator of Hsp90α ATPase) and/or client proteins (McLaughlin, Smith, and Jackson 2002; Panaretou et al. 2002). However, a multicomponent assay demands tremendous work on protein expression and purification and may also complicate the data interpretation.

To assess the abilities of compounds to modulate pure Hsp90α ATPase activity, we developed a relatively sensitive assay based on a fluorescent probe, MDCC–PBP, which can detect micromolar amounts of inorganic phosphate (P_i). The MDCC–PBP (Brune et al. 1998) is maximally excited at 425 nm, producing an emission maximum at 474 nm in the absence of P_i. The maximum emission shifts to 464 nm upon binding of MDCC–PBP to P_i. The same assay format can be easily adapted to TRAP1, which has much stronger intrinsic ATPase activity than Hsp90α, and for which we failed to develop a suitable biochemical binding assay. This allowed us to evaluate the relative potency of compounds to inhibit TRAP1.

5.5.2 PROTOCOL

In this assay, recombinant full-length Hsp90α and TRAP1 containing C terminal His tags purified from SF9 cells were employed. ATP and MDCC–PBP were diluted to two times their final concentrations in the assay buffer and dispensed at 10 μL per well. The buffer consisted of 50 mM Tris, pH 7.5, 6 mM MgCl$_2$, 20 mM KCl, and 1 mM freshly added DTT. Low volume plates were used to conserve MDCC-PBP, the most limited reagent. Full-length Hsp90α and TRAP1 were diluted to two times their final concentrations in assay buffer and 50 µL per well were dispensed into 96-well polypropylene plates. Compounds tested were diluted to 40 times their final concentrations in a separate 96-well polypropylene plate using 100% DMSO. Serial dilutions were made in 100% DMSO; 2.5 µL of 40 times diluted compounds or DMSO alone were added to the 50 µL of Hsp90α and TRAP1.

The proteins were first set up in this larger volume to accurately add compound and keep the final DMSO concentration at 2.5%. After mixing, 10 µL of the protein–compound–DMSO solutions were combined with the 10 µL of ATP–MDCC–PBP in the low volume plates. An initial reading was taken before the plates were sealed and placed in a 37°C incubator. Background wells consisted of MDCC–PBP plus ATP only and no protein. Plates were read in a VICTOR^2 equipped with a 405-nm excitation filter and a 460-nm emission filter. The increase in MDCC–PBP fluorescence at 460 nm was monitored every 30 to 60 min. The plates were unsealed before each reading and resealed before returning to 37°C.

5.5.3 OPTIMIZATION AND VALIDATION

The optimal concentrations of Hsp90α and TRAP1 used in the assays were determined by con-centration-activity titration experiments (Figure 5.4A). Hsp90α at 1 µM and at 0.5 µM produced P$_i$

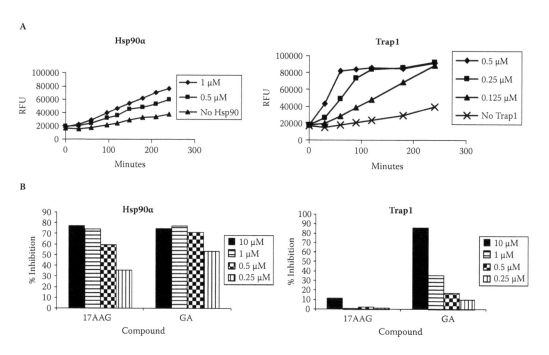

FIGURE 5.4 Hsp90 ATPase assay. (A) Hsp90α and TRAP1 concentration–activity titration. ATPase assays were set up as described in Section 5.5. Hsp90α was at final concentrations of 1 µM and 0.5 µM, and TRAP1 was at final concentrations of 0.5, 0.25, and 0.125 µM. ATP and MDCC–PBP were at 500 µM and 14 µM, respectively, for both proteins. Plates were read out through 4 hr. (B) Inhibition of Hsp90α and TRAP1 ATPase activity by GA and 17AAG. Compounds were diluted serially into DMSO before adding to the assays. The final DMSO concentration was 2.5%. For all experiments, 0.5 µM Hsp90α and 0.125 µM TRAP1 were used with 500 µM ATP and 14 µM MDCC–PBP. Readings were taken through 4 hr. Percent inhibitions were calculated from the rate (slope) of P_i production in the presence and absence of inhibitors. Background samples contained ATP and MDCC–PBP but no protein.

within the detection limits of the MDCC–PBP through the course of the assay. TRAP1 showed a stronger ATPase activity than Hsp90α. At 0.5 and 0.25 µM, TRAP1 quickly produced P_i at levels beyond the maximum detection limit of the added MDCC–PBP. In order to obtain the most sensitive assay possible, the lowest concentration of protein that produced a two- to threefold signal-to-background ratio was chosen for the finalized protocol. For subsequent assays, Hsp90α was used at final 0.5 µM and TRAP1 at final 0.125 µM concentrations.

Using this assay, we determined the potencies of GA and 17AAG to inhibit the ATPase activity of Hsp90α and TRAP1. As shown in Figure 5.4B, both 17AAG and GA significantly inhibited ATPase of Hsp90α, approaching 80% inhibition at 10 µM. In contrast, 17AAG was much less potent on TRAP1 compared with GA, achieving only 12% inhibition at 10 µM (versus >80% inhibition by GA). This was the first demonstration that the two closely related analogs exerted differential activities against TRAP1. Whether this will translate into differences in cellular activity remains to be investigated.

Under similar conditions, we also readily assessed the ATPase activity of Hsp90β in the absence or presence of compound treatment (data not shown). We were not, however, able to detect any ATPase activity with purified full-length Grp94 despite the clear demonstration of an ATP binding pocket in the crystal structure (Soldano et al. 2003). The data supported the prevailing notion that Grp94 was capable of binding but not able to hydrolyze ATP.

5.5.4 OTHER CONSIDERATIONS

In order to obtain the best ratio of signal to background, it was important to start with an ATP solution with minimal contaminating P_i. Therefore, the ATP solution should be made from the purest grade of ATP available.

5.6 CELL-BASED MECHANISTIC ASSAYS

5.6.1 OBJECTIVES AND RATIONALE

A number of cell-based assays were developed to evaluate the cellular effect of Hsp90 inhibition. While intracellular Hsp90 ATPase activity cannot be readily measured, several surrogate response markers have been identified based on unique cellular reactions to Hsp90 inhibition. The chaperone cycle of Hsp90–client protein interaction is driven by ATP loading and hydrolysis and requires recruitment of various co-chaperones such as p23 at different stages of the cycle (Kamal, Boehm, and Burrows 2004). The co-chaperone p23 is important for catalytic Hsp90 activity and its association with Hsp90 is mandatory for overall chaperoning activity. The Hsp90–p23 interaction requires ATP binding but not ATP hydrolysis. In addition, p23 specifically recognizes the Hsp90–ATP complexes and not Hsp90 alone. An ATP-competitive small molecule inhibitor prevents binding of ATP to Hsp90 and causes rapid dissociation of p23 from hsp90.

Consequently, client proteins become unstable and eventually undergo proteasome-dependent degradation. On the other hand, an Hsp90 inhibitor also elicits a compensatory stress response via induction of Hsp70. This is accomplished by Hsp90 inhibitor-mediated activation of heat shock factor (HSF) either directly via dissociation of HSF from the inhibitory Hsp90, and/or indirectly via accumulation of destabilized client proteins. Collectively, Hsp90–p23 dissociation, client depletion, and Hsp70 upregulation are considered signature responses indicative of Hsp90 inhibition, and form the molecular basis of target modulation assays for Hsp90 inhibitors in cells.

Using TRF in-cell Western technology, we developed cell-based assays for routine determination of the potencies of compounds to downregulate client proteins (Raf, c-Met, etc.) and induce Hsp70. The traditional Western blot analysis is time consuming, labor intensive, low throughput and at best, semi-quantitative. The in-cell Western (Figure 5.5A) detects target proteins in their cellular environment, and thereby eliminates the need for cell lysate preparation, electrophoresis, transfer

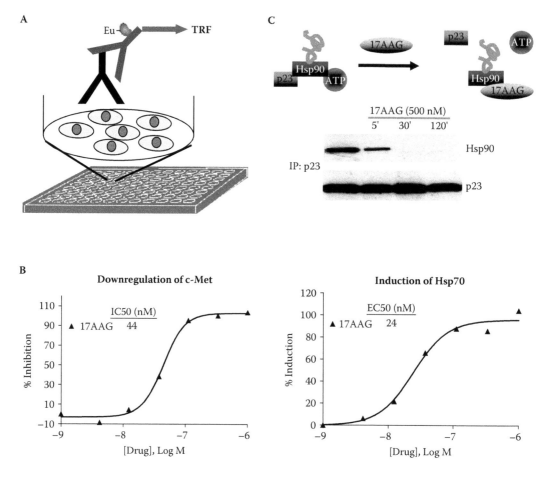

FIGURE 5.5 (A) In-cell Western target modulation assay. (B) Potency of 17AAG to downregulate client protein (c-Met) and induce hsp70. GTL-16 gastric tumor cells were treated with various concentrations of 17AAG for 20 hr. IC_{50} and EC_{50} values for 17AAG to cause c-Met degradation or Hsp70 induction were determined by in-cell Western analysis. (C) Dissociation of Hsp90–p23 complex by 17AAG. GTL-16 cells were treated with 17AAG (500 nM) for various times. Whole cell lysates were collected after treatment and subjected to immunoprecipitation (IP) of p23. The amount of p23 bound to Hsp90 was determined by Western blot analysis of the immunoprecipitate.

of protein to membrane, and film development. Moreover, the in-cell Western performed in 96-well microplates allows quantitative and accurate determination of cellular potency (IC_{50} or EC_{50}) of compounds in a medium or high-throughput manner. A large number of compounds can be evaluated in multiple cell lines for effects on the same or different target proteins in a single experiment. For selected inhibitors, we also evaluated their ability to disrupt Hsp90–p23 interaction using IP (immunoprecipitation) Western blot analysis (a medium-throughput ELISA assay is currently in development but will not be discussed here).

5.6.2 PROTOCOL

Cells were seeded at ~10,000 per well and cultured in clear 96-well flat bottom TC plates for 24 to 36 hr, followed by treatment with compounds for an additional 20 hr. All compounds (stock solution in DMSO) were diluted in 10% FBS culture medium with a final DMSO concentration of 0.5%.

Cells were then fixed with 75 µL 4% paraformaldehyde–PBS at 37°C for 20 to 30 min, washed (three times with 300 µL) and permeabilized with PBS–0.1 % TritonX-100 (PBST).

The wells were blocked with 100 µL of 1% BSA–PBST at 37°C for 30 min. After removal of the blocking buffer, 75 µL of primary antibody in PBST–1% BSA were added and incubated for 1.5 hr at 37°C (or overnight at 4°C). The wells were then washed three times with 300 µL DELFIA wash buffer, and 100 µL of secondary europium-labeled anti-rabbit (1:4000) or anti-mouse (1:500) antibody in DELFIA assay buffer added and incubated at 37°C for 1 hr in darkness. Following three washings with 300 µL DELFIA wash buffer, 100 µL DELFIA enhancement solution were added into each well and plates were incubated on a shaker at room temperature for a minimum of 15 min. The plates were read using the Victor^2.

5.6.3 OPTIMIZATION AND PERFORMANCE

A robust in-cell Western assay requires highly specific antibodies for target proteins. The first step was, therefore, to screen and identify antibodies with low backgrounds on traditional Western blots. The assay development then consisted of selecting appropriate buffers for fixing and permeabilizing the cells (depending on cellular locations of target proteins), blocking the non-specific binding sites, and optimizing the concentrations of primary and secondary antibodies as well as the cell numbers. Finally, under optimal conditions, lowering the cell numbers should produce a corresponding decrease of the signal, demonstrating the linearity of the assay response. Ideally, this should be evaluated with each well in microplates containing the same numbers of cells (to keep the background signal constant) but expressing decreasing amounts of target proteins. Practically, however, this is difficult to achieve.

Using the TRF in-cell Western assay, we determined the potency of 17AAG to inhibit Hsp90 in cells. As shown in Figure 5.5B, 17AAG caused a concentration-dependent decrease of c-Met and induction of Hsp70 in GTL-16 gastric tumor cells. With EC_{50} values of 44 nM and 24 nM, respectively, 17AAG induced a maximal four- to fivefold decrease of c-Met and a three- to fourfold increase of Hsp70. The in-cell Western assays for c-Met and Hsp70 in GTL-16 cells were very robust, with Z′ values of 0.8 and 0.9, respectively.

We also evaluated the ability of 17AAG to destabilize the Hsp90–p23 complex. IP Western blot analysis (Figure 5.5C) indicated that 17AAG induced a rapid dissociation of p23 from Hsp90 in a time-dependent manner. Collectively, these data demonstrated that 17AAG was capable of eliciting signature responses of Hsp90 inhibition in cells. It displayed highly potent cellular activities despite weak *in vitro* binding affinity for Hsp90. Possible explanations for this discrepancy include the reduction of 17AAG to the high affinity dihydroquinone ansamycin in the cells (Maroney et al. 2006) and the much higher intracellular concentration of 17AAG due to its cellular uptake and accumulation (Chiosis and Huezo et al. 2003)

5.7 SUMMARY AND FUTURE DIRECTION

In this chapter, we have discussed a series of Hsp90 biochemical and cell-based assays that constituted the major components of an *in vitro* testing funnel for hit identification and lead optimization. These or similar assays developed by others have proven successful for driving the Hsp90 drug discovery program and led to the identification of several novel small molecule Hsp90 inhibitors currently under evaluation at various stages of preclinical and clinical development.

While most drug development efforts have centered on the inhibitors of the N terminal binding domain of Hsp90, basic research has opened new avenues and opportunities of inhibiting Hsp90 for cancer treatment. These include proposals to develop compounds targeting (1) a second ATP binding site located at the C terminus of Hsp90 (Marcu et al. 2000), (2) one or more particular Hsp90 isoforms, (3) tumor-derived Hsp90 (Kamal et al. 2003; Workman 2004), and (4) the interaction of Hsp90 and a specific client protein or co-chaperone (e.g. survivin or kinase-specific co-chaperone cdc37) (Fortugno et al. 2003; Suriawinata 2004; Pearl 2005). These novel strategies of interfering with Hsp90 functions hold the promise of more effective therapeutic agents with better toxicity profiles. Future efforts are needed to develop suitable and perhaps more sophisticated biochemical assays to allow rapid identification of such inhibitors.

While this chapter focused on assay development for Hsp90 inhibitors, the same principles and concepts will be generally applicable to drug discovery programs targeting other heat shock proteins. Among the promising candidates, Hsp70, which possesses powerful anti-apoptotic properties, is predicted to be the next high-profile small molecule target for more effective cancer treatment.

ACKNOWLEDGMENTS

We would like to thank Dr. Alex Harris, Dr. Jaime Escobedo, and Dr. Mike Doyle for their insightful comments and feedback.

REFERENCES

Agatsuma, T. et al. 2002. Halohydrin and oxime derivatives of radicicol: synthesis and antitumor activities. *Bioorg. Med. Chem.* 10, 3445–3454.

Brune, M. et al. 1998. Mechanism of inorganic phosphate interaction with phosphate binding protein from *Escherichia coli. Biochemistry* 37, 10370–10380.

Chiosis, G., Huezo, H. et al. 2003. 17AAG: low target binding affinity and potent cell activity: finding an explanation. *Mol. Cancer Ther.* 2, 123–129.

Chiosis, G., Lucas, B. et al. 2003. Development of purine-scaffold small molecule inhibitors of Hsp90. *Curr. Cancer Drug Targets* 3, 371–376.

Du, Y. et al. 2007. High-throughput screening fluorescence polarization assay for tumor-specific Hsp90. *J. Biomol. Screen.* 12, 915–924.

Fortugno, P. et al. 2003. Regulation of survivin function by Hsp90. *Proc. Natl. Acad. Sci. USA* 100, 13791–13796.

Goetz, M.P. et al. 2003. The Hsp90 chaperone complex as a novel target for cancer therapy. *Ann. Oncol.* 14, 1169–1176.

Howes, R. et al. 2006. A fluorescence polarization assay for inhibitors of Hsp90. *Anal. Biochem.* 350, 202–213.

Kamal, A., M.F. Boehm, and F.J. Burrows. 2004. Therapeutic and diagnostic implications of Hsp90 activation. *Trends Mol. Med.* 10, 283–290.

Kamal, A. et al. 2003. A high-affinity conformation of Hsp90 confers tumour selectivity on Hsp90 inhibitors. *Nature* 425, 407–410.

Kim, J. et al. 2004. Development of a fluorescence polarization assay for the molecular chaperone Hsp90. *J. Biomol. Screen.* 9, 375–381.

Kwon, H.J. et al. 1992. Potent and specific inhibition of p60v-src protein kinase both *in vivo* and *in vitro* by radicicol. *Cancer Res.* 52, 6926–6930.

Marcu, M.G. et al. 2000. The heat shock protein 90 antagonist novobiocin interacts with a previously unrecognized ATP-binding domain in the carboxyl terminus of the chaperone. *J. Biol. Chem.* 275, 37181–37186.

Maroney, A.C. et al. 2006. Dihydroquinone ansamycins: toward resolving the conflict between low *in vitro* affinity and high cellular potency of geldanamycin derivatives. *Biochemistry* 45, 5678–5685.

McLaughlin, S.H., H.W. Smith, and S.E. Jackson. 2002. Stimulation of the weak ATPase activity of human hsp90 by a client protein. *J. Mol. Biol.* 315, 787–798.

Owen, B.A. et al. 2002. Regulation of heat shock protein 90 ATPase activity by sequences in the carboxyl terminus. *J. Biol. Chem.* 277, 7086–7091.

Pacey, S. et al. 2006. Hsp90 inhibitors in the clinic. *Handb. Exp. Pharmacol.* 172, 331–358.

Panaretou, B. et al. 2002. Activation of the ATPase activity of hsp90 by the stress-regulated co-chaperone aha1. *Mol. Cell* 10, 1307–1318.

Pearl, L.H. 2005. Hsp90 and Cdc37: a chaperone cancer conspiracy. *Curr. Opin. Genet. Dev.* 15, 55–61.

Pratt, W.B. and D.O. Toft. 2003. Regulation of signaling protein function and trafficking by the hsp90/hsp70-based chaperone machinery. *Exp. Biol. Med.* 228, 111–133.

Rowlands, M.G. et al. 2004. High-throughput screening assay for inhibitors of heat-shock protein 90 ATPase activity. *Anal. Biochem.* 327, 176–183.

Sausville, E.A., J.E. Tomaszewski, and P. Ivy. 2003. Clinical development of 17-allylamino, 17-demethoxygeldanamycin. *Curr. Cancer Drug Targets* 3, 377–383.

Soldano, K.L. et al. 2003. Structure of the N-terminal domain of GRP94: basis for ligand specificity and regulation. *J. Biol. Chem.* 278, 48330–48338.

Sreedhar, A.S. et al. 2004. Hsp90 isoforms: functions, expression and clinical importance. *FEBS Lett.* 562, 11–15.

Suriawinata, A. 2004. Survivin: apoptosis inhibitor and its regulation by Hsp90. *Lab. Invest.* 84, 395–403.

Whitesell, L. et al. 1994. Inhibition of heat shock protein 90–pp60v–src heteroprotein complex formation by benzoquinone ansamycins: essential role for stress proteins in oncogenic transformation. *Proc. Natl. Acad. Sci. USA* 91, 8324–8328.

Workman, P. 2004. Altered states: selectively drugging the Hsp90 cancer chaperone. *Trends Mol. Med.* 10, 47–51.

Workman, P. et al. 2007. Drugging the cancer chaperone HSP90: combinatorial therapeutic exploitation of oncogene addiction and tumor stress. *Ann. NY Acad. Sci.* 1113, 202–216.

Zhou, V. et al. 2004. A time-resolved fluorescence resonance energy transfer-based HTS assay and a surface plasmon resonance-based binding assay for heat shock protein 90 inhibitors. *Anal. Biochem.* 331, 349–357.

6 Assay Development for Cell Viability and Apoptosis for High-Throughput Screening

Terry L. Riss, Richard A. Moravec, and Andrew L. Niles

CONTENTS

6.1 INTRODUCTION

Cell-based assays are experimental tools constructed to serve as predictive models for *in vivo* biology. Although cell-based assays are far from perfect, by choosing physiologically relevant model systems they demonstrate utility for testing large numbers of compounds during drug discovery and investigating signaling pathways (Ekwall et al. 1998). The predominant applications of cell viability assays in high-throughput screening (HTS) are to indicate potential *in vivo* toxicity of candidate compounds or to serve as internal controls in a multiplex format to normalize the results of other assays to the number of viable cells remaining at the end of the experimental treatment.

In addition to determining whether cells are alive or dead, it is often of interest to determine whether cell death was due to apoptosis. An apoptosis assay is the logical choice to screen libraries to identify candidate compounds for development of cancer therapeutics where inducing apoptosis is often the clinical goal. Screening to detect apoptosis is more likely to rule out problematic compounds that induce undesirable outcomes such as necrotic cell death.

Many factors must be considered during development of assays to measure cell viability and apoptosis. Several special features contribute to the ability to miniaturize an assay for HTS. Miniaturization became possible with the development of high density multiwell plates, devices for reproducibly dispensing small volumes of liquids, instrumentation for recording fluorescent or luminescent signals, and powerful computing systems for handling and analyzing large data sets.

Detecting the results of cell-based assays during HTS is most often done by measuring the signal produced by a population of cells rather than detecting individual events. High content screening (HCS) involving measurement of individual events using instrumentation based on automated microscopy (O'Brien and Haskins 2006) or flow cytometry (Ivnitski-Steele et al. 2008) will be covered in detail in Chapter 8.

Although automated imaging instruments are extremely useful for measuring multiple events from individual cells, their throughput is often far less than microplate reader assays due to data acquisition rate limitations of the instruments and limited ability to analyze huge amounts of imaging data. Because of these factors, HCS may not be the best choice for efficient measurement of markers of cell viability or apoptosis in large numbers of samples. In most cases the assay chemistries and detection instruments are different for HCS; but, many of the same issues must be considered for developing a relevant cell culture model system. A cell-based assay for screening has three essential elements: the cell culture model system, the assay chemistry, and instrumentation components such as automated liquid handlers and detectors.

6.2 CELL CULTURE MODEL SYSTEM

Although a successful cell-based assay depends on the integration of several essential elements, it can be argued that the cell culture model system is the most important component. If the model system does not reflect the relevant biological event, the assay chemistry and instrumentation cannot compensate. If the culture model is a good representation of events to be measured, it will provide more flexibility in the choices of assay chemistry and instrumentation used. A number of questions should be addressed during development of cell-based assays to measure viability or apoptosis to ensure the model system represents a physiologically relevant event and will provide meaningful results. The relevant questions include:

- What type of cells are the best predictors of human toxicity?
- What culture conditions can produce enough cells for screening?
- Should antibiotics be used for stock culture of cells or during assay?
- What sample density (96-, 384-, or 1536-well) should be used?
- How many cells per well are used for different plates?
- Which robotic liquid handlers can dispense live cells?
- When should the test compound be added relative to plating cells?
- What concentration of compound should be tested?
- Should more than one concentration of test compound be assayed?
- How much DMSO can be tolerated by cells in the assay?
- Does each assay plate need positive and negative controls and is placement on the plate critical?
- How long should cells be exposed to the test compound before recording viability?
- Will the assay chemistry affect the results?
- Should the same assay be repeated to confirm "hits" or should a different method be used?
- Can multiplexing more than one assay be done simultaneously to increase efficiency?

Initially, the list may seem overwhelming; but investment of time up front to consider these issues during assay development will decrease the occurrence of artifacts and false hits. The answers to all of these questions will depend on the *in vitro* model system chosen.

6.2.1 CELLS

The choice of cell type for testing cytotoxicity is often dictated by individual project goals. Candidate cell types may contain tissue-specific enzymes or signaling pathways that are relevant to a specific target. Special biological features such as the presence of drug transporters or abnormal growth characteristics should be considered for their potential influence on results. The goal for high-throughput cytotoxicity screening is to select the most physiologically relevant *in vitro* model system predictive of *in vivo* toxicity. In some cases, the strategy may be screening via multiple cell lines such as those known as the NCI60 human tumor cells (Shoemaker 2006). Established cell lines, fresh primary cells, frozen cells, stem cells, multicellular aggregates, and division-arrested are among the options as tools for screening (Digan et al. 2005).

While stem cells, mixed populations of cells, or multicellular aggregates may represent more physiologically relevant models to study cytotoxic events compared to continuous cell lines, the difficulty in reproducibly handling and using more complex model systems may be a barrier for use in high-throughput screening. Primary cells, especially human hepatocytes, are considered more physiologically relevant for ADMETox studies, but the limitations in supply, cost, and donor history may restrict the ability to screen large numbers of compounds.

The use of cryopreserved cells for screening is gaining popularity and has been validated for a variety of cell types to provide adequate results in many screening systems (Wigglesworth et al. 2008; Zaman et al. 2007). The availability of large blended pools of primary cells from different donors or identical frozen aliquots of cells from the same passage can provide consistency for studies that may take place over a long period. For comparative purposes, it is essential to have a consistent cellular response during the entire project. Additional advantages of using frozen cells include a reduction in labor compared to continuously maintaining large cultures for screening and the ability to use identical stocks in different laboratories. Many problems associated with coordinating scale-up of fresh cultures for a screening campaign are eliminated.

6.2.2 CULTURE CONDITIONS

While cultured cells are arguably the most important components of cell-based assays, they are probably also the largest contributor to variations among replicate samples. The ability to reproducibly prepare a uniform suspension of cells, deliver a consistent number of cells per well, and incubate assay plates using conditions to avoid edge effects are all critical for reproducibility. The physical problems of dispensing cells, avoiding uneven evaporation of medium, and maintaining consistent temperature and incubation conditions when moving plates to different locations can be identified. These sources of variability can be improved by adjustments to equipment, using specially designed assay plates, or controlling the environment. To reduce variability from the population of cells, the logical place to start is to establish and strictly follow standard operating procedures for maintaining growing stock cultures. The goal is to have the cultured cells become as consistent and dependable as standard biochemical reagents.

Using consistent trypsinization procedures and seeding the same number of cells per unit of surface area is more reliable than splitting cultures at a defined ratio. Passage number is one parameter easily documented, but often misleading and not always an accurate representation of how many population doublings the cells have undergone. Normal trypsinization procedures can select for the most rapidly growing subpopulation and select against strongly adherent cells and may eventually result in drift of cell responsiveness. The responsiveness of populations of cells also will change, depending on the recent history of the stock culture. For example, cells harvested from two different flasks grown to different densities have been shown to exhibit different tolerances to toxic compounds (Riss and Moravec 2004).

Although culture medium is often overlooked as a source of error, storage at a recommended temperature shielded from light is necessary to maintain optimum quality and shelf-life. The performance of each lot of animal serum used to supplement culture medium should be verified before routine use. The use of antibiotics for culturing cells is unnecessary. It can hide poor culture technique and may mask bacterial contamination of stock cultures that can interact with the chemical compounds screened. Antibiotics also may affect the biology of the cells and ultimately the interpretation of results.

Mitigating risk of contamination should be an integral part of standard cell culture laboratory procedures. Routine periodic testing of eukaryotic cell cultures for the possibility of mycoplasma contamination should be mandatory. Even if cells are obtained from a source certifying they are mycoplasma-free, subsequent contamination can be introduced by aerosols, unmasked laboratory staff, or other cell lines. Contamination can go unobserved throughout routine low power microscopic examination during maintenance of cultures. Undetected contamination can have adverse effects on the general health and metabolism of a culture, leading to altered cell responsiveness and assay artifacts.

Verifying the identity of cell lines chosen for use should also become a routine task in cell culture laboratories. The number of documented cases of misidentification and cross-contamination of cell lines is growing (Chatterjee 2007). The National Institutes of Health (NIH) issued a notice on November 28, 2007 (NOT-OD-08-017) recognizing the importance of this issue and suggesting that peer reviewers of grant applications and manuscripts submitted for publication should assure that scientists employ available authentication procedures. As the pressure to authenticate cell lines continues, we can expect core facilities and contract research organizations to expand their offerings of such services.

We should keep in perspective that eukaryotic cell cultures are models with limitations. It is ironic that a growth rate representing a doubling of the population of cells once a day on a plastic surface has become accepted as a desirable characteristic representative of a healthy culture. That rapid rate of growth would be considered abnormal for most human cell types. The validity of rapidly growing *in vitro* cultures as a representative model has been questioned and is leading to increased efforts to develop more physiologically relevant screening systems.

6.2.3 ASSAY PLATES

Arguments have questioned the biological relevance of using a single layer of cells growing at the bottom of a plastic chamber that is tissue culture-treated to enhance cell attachment. This is in contrast to the *in vivo* situation in which cell types interact with neighbors in a three dimensional arrangement. To mimic the *in vivo* environment, culture surfaces may be coated with protein mixtures such as collagen or Matrigel, but the multistep procedures required to prepare coated plates can act as sources for variability.

Alternatively, mass produced protein-coated plates can be purchased from commercial sources but may add significant cost to an assay. Advances in producing modified plastic surfaces to mimic the structure of collagen also have been used to provide an improved environment to culture cells. This approach can eliminate the use of animal-derived products that may contain undefined ingredients (Vukicevic et al. 1992) and provide a more consistent surface. These surfaces may prove to be a useful alternative for screening with cells that do not exhibit the desired responsiveness on a standard tissue culture treated polystyrene surface.

The selection of the type of assay plate is also dependent on the type of signal(s) to be detected and whether a clear bottom is required to observe cells under a microscope. Opaque black plates are often chosen for fluorescent assays and white plates are best for luminescent assays because of differential light scattering properties affecting signal-to-background ratios. The choice of clear-bottom assay plates enables morphological observation using a microscope or recording signals through the bottoms of the wells. Observation of the cell population is extremely useful during assay development, but may not be necessary after the assay protocol has been validated or during HTS.

Despite the limitations of *in vitro* culture models, the use of cell-based assays for screening is continuing to grow as the proportion of biochemical assays is declining. This trend to use defined cell lines for screening is likely to continue until a cost-effective alternative can demonstrate greater physiological relevance. Advances in technology for culturing stem cells will likely result in improved model systems for future use.

6.2.4   ASSAY CHARACTERIZATION

After a cell type (*in vitro* model system) has been chosen for cytotoxicity or apoptosis assays, several key parameters must be determined to characterize assay performance: (1) optimal cell seeding density, (2) volume of culture medium per well, (3) equilibration period after dispensing cells, (4) concentration of compound to be tested, (5) length of exposure of cells to the test compound, and (6) selection of appropriate assay chemistry.

6.2.5   SEEDING DENSITY

The number of cells used per well should be high enough to ensure reproducible dispensing of a consistent number, but not so high that nutrients are depleted from the medium, pH is changed, or metabolic wastes accumulate during the assay period. The responsiveness of cells can be density-dependent (Riss and Moravec 2004). As cells transition from logarithmic growth to confluence, metabolic changes will occur. Increasing numbers of cells per well by seeding may not be the best approach to achieve increased assay sensitivity. Using too many cells can result in exceeding the linear range of a particular assay or the instrument used to capture the signal. As described below, assay chemistries are available with adequate detection sensitivity to enable use of fewer than 1000 cells per well.

6.2.6   VOLUME OF CULTURE MEDIUM PER WELL

The volume of culture medium used per well should be adequate to nourish the cells for the entire duration of the experiment. The cost of culture medium is usually not a driving factor for experimental design. If the experimental design uses long incubation periods, larger volumes may be required to provide adequate nutrients and avoid any problems due to evaporation.

An important consideration when determining volume is to ensure adequate space to accommodate the suspension of cells, the test compound, and one or more assay reagent. For assay protocols that use motion-induced mixing, the total volume may need to be lower to prevent loss of fluid and cross-contamination of wells. Changes in the surface-to-volume ratio of the culture medium may be important to consider when miniaturizing assays. For some fluorescence plate readers, the depth of medium may also be important and adjustments to the focal plane of the excitation light source may be necessary.

6.2.7   EQUILIBRATION PERIOD AFTER DISPENSING CELLS

An equilibration period is often used after dispensing cells to allow uniform attachment before addition of test compound. The length of equilibration can be easily determined and optimized to reach an acceptable window between positive and negative control responses. In some cases, the equilibration step is eliminated when compounds are dispensed immediately after seeding cells and in other cases the test compounds may have already been dispensed into the assay plates prior to dispensing cells.

6.2.8 CONCENTRATION OF COMPOUND TESTED

The available concentration of a compound may determine whether it is toxic to cells. The concentration of a test compound available to cells may not be the same as the final concentration added to the sample if the toxin is bound by albumin present in serum used to supplement the medium. Chemical compounds also may bind to plastic surfaces, become concentrated inside cells, be chemically converted by modifying enzymes in the cytoplasm, or actively pumped out of cells.

Because chemical libraries contain compounds with diverse physiochemical properties, it may be a challenge to choose an initial concentration to test. Many screening campaigns testing small molecule libraries typically use 10 μM compound delivered in 0.1% to 0.2% DMSO. Although this represents a good starting point, using quantitative HTS (qHTS; Inglese et al. 2006; Xia et al. 2008) presents a tremendous advantage for testing a range of concentrations during primary screening. The qHTS approach that typically uses the 1536-well format provides much additional data and can overcome many of the limitations of screening at a single toxin concentration.

6.2.9 LENGTH OF EXPOSURE TO TEST COMPOUND

Determining the appropriate length of exposure of cells to test compounds can be a balance between the limits of nutrient depletion and the time needed by cells to respond to an experimental treatment. The toxicity of a compound to a population of cells and the magnitude of response can be related to both the concentration and length of exposure (Riss and Moravec 2004). Figures 6.1 and 6.2 show that the duration of exposure of cells to test compounds can directly affect whether the population of cells will be scored as viable, dead or apoptotic. For instance, some test compounds can be exposed for a few hours with no apparent effect on viability, but induce markers of apoptosis.

One of the main challenges in designing cytotoxicity and apoptosis assays for screening is determining an appropriate length of compound exposure. Difficulties arise because different classes of

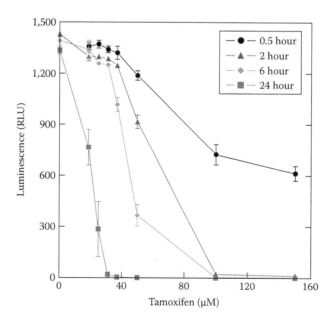

FIGURE 6.1 Characterization of toxic effects of tamoxifen on HepG2 cells measured using a luminescent ATP assay as an indicator of cell viability. Times in the legend indicate durations of tamoxifen exposure. Loss of viability is dependent on concentration of toxin and duration of exposure. (*Source:* Modified from Riss, T.L. et al. 2005. Selecting cell-based assays for drug discovery screening. *Promega Cell Notes* 13, 16–21.)

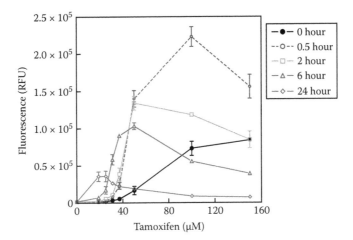

FIGURE 6.2 Characterization of toxic effects of tamoxifen on HepG2 cells measured with the *bis*-Z-DEVD-R110 substrate in a fluorescent caspase-3/7 assay as an indicator of apoptosis. Times in the legend indicate duration of tamoxifen exposure. Activation of apoptosis is dependent on concentration of toxin and duration of exposure. Activation of caspase activity in time zero sample occurred during incubation with fluorogenic reagent. (*Source:* Modified from Riss, T.L. et al. 2005. Selecting cell-based assays for drug discovery screening. *Promega Cell Notes* 13, 16–21.)

compounds in a diverse library will exhibit a broad range of effects on the model system chosen and may require longer exposure periods to elicit responses. For instance, long term exposure to sublethal doses of compounds may eventually result in toxicity (O'Brien and Haskins 2006) that would not be detected using a shorter incubation period.

Time course experiments can be performed using different classes of control toxins to determine the length of exposure necessary to induce apoptosis or result in necrotic cell death. A toxin with properties that disrupt cell membranes will result in rapid necrotic cell death. Other chemicals may not become toxic until after conversion by modifying enzymes in the cytoplasm. In many cases, sub-populations of cells at different stages of the cell cycle may undergo cell death at different times.

All cells undergoing apoptosis *in vitro* eventually undergo secondary necrosis. This results in disruption of the cell membranes and spilling of cytoplasmic components into the surrounding medium. This is in contrast to the *in vivo* situation in which early engulfment of apoptotic cells by macrophages occurs. Caspase activity (as a marker of apoptosis) may increase initially in a population of cells induced to undergo apoptosis, but eventually decline after secondary necrosis occurs as the enzymes become degraded. One of the benefits of understanding the sequence of apoptotic events *in vitro* is knowing that assays to measure markers of cell viability or necrosis can be used to detect the end result of apoptosis. Regardless of whether the mechanism of cell death is apoptosis or necrosis, after long term exposure all the cells will be scored as dead with a cell viability assay.

Two general approaches may overcome uncertainty in establishing the duration of incubation. One approach that can be useful when testing a single concentration of test compound is to use multiplexing to measure more than one assay endpoint. Multiplexing can help confirm or rule out a positive response (hit) by measuring two independent markers of the same event or two different events (cell viability and apoptosis). Another and perhaps more powerful approach is to use qHTS to test a broad range of compound concentrations. The rationale is that a range of concentrations is more likely to demonstrate toxicity during a defined incubation period, whereas testing only a single standard dose may not show an effect.

For example, if a cell viability marker assay indicated a population of cells was dead after 48-hr incubation with 10 μM of a test compound, it would be impossible to determine whether cell death resulted from apoptosis. It is possible the cell population underwent apoptosis soon after chemical treatment followed by secondary necrosis, and the apoptosis marker (caspase activity) subsequently degraded in a pool of cell debris and culture medium. A relatively high concentration of test compound may induce apoptosis in a population of cells in less than an hour, but a much lower concentration of the same compound may not induce apoptosis for several hours. In cases where lower concentrations of a chemical induce apoptosis only after longer incubation periods, testing a broad range of concentrations of the same compound may indicate that 0.01 µM induces apoptosis not observed with higher concentrations. Dosage and length of exposure clearly impact the result of endpoint assays.

6.3 METHODS FOR MEASURING CELL VIABILITY

A variety of different assay chemistries have been developed for screening using cell-based assays. Making the most logical choice among the different methods will depend on what you want to measure at the end of the experiment. The number of viable cells, the number of cells that died as a result of treatment, and investigating whether cell death occurred via apoptosis or necrosis can all be determined with homogeneous assays that have adequate sensitivity for use in 1536-well formats.

Measuring the effect of chemical or growth factor treatment on the viability of cells in culture historically was based on the ability of the cell population to continue to divide and form new colonies. Directly counting the number of cells using a microscope or an electronic particle counter were reliable methods, but very tedious and time consuming for multiple samples. Improved procedures to handle more samples were developed based on measurement of surrogate markers that reflected cell growth or total biomass present at the end of a treatment period. Most methods were based on measuring cell membrane integrity, events related to DNA synthesis, selective staining, or markers indicating active metabolism. While many methods are useful for specific applications, only a few have been adopted for high density HTS. The following section provides a brief overview of the history and evolution of various methods for measuring cell viability and highlight the advantages and disadvantages of the most frequently used methods.

6.3.1 STAINING METHODS

Several direct dye staining procedures can measure total protein biomass, nucleic acid content, or selective uptake by organelles such as mitochondria or lysosomes. Many staining methods are appropriate for small numbers of samples or for image-based high content screening; but in most cases, the multistep procedures exhibit limitations for HTS.

During routine maintenance of cell cultures, the most common method of determining viability is assessing membrane integrity by using a hemacytometer and a microscope to observe samples of Trypan blue-stained cells. Cells that have lost membrane integrity stain positive with Trypan blue and are considered non-viable. Viable cells with intact membranes exclude Trypan blue and appear colorless. Various instruments and semi-automated methods using Trypan blue have been developed and may prove useful for reducing variability resulting from different operators or biased observation during routine examination or daily maintenance of cultures, but the throughput is inadequate for HTS applications.

A large number of dye molecules have been demonstrated to increase fluorescence (10- to 1000-fold) upon binding to and intercalating into the structure of DNA. Many have been used successfully for quantitation of eukaryotic and prokaryotic cell numbers. Examples include propidium iodide, ethidium homodimer, and several different monomeric and dimeric cyanine dyes. In general, these are polar dyes that are not permeable to viable cells but can penetrate and stain dead cells with compromised membranes. A common use of these dyes is to stain only dead cells by adding a reagent solution that does not damage the viable cells present. However, nucleic acid binding dyes also can be used to estimate total cell numbers if a sample is treated with a lysis solution to permeabolize viable cells.

Many of the non-permeable dyes can be combined in a multiplex format with other fluorescent methods that measure viable cell numbers as long as adequate spectral discrimination between the detection wavelengths of the two methods is possible and appropriate instrumentation and filter sets are used. A disadvantage of staining with these dyes or using any fluorescent method is the possibility of fluorescence interference with small molecule compounds in chemical libraries that may either quench or contribute to the fluorescent signal. Another disadvantage with some dyes is a lack of specificity for DNA. In some cases, double-stranded RNA can be stained; this is problematic because levels of RNA may change during different physiological states.

The acetoxymethyl ester form of the calcein dye (calcein-AM) has been widely used as an assay to measure viable cells for microscopy, flow cytometry, and some plate-based methods. Calcein-AM is a cell-permeable non-fluorescent compound added to cells as a solution in DMSO. Non-specific esterase activity present in viable cells removes the lipophilic AM group, resulting in conversion of the non-fluorescent calcein-AM into a green fluorescent calcein molecule that is generally retained inside viable cells.

Calcein-AM staining of viable cells is often combined with staining dead cells with a non-permeable nucleic acid binding dye that emits fluorescence at a different wavelength (Garner et al. 1994). The major limitation of this method results from esterase activity in serum used to supplement culture medium. Cell washing and removal of culture medium are required prior to the assay for optimal performance in a multiwell plate format. In addition, dead cells may release or retain residual esterase activity that can cleave calcein-AM and contribute to background fluorescent signal.

Another complication of this staining method is that some cell types are known to actively pump calcein from the cytoplasm through the action of the multidrug resistance pump (Essodaigui, Broxterman, and Garnier-Suillerot 1998). In fact, calcein-AM has been used as a multidrug resistance transporter assay because calcein is a known substrate to measure efflux activity of p-glycoprotein (Karászi et al. 2001). Although calcein-AM is useful for some imaging and flow cytometry applications to measure viability, it is not generally used for HTS assays in multiwell plates because of the liquid handling steps to remove non-specific esterase activity required for optimal performance.

Sulforhodamine B [SRB; 2-(3-diethylamino-6-diethylazaniumylidene-xanthen-9-yl)-5-sulfobenzenesulfonate] dye binds electrostatically in a pH-dependent manner to basic amino acid residues on proteins. This staining method is used to measure accumulation of cellular protein biomass in assay wells. The procedure involves washing, trichloroacetic acid fixing, and staining monolayers of cells with SRB followed by additional wash steps and elution of dye before recording absorbance. Mild basic conditions are used to extract the dye from cells for measurement. This procedure was developed and has been used effectively by the National Cancer Institute to screen for cytotoxic effects using a large panel of tumor cell lines derived from different kinds of cancers (Rubinstein et al. 1990; Skehan et al. 1989).

Neutral red (3-amino-7-dimethylamino-2-methylphenazine hydrochloride) is an example of a selective organelle stain. It is a cell-permeable dye that passes through intact cell membranes and becomes concentrated in lysosomes (Borenfruend and Puerner 1985). When cells die, neutral red uptake will not occur. This lysosomal uptake and staining approach has been used with many cell types to measure toxic effects of chemicals. This method has been validated for use on 3T3 fibroblasts as an *in vitro* cytotoxicity test for estimating starting doses for acute oral systemic toxicity tests by the Interagency Coordinating Committee on the Validation of Alternative Methods (ICCVAM). The procedure requires multiple steps to remove medium, apply dye, remove excess dye, wash cells, and elute dye before recording absorbance at 540 nm which is proportional to viable cell number (Cavanaugh et al. 1990).

Both the neutral red and SRB staining methods present the advantages that they avoid some potential for artifacts related to chemical interference with the tetrazolium and resazurin assay chemistries used to measure cell metabolism and the signals are stable. The obvious and major disadvantage of these dye binding assays is the need for multiple steps for washing cells and elution of dye. Although these methods have been used for a large body of work, the requirement for multiple liquid handling steps, limitations in sensitivity to 1000 to 2000 cells, and challenges working with suspension cultures have all contributed to preventing these methods from wide adaptation for HTS.

6.3.2 DNA Synthesis

The incorporation of tritiated thymidine into DNA was frequently used as an assay to identify growth factors and study their effects on cell proliferation. The method involves addition of a radioactive precursor that is incorporated into newly synthesized DNA during the S phase of the cell cycle. Cells are cultured to allow incorporation of the tracer, followed by acid treatment of samples to precipitate DNA contained in cells, and subsequent separation of unincorporated tracer. Special equipment was designed to process increasing numbers of radioactive samples.

Cell harvesters were developed to capture multiple samples of cells on membrane filters, wash away unincorporated isotopes, and prepare samples for liquid scintillation counting on special equipment developed to process and count multiple samples. Despite miniaturization and improvements in efficiency of this technique, the disadvantages of multiple liquid handling steps and increasing costs for disposal of radioactive waste materials severely limit its usefulness. Although specific applications require measuring DNA synthesis as a marker for cell proliferation, much better choices are available for detecting viable cell number for HTS.

6.3.3 Metabolic Markers of Cell Viability

The design principle for assays that detect active cell metabolism as a marker for viable cells is to provide a substrate molecule that only viable cells will convert into a product. This conversion can be measured based on a change in absorbance of light, changes in fluorescence, or emission of a luminescent signal. The approach relies on the concept that metabolic activities stop when a cell dies, the ability to convert the substrate into an indicator is lost, and no signal is generated. Despite minor differences in assay procedures, tetrazolium reduction, resazurin reduction, alkaline phosphatase activity, protease activity, and ATP content all represent methods of monitoring some aspect of general metabolism or enzymatic activity present in viable cells. Except for the ATP detection method (described below), these metabolic marker assays require incubation of the substrate with a population of viable cells for an adequate period to generate a sufficient amount of product that can be detected above background.

6.3.3.1 Tetrazolium Reduction Assays

Among the first cell viability assays developed for HTS was the MTT [3-(4,5-dimethylthiazol-2-yl)-2,5-diphenyltetrazolium bromide] tetrazolium reduction assay (Mosmann 1983) that served as a milestone for this type of study. The assay offered a non-radioactive alternative to tritiated thymidine incorporation into DNA as a method of measuring cell proliferation. In many cases, the MTT assay can directly substitute for the tritiated thymidine incorporation assay (Figure 6.3). The MTT tetrazolium compound is prepared in a physiologically balanced solution, added to cells in culture, and incubated for approximately 4 hr. Viable cells convert MTT into an intensely colored formazan product that can be quantitated by recording changes in absorbance at specific wavelengths.

The formazan product resulting from reduction of MTT is deposited as a precipitate both inside and outside of cells. The precipitate must be solubilized before recording 570 nm absorbance that requires addition of a second reagent to generate a uniformly colored solution within the assay well. A variety of different combinations of organic solvents and detergents were developed as reagents to solubilize the formazan product, stabilize the color, avoid evaporation, and reduce interference by phenol red often present in culture medium (Tada et al. 1986; Hanson, Nielsen, and Berg 1989; Denizot and Lang 1986).

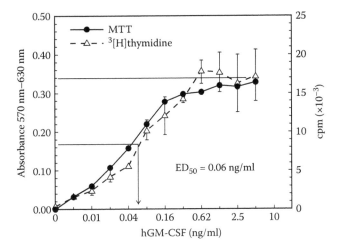

FIGURE 6.3 Comparison of MTT tetrazolium and tritiated thymidine incorporation assays to measure effects of hGM-CSF on proliferation of TF-1 cells. A no-cell blank absorbance of 0.065 was subtracted from all MTT values before plotting. Similar ED_{50} values are shown for both assays. (*Source:* Modified from Promega Corporation Technical Bulletin 112. *CellTiter 96® Non-Radioactive Cell Proliferation Assay.*)

Tetrazolium assay technology is widely established for use in 96-well plates as evidenced by hundreds of published articles about the technique. Although the formazans are intensely colored, spectrophotometric detection has practical limits of sensitivity. The lack of sensitivity restricts the ability to miniaturize the assay into a high density plate format and has limited its adoption for HTS. The limit of detection above background using this light absorbance method is generally about 1000 cells; however, the ability to convert tetrazolium to formazan and thus sensitivity is highly dependent on metabolic activity of the cell type measured. Incubating cells for longer periods results in an accumulation of color and increased assay sensitivity up to a point, but incubation time is limited because of the toxicities of the detection reagents that utilize energy (reducing equivalents such as NADH) from the cell to generate a signal.

A significant advance in the convenience of performing tetrazolium assays occurred with the development of reagents that form aqueous soluble formazan products when chemically reduced by viable cells. This group of tetrazolium compounds includes MTS, XTT, and WST (Cory et al. 1991; Barltrop and Owen, 1981; Paull et al. 1988; Ishiyama et al. 1993; Tominaga et al. 1999). These improved reagents eliminated the need to add a second solubilization reagent and thus eliminated a liquid handling step.

The molecular charge properties of these new tetrazolium reagents that enable the formation of water-soluble formazan products also restrict membrane permeability and entry into viable cells. To overcome this problem, cell-permeable electron transfer reagents such as phenazine methosulfate, phenazine ethosulfate, and menadione are used to facilitate shuttling of electrons from cytoplasmic reducing compounds (e.g., NADH) to the tetrazolium compound in the surrounding culture medium. Although a liquid handling step is eliminated with this newer class of tetrazolium reagents, the addition of the electron transfer reagent can be considered a disadvantage because of the increased risk of adverse effects on cell physiology.

6.3.3.2 Resazurin Reduction

Resazurin is a chemical redox indicator that functions like the tetrazolium compounds. The resazurin reduction assay approach has successfully replaced tritiated thymidine incorporation for some HTS applications (Ahmed, Gogal, and Walsh 1994; Shum et al. 2008). Resazurin can be dissolved in physiological buffers resulting in an intense blue solution that is added directly to growing cultures in a homogeneous assay format. After addition of reagent, the assay plates are returned to an incubator to enable viable cells to convert resazurin into a pink fluorescent product (resorufin) that is quantitated using a spectrophotometer and a microplate fluorometer equipped with a 560-nm excitation/590-nm emission filter set.

For HTS applications, the incubation period required to generate adequate fluorescent signal above background is usually about 2 to 4 hr, depending on the number of cells per well and their metabolic activity. Although longer incubation periods have been used, caution should be observed to avoid artifacts caused by the potential for secondary chemical reduction of resorufin into a colorless non-fluorescent compound and the known toxic effects of resazurin exposure to cells.

The major advantages of this approach of estimating the number of viable cells for HTS include the: (1) homogeneous add-incubate-measure protocol, (2) ability to measure the signal using fluorescence, (3) relatively inexpensive reagents, and (4) capability of multiplexing with other methods such as caspase assays (Wesierska-Gadek et al. 2005) to gather more data about the mechanisms of toxicity. The fluorescent resazurin reduction assay has been demonstrated to have adequate sensitivity for miniaturization into the 1536-well format in some cases (Shum et al. 2008); however, for most applications, the culture conditions required to maintain the number of cells necessary to generate an adequate signal has limited miniaturization to the to 384-well format.

6.3.3.3 Disadvantages of Metabolic Reduction Assays

Some of the major disadvantages of both the tetrazolium and resazurin reduction assays are related to the requirement to incubate a substrate with viable cells for a sufficient time to generate a measurable signal. This is a very important feature to consider when designing assays for HTS. In addition to the extra plate handling steps needed to return the cells to a 37°C incubator, the extended incubation of the detection reagents with viable cells increases the possibility of undesirable artifacts resulting from chemical interactions among the assay chemistries, the compounds tested, and the biochemistry of the cells.

One example is the known interference by reducing compounds that affect the chemical conversion of substrate to a colored indicator. This is especially true for the tetrazolium assays (Ulukaya, Colakogullari, and Wood 2004; Chakrabarti et al. 2000; Pagliacci et al. 1993; Collier and Pritsos 2003). The growing list of interfering compounds includes ascorbic acid and sulfhydryl reagents such as glutathione, coenzyme A, dithiothreitol, etc. Similar interferences by compounds that affect the oxidation and reduction chemistry of cells are likely to cause artifacts with the resazurin reduction assay. Assays that measure markers of metabolism also can be influenced by the pH of the culture medium and other factors that may stimulate or stress the metabolic rates of cells.

One of the major disadvantages of the tetrazolium and resazurin assays often overlooked is the toxic effects of the detection reagents on cells (Squatrito, Connor, and Buller 1995). Exposing cells to resazurin or tetrazolium reagents for long periods (or elevated concentrations for shorter periods) will result in cytotoxicity that has the potential to mask or interfere with the experimental outcome. The concentration of reagent and incubation time must be optimized to reduce cytotoxic effects to ensure avoidance of artifacts. Determining toxic effects can be accomplished by treating a population of cells with resazurin for various durations followed by determining viability using a different method such as measuring ATP. Figure 6.4 shows the toxic effects of Alamar blue on cells measured using ATP as a marker of viability.

Additional general limitations of all cell-based assays that utilize fluorescence detection methods include the possibility of fluorescence interference from small molecule compounds in chemical libraries and color quenching of signals from multiplex assays. Fluorescence interference can occur if the compound tested has properties that alter the normal detection of signal. Compounds may contribute to an artificially high fluorescent signal or may serve to quench normal signals from fluorescent indicators. Depending on the nature and size of the library tested, the effects of fluorescence interference may result in a substantial increase in the number of repeat assays to confirm hits or rule out interfering chemicals. The concentration of resazurin used for cell viability assays

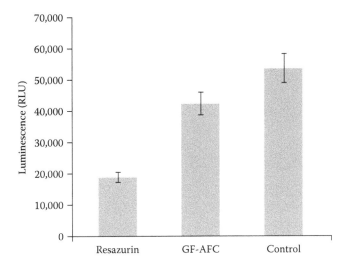

FIGURE 6.4 Comparison of effects of Alamar blue (resazurin) and GF-AFC reagents on viability of cells measured using a luminescent ATP assay. Resazurin or GF-AFC was incubated with 10,000 DU145 cells per well for 18 hr prior to measuring ATP as an indicator of cell viability. Alamar blue reagent is more toxic to cells.

produces solutions that are intensely colored and can result in quenching of secondary signals from other multiplexed fluorescent or luminescent assays. Despite several disadvantages that may lead to false hits and the need to repeat, the lower reagent cost per well has often been a deciding factor for choosing an assay method.

6.3.3.4 Aminopeptidase Markers

The recent identification of selective protease substrates to simultaneously measure markers of both viable and dead cells led to the development of optional methods for HTS that provide flexibility and added advantages (Niles et al. 2007a). The assay to measure viable cells is based on a cell-permeable protease substrate called glycyl-phenylalanyl-aminofluorocoumarin (GF-AFC). The procedure is a homogeneous add-incubate-measure method that is faster, more sensitive, and less toxic to cells than the tetrazolium and resazurin reduction assays. The substrate can be prepared in an aqueous buffer and is added directly to samples containing cells. The substrate permeates viable cells where constitutive protease activity in the cytoplasm rapidly removes the amino acids, yielding free AFC. The amount of AFC released is directly proportional to viable cell numbers and shows improved sensitivity compared to the resazurin assay (Figure 6.5). The AFC is detected via a microplate fluorometer equipped with a (380- to 400-nm excitation/505-nm emission) filter set.

The selective detection of viable cells by this method arises from the rapid loss of the protease activity that cleaves GF-AFC upon cell death, so only the viable population of cells contributes substantially to generating signals from free AFC. A 30-min incubation of the protease substrate with viable cells is generally sufficient to generate adequate signal for use in 384- or 1536-well format for HTS. The single liquid handling step contributes to improved assay performance and acceptable Z′ factor values (Figure 6.6).

A major advantage of this method is the ability to multiplex with other assays. The GF-AFC substrate used to detect viable cells was designed for use in combination with another substrate that selectively detects protease activity from dead cells (Niles, Moravec, and Riss 2008). The method used to measure dead cells is based on the *bis*-Ala-Ala-Phe-rhodamine 110 (AAF-R110) protease substrate. This substrate is non-permeable; thus viable cells do not substantially contribute to signal. Dead cells with compromised membranes leak protease activity into the surrounding medium

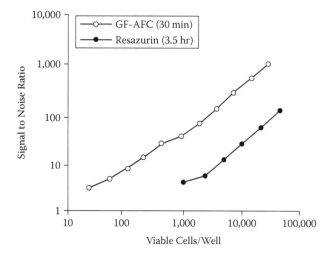

FIGURE 6.5 Comparison of sensitivity of resazurin reduction and GF-AFC cleavage assays for detection of numbers of cells. Values are plotted as signal-to-noise (S:N) ratios. Comparison of data at S:N = 3 indicates approximately 30-fold better sensitivity of the GF-AFC assay after 30 min of incubation compared to resazurin incubated for 3.5 hr.

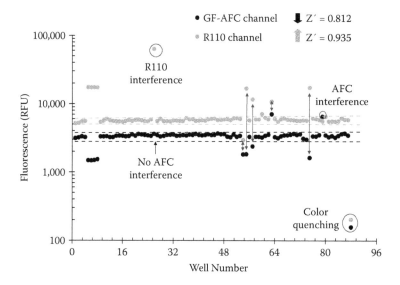

FIGURE 6.6 Multiplexing cell viability and cytotoxicity assays to identify artifacts from various sources. HeLa cells (10,000 per well in 50 μL medium) were added to white, clear bottomed plates. Test compounds or controls diluted in 50 μL medium were added to appropriate wells. MultiTox-Fluor reagent (containing GF-AFC and AAF-R110) was prepared, added in 100 μL, and incubated at 37°C for 30 min. Fluorescence was measured at 400 nm excitation/505 nm emission and 485 nm excitation/520 nm emission. Wells 1 to 4 were untreated viability controls. Wells 5 to 8 contained 30 μg/mL digitonin as a positive control to kill cells. Well 26 shows that R110 fluorescence interference at 485/520 scored as cytotoxic whereas the AFC fluorescence in well 26 was unaffected. The opposite scenario was demonstrated in well 79 where AFC fluorescence at 400/505 suggests an increase in viability. Wells 55, 57, and 75 containing cytotoxic compounds demonstrated an inverse relationship between the two biomarker values (long double-headed arrows). Well 88 demonstrated "flagging" of color quenching compounds not possible by a single parameter measure. Wells 54 and 63 received 5,000 and 20,000 cells, respectively, to demonstrate effects on magnitude of fluorescent values and how ratiometric measurements can be used to identify pipetting errors. (*Source:* Modified from Niles, A.L. et al. 2007b. Using protease biomarkers to measure viability and cytotoxicity. *Promega Cell Notes* 19, 16–20.)

where it can cleave amino acids from the AAF-R110 substrate and generate fluorescent rhodamine 110. The protease activity released from dead cells that cleaves AAF-R110 is relatively stable in culture medium, with a half-life of ~9 to 10 hr. The combination of the non-permeable nature of AAF-R110 and the rapid loss of the protease activity that cleaves GF-AFC upon cell death results in the ability to selectively detect both viable and dead populations of cells in the same sample.

The multiplex protocol to measure viable and dead cells in one sample utilizes a single reagent solution containing both the GF-AFC and AAF-R110 substrates. The combined substrates are added directly to samples of cells, incubated for 30 min, and fluorescence is recorded using two different filter sets to quantify the protease markers from the viable and dead populations. This homogeneous add-incubate-measure protocol to measure two endpoints can be used as an internal normalization control or provide additional useful information about the status of the population of cells treated (Figure 6.6). For example, if a treatment results in a decrease in viability measured with the GF-AFC substrate, you would expect to see an increase in dead cell marker activity in the same well. If fluorescence in the R110 channel increases (suggesting cell death) but the expected decrease in AFC fluorescence is not observed in that sample, the data may suggest that the tested compound did not reduce viable cell numbers but possibly contributed fluorescence in the R110 channel.

Measuring multiple parameters can be very useful for characterization of a cell culture model system during initial assay development and during subsequent optimization or miniaturization to a microwell plate format. The use of multiplexed internal controls also can be useful to identify false hits.

6.3.3.5 Luminogenic ATP Assay

Measurement of ATP is widely accepted as a valid method for estimating the number of viable cells from a variety of prokaryotic and eukaryotic sources. ATP levels are closely regulated by viable cells. When cells die, they lose the ability to synthesize ATP and endogenous ATPases rapidly deplete cytoplasmic stores to form ADP. ATP is a substrate for beetle luciferase. Although the chemical reaction is far more complex, a simplistic form to help explain the ATP assay for cell viability can be shown as:

$$ATP + luciferin + luciferase \rightarrow light$$

Early protocols to measure ATP from cell cultures involved an acid or alkaline sample extraction step to precipitate proteins and inhibit ATPases (Lundin et al. 1986; Crouch et al. 1993). This was followed by removal of debris by centrifugation or filtration and a step to neutralize the pH of the sample prior to addition of luciferin and firefly luciferase to generate a flash of light proportional to the amount of ATP. Early reagents that created "flash" kinetics required luminometers with injection capabilities that hindered their use for HTS.

Years of effort have been dedicated to optimizing reagent systems to capitalize on the advantages of using luminescence as a signal in biological samples. Several technological breakthroughs and advances resulted in homogeneous procedures that require only a single reagent addition step to measure ATP in a variety of sample types. The elimination of the liquid handling steps provided flexibility to measure ATP in suspension or attached cell types.

Probably the most important advance enabling the development of the modern homogeneous ATP assay was the engineering of improved luciferase molecules. Directed evolution techniques were used to select luciferase mutants based on thermal stability (Hall et al. 1998). The result was identification of luciferases that could remain enzymatically active in the presence of ATPase inhibitors and harsh detergents used for cell lysis. These properties were critical for the development of robust ATP assays performed with a single reagent addition step and yielding extended signal stability.

The modern ATP assay has become the HTS method of choice to measure cell viability. It is the fastest and most sensitive assay and is less prone to artifacts than fluorescent methods. A growing list of publications (Rossi et al. 2007; Severson et al. 2007; Melnick et al. 2006) and examples posted to the PubChem Bioassays web site (http://pubchem.ncbi.nlm.nih.gov/assay/) make available data resulting from screening supported by the Roadmap Initiative of the National Institutes of Health.

The assay is performed by addition of a single reagent that contains a detergent to lyse cells immediately, ATPase inhibitors to stabilize the ATP present, and luciferin and luciferase contained in an optimized buffer formulation to allow the prolonged generation of photons of light. Within 10 min after the addition of reagent, the signal stabilizes and will glow for several hours, typically with a half-life of about 5 hr.

The ATP assay is faster because it uses a fundamentally different approach from tetrazolium and resazurin assay protocols that usually require 2- to 4-hr incubation when the plates must be returned to a 37°C incubator to allow viable cells to generate measurable product. The first step of the ATP assay includes cell lysis that eliminates co-incubation of the reagent and test compound in the presence of living cells. The homogeneous add-mix-measure protocol of the ATP assay also eliminates handling steps required to return assay plates to a cell culture incubator for development of a signal dependent on viable cells. In addition to making the assay more convenient, elimination of an incubation step reduces possible artifacts caused by interaction of the reagent chemistry with viable cells as occurs with tetrazolium and resazurin reduction assays.

The luminescent ATP assay is the most sensitive HTS method available for measuring the viability of cell populations in microwell plates. The limits of detection determined in samples of eukaryotic cells serially diluted from a known concentration may fall below 10 cells per well (Figure 6.7). This enables miniaturization to a 1536-well format.

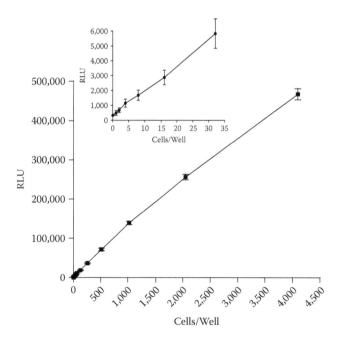

FIGURE 6.7 Sensitivity of detection of Jurkat cells using luminescent ATP assay. Serial twofold dilutions of cells were made in a 384-well solid white plate as 25 μL per well samples. An equal volume of ATP detection reagent was added and luminescence recorded after 10 min. Values represent mean ± standard deviation of eight replicates. Direct relationship between luminescence and cell number ($r^2 = 0.99$) is demonstrated. A Student's T-test indicates luminescence from a dilution containing four cells is significant over background. (*Source:* Modified from Hanna et al. 2001. CellTiter-Glo™ luminescent cell viability assay: a sensitive and rapid method for determining cell viability. *Promega Cell Notes* 2, 11–13.)

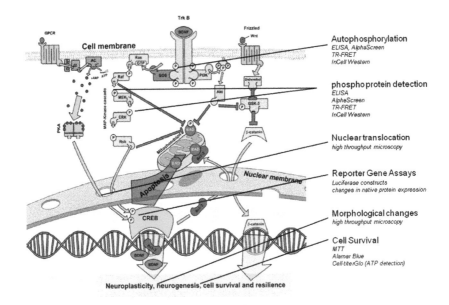

FIGURE 1.2   Cell-based assay concepts for kinase and phosphatase HTS.

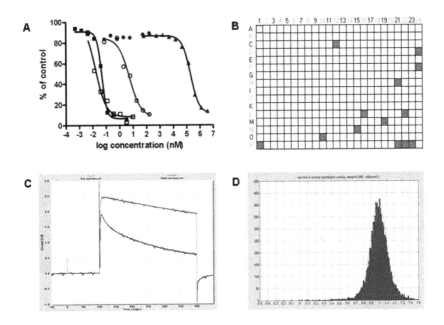

FIGURE 4.6   Pharmacology and screening results for $K_V1.3$ assay. (A) Concentration–response curves for inhibition of K+ currents by 4-AP (▲) and ShK-Dap[22] (□), ShK (■), charybdotoxin (○), and dendrotoxin (●, data not fitted with curve) peptides. The estimated IC_{50} values are listed in Table 4.1. Dendrotoxin is a selective $K_V1.1$ blocker and did not inhibit the $K_V1.3$ current in our assay. (B) View of typical compound plate from screening campaign. Thirteen wells failed to form seals (blue squares). The yellow squares identify wells in which K+ current was inhibited by >30%. All successful wells in columns 23 and 24 are yellow because they contain 3.33 mM of 4-AP, a standard blocking agent for $K_V1.3$; randomly located yellow wells represent unconfirmed hits. (C) Representative pre-compound (red) and post-compound (blue) K+ current traces from well H6 of the plate in (A). The test compound appears to be an open channel blocker, as evidenced by the increasing current inhibition at the later times during the pulse. (D) Frequency histogram of normalized K+ current amplitudes (post- and pre-compound) for the ion channel-focused library. The histogram shows a normal distribution curve with an average value of 0.995 and a standard deviation of 0.1. Therefore, a hit was defined in POC terms as a compound with POC <70 i.e., an inhibitory effect of >30%.

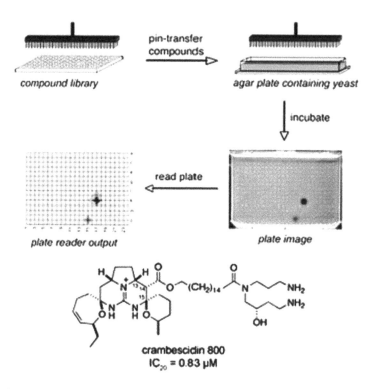

FIGURE 7.9 High-throughput yeast halo assay strategy (top) and structure of a potent antifungal agent (crambescidin 800) identified (bottom) (Source: Gassner, N.C. et al. 2007. *J. Nat. Prod.* 70, 383. Copyright 2007 American Chemical Society.)

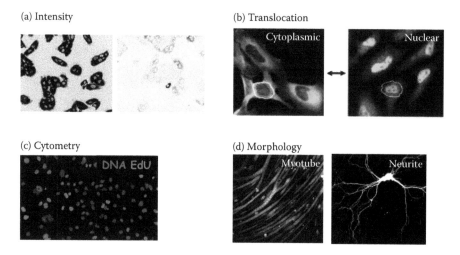

FIGURE 8.2 HCS assay formats. (a) Intensity measurement of phospho-S6. (b) Cytoplasmic–nuclear translocation of GFP-Foxo3a. Ratio of GFP-Foxo3a intensity in the nuclear mask region (red circle) and in the cytoplasmic mask region (yellow ring) was used as quantitative measurement. (c) Cytometry assay measuring percentage of S-phase cells. S-phase cells were labeled with ethynyl-dU incorporation and azido-rhodamine label (red) and all cells were labeled with Hoechst 33342 dye (blue). (d) Morphology assays on myotube formation and neurite formation. Myotube length and width and neurite length and branches were among the parameters of interest.

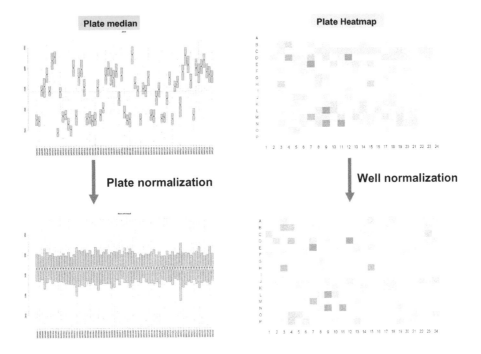

FIGURE 8.3 Pre-processing data normalization. Normalization reduces plate-to-plate and well-to-well variations, allowing uniform analysis of entire HCS dataset in other modules.

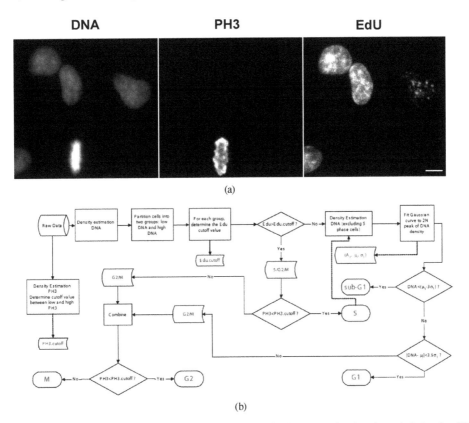

FIGURE 8.4 Decision tree cell cycle phase classification. (a) Cells were stained to show their levels of DNA, phospho-H3, and EdU in each nucleus. (b) Every cell was classified as G1 (2N DNA), G2 (4N DNA), M (pH3, positive) or S (EdU positive) phases using the automated four-parameter decision tree model. (c) The results of the classification were shown in two scatter plots where cells in different phases were labeled with different colors, G1 (orange), G2 (green), M (red), and S (blue), respectively.

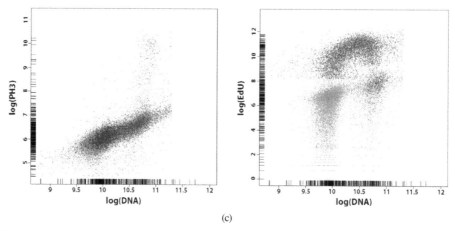

(c)

FIGURE 8.4 (*Continued*).

RNAi screening protocol

1. Infect shRNA virus or transfect siRNA into cells

2. Allow cells to reach desired knockdown (48h-120+ h)

3. Sensitize cells with agonist / antagonist

4. Read absorbance or acquire high content images from assay

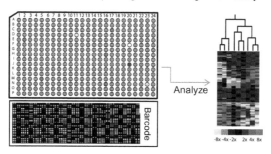

Analyze

FIGURE 9.1 RNAi screening protocol. Custom libraries or genome-wide RNAi libraries are designed, constructed, and synthesized with up to five duplexes for each gene. In gene-by-gene screening RNAi duplexes for each target are combined and arrayed in a 96- or 384-well format. Viral shRNAs constructs or liposome mediated siRNA oligonucleotides are forward (or reverse) transfected onto disease-relevant cell types. On the basis of validation data, one should achieve greater than 95% infection or transfection efficiency and, on average, greater than 80% knockdown for each mRNA target species tested. Infection or transfection proceeds for 48+ hr to allow for sufficient knockdown before cells are sensitized with agonist or antagonist compounds before reading the assay. RNAi barcode pooling approaches also involve determining the precise barcode that generated the desired phenotypic results.

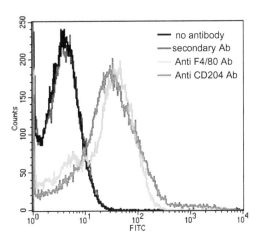

FIGURE 10.1 Primary mouse macrophages were analyzed by FACS after treatment with two different rat antibodies: anti-CD204 (dark green) and anti-F4/80 (pale green). As controls, untreated macrophages (black line) and the same cells treated with only secondary anti-rat FITC antibody (red line) were analyzed. The positive cell population clearly shifted to the right.

FIGURE 12.2 Example of large-scale automated storage and retrieval system. Left: environmentally controlled store with tube-filled trays inserted in slots of shelves. A transporting robot pulls a tray and sends it to a picking robot (upper right) where tubes are picked and placed in a rack with SBS standard plate footprint. The rack with tubes is sent for processing via a conveyance (lower right).

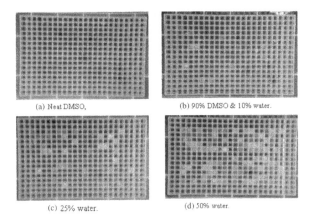

FIGURE 12.3 Effect of water on compound precipitation. A set of selected samples that previously showed signs of precipitation was placed in a 384-well plate. Four identical copies were made and the copies were diluted to sample volume; each had a different DMSO-to-water ratio. After 1 week in an ambient environment, the number of wells showing precipitates increased with the increase of water.

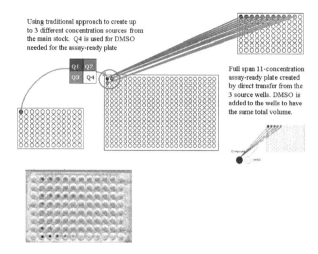

FIGURE 12.7 New methods of creating IC$_{50}$ assay-ready plate via direct transfer. Top: achieving 11 point, half log serial dilution by first creating a three-concentration source plate with a traditional approach and obtaining a full span assay-ready plate. Bottom: visual inspection of the plate.

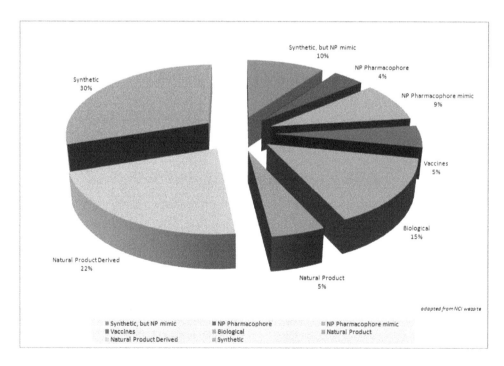

FIGURE 13.1 Pie chart showing genesis of all drugs approved ($n = 1272$) from January 1, 1981 through October 12, 2008. (*Source:* Adapted from Newman, D.J. and Cragg, G.M. 2007. *J. Nat. Prod.* 70, 461–477.)

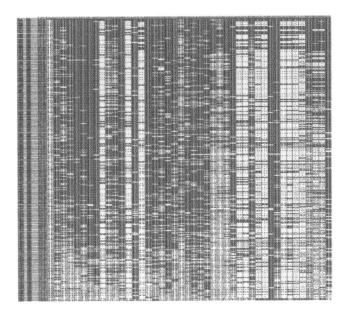

FIGURE 13.5 Partial view of screen in New Leads Generation (combinatorial chemistry and HTS) generating a huge quantity of hard-to-analyze results.

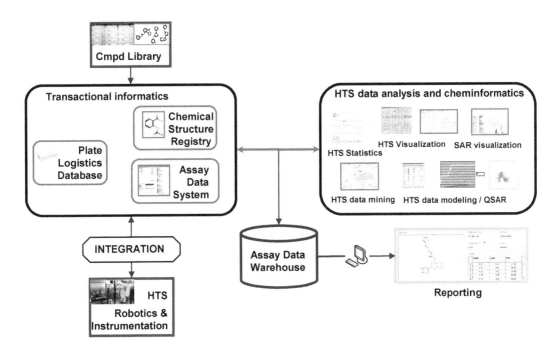

FIGURE 14.2 Components of transactional and data analysis (statistical and cheminformatics) environment in the context of HTS campaign.

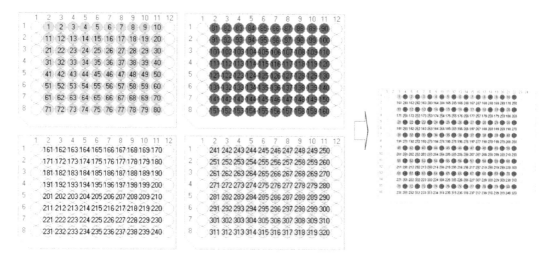

FIGURE 14.8 Simple plate remapping operation: 4×96 (1, 12 empty) $\rightarrow$ 384 (1, 2, 23, 24 empty).

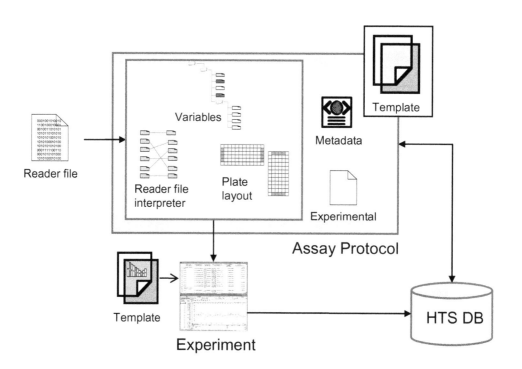

FIGURE 14.10 Components of assay data management system.

Modification of the assay chemistry designed for measuring bacterial cells can achieve detection of lower concentrations of ATP (Junker and Clardy 2007; Fan et al. 2005). At this sensitivity, the assay performance becomes less limited by the ATP detection chemistry than the ability to reproducibly deliver and culture cells in microwell plates. The sensitivity is based on the high signal-to-background ratios that result from extremely low background luminescence. This is in contrast to fluorescent detection methods that require excitation of a fluorophore by illuminating a sample with an incident beam of light. The result is excitation of many other fluorescent molecules present in cells and serum supplemented culture media that can contribute to background fluorescence (Auld et al. 2008; Fan and Wood 2007).

6.3.3.6   Disadvantages of ATP Assay

One disadvantage of the ATP assay is the need for constant temperature to achieve consistent readings among different assay plates and within different locations within the same plate. The luminescent signal is dependent on the rate of the luciferase reaction that is related to temperature. For that reason, assay plates are usually allowed to equilibrate to a constant ambient temperature before addition of reagent. Complete equilibration is necessary because even slight temperature differences across the interiors, edges, and corners of a plate will result in altered luminescence readings.

With all assays that measure metabolic markers, cells with different metabolic capacities will produce different signal intensities. Factors that alter cell metabolism may influence the amount of ATP per cell and result in misleading data. The stimulation of the amount of ATP per cell has been used as an advantage in some applications such as measuring lymphocyte activation (Bulanova et al. 1995; White et al. 1989; Augustine, Pasi, and Hill 2007). Because ATP levels are tightly regulated, a decrease in ATP often results in cell death. Another disadvantage of the ATP assay (or any assay based on luciferase) is the possibility of interference by luciferase inhibitors present in small molecule chemical libraries. Molecules that interfere with luciferase measurements have been identified and annotated (Inglese et al. 2006; Kashem et al. 2007). Reagent formulations may affect the occurrence and extent of inhibition.

Many of these disadvantages can be overcome by implementing appropriate internal assay controls or by confirming results via an independent multiplexed measurement. One limitation of the ATP assay for multiplexing is that the protocol kills cells and produces a sample with high detergent and luciferin content. Sequential multiplexing of assays is still possible as long as the assay chemistries are compatible; however, other assays that use viable cells should be performed first in the sequence, before the addition of the ATP assay reagent. Examples of multiplexing with the ATP assay include genetic reporters, glutathione, cytochrome P450, resazurin reduction, GF-AFC aminopeptidase, etc.

6.4   APOPTOSIS ASSAYS

Programmed cell death was originally described based on morphological observations of developing tissues. Apoptosis is an active and well defined form of programmed cell death that plays an important role in regulating population dynamics of cells in multicellular organisms under normal and pathological conditions (Locksin and Zakeri 2001; Kerr, Wyllie, and Currie 1972; Wyllie, Kerr, and Currie 1980).

Apoptosis has served as an increasingly popular research topic over the past two decades because we recognize that elements in the apoptotic signaling pathways may be important therapeutic targets for treating cancer and other diseases. Interest in screening large libraries of small molecules has been growing in attempts to identify modulators of the apoptotic pathway and this led to the need for easy-to-use HTS assay methods.

A variety of methods are available to measure different processes occurring during apoptosis, including direct observation of morphological events, changes in mitochondrial membrane potential, movements of proteins to different subcellular compartments, DNA fragmentation, caspase-specific cleavage of target proteins, covalent binding and staining with fluorescent caspase inhibitors, and binding of fluorescently labeled Annexin V to phosphatidyl serine that becomes exposed on the outer leaflets of cell membranes.

Many elements of complex intrinsic and extrinsic signaling pathways leading to apoptosis have been identified. During apoptosis, a cascade of proteases termed caspases is involved with upstream signaling events and downstream executioner events. Caspases are cysteine-dependent, aspartate-specific proteases that contain highly conserved cysteine residues in their active sites and cleave substrates leaving C terminal Asp residues. Caspase-3 is one of the main effector molecules of the apoptotic process. It cleaves several target proteins and serves as one of the executioner caspases that implement apoptosis. Despite reports of caspase-independent apoptosis (Bröker et al. 2005), caspase-3 has become the most widely accepted and most frequently measured apoptosis marker for HTS.

Analogous to cell viability measurements, many of the same basic concepts apply to the development of cell-based assays for apoptosis. The length of incubation of cells with the test compound is among the most important issues to address and optimize. The length of incubation is important because the markers of apoptosis may be present for relatively brief transient periods and subsequently disappear as the population of cells undergoes secondary necrosis. The induction of measurable caspase activity can occur in only a few minutes or can take days, depending on the model cell line, type of inducer, and effective concentration inside the cells.

The kinetics of apoptosis induction depend on the specific cell type used and the culture environment, so it is advisable to optimize assay parameters for each cell line and also when changing culture conditions during miniaturization of assays.

To measure apoptosis, small peptide substrates selective for different caspase family members were developed based on technology used for other proteases (Gargiulo et al. 1981). Substrate design evolved from a colorimetric approach using paranitroanalide as the leaving group on the C terminal side of aspartic acid to fluorogenic substrates containing different coumarin derivatives or R110.

Although the peptide aminomethylcoumarin (AMC) substrate approach has been widely used for measuring the activities of many proteases, a disadvantage of this approach arises from the ultraviolet excitation and emission wavelengths required for detecting free AMC where there is a possibility for fluorescence interference. R110-labeled substrates are less encumbered by interference due to a spectral red shift, but many compounds in small molecule chemical libraries exhibit fluorescent properties that may interfere with assays based on fluorescence detection (Simenov et al. 2008).

Luminogenic substrates containing peptides linked to aminoluciferin overcome the problem of fluorescence interference while providing much better detection sensitivity (Monsees, Miska, and Geiger 1994; O'Brien et al. 2005). Figure 6.8 shows coupled enzymatic reactions in which caspase-3 cleaves the peptide substrate to liberate aminoluciferin that then becomes available as a substrate for the luciferase reaction to generate light. The luminogenic caspase-3 assay is the fastest and most sensitive method available for miniaturized HTS. The homogeneous add-incubate-read assay procedure involves adding a single reagent mixture directly to cells growing in multiwell plates. The samples are incubated for 30 to 60 min to achieve a steady state between the protease and luciferase enzymes, then luminescence is recorded to quantify the caspase-3 activity present. The reagent contains detergent, a stable form of luciferase, and ATP in a buffered formulation optimized to efficiently lyse cells and stabilize both caspase-3 and luciferase activities for hours. The half-life of the luminescent signal is typically >5 hr.

Direct comparison of the sensitivity of fluorogenic and luminogenic caspase-3 assays using R110 or aminoluciferin conjugated with the same DEVD peptide sequence has been reported to yield 20-fold greater sensitivity by the luminogenic method (O'Brien et al. 2005). As with any assay based on luciferase, the compounds may interfere with the luciferase activity. A decrease in luminescent signal may result from inhibition of caspase-3 or luciferase. A direct comparison of caspase-3 inhibition using fluorogenic and luminogenic assays to screen a library of 640 pharmacologically active compounds resulted in a lower false hit rate with the luminescent assay (O'Brien et al. 2005). Temperature is another factor that can affect the results of luminescent caspase-3 assays. To ensure consistent plate-to-plate results, samples are usually allowed to equilibrate to a constant ambient temperature before addition of reagent, similar to the procedure recommended for the ATP assay.

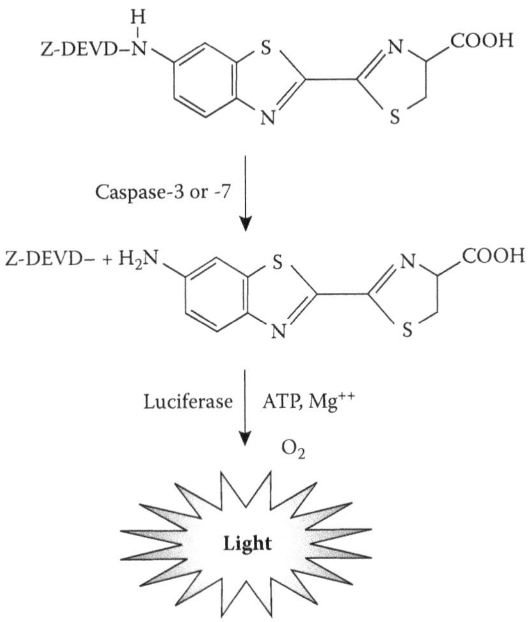

FIGURE 6.8 Dual enzymatic reactions of luminogenic protease assay to measure caspase-3 or -7 as an indicator of apoptosis. Caspase-3 ot -7 cleavage of the luminogenic protease substrate containing Z-DEVD-aminoluciferin releases aminoluciferin that can be used as a substrate for luciferase, resulting in production of light. (*Source:* Modified from Promega Corporation Technical Bulletin 323. *Caspase-Glo® 3/7 Assay.*)

6.5 MULTIPLEXING CELL-BASED ASSAYS

False hits caused by assay artifacts represent a major concern for HTS. The risk of collecting flawed or misleading data can be partially mitigated by repeating an assay or using a confirmatory assay that detects a different parameter to measure the same event. Regardless of the type of cell-based assay used, knowing the number of viable cells at the end of the treatment period is practical information that can be utilized to normalize results. For example, if a single genetic reporter assay shows a decrease in signal, it is impossible to tell whether the decrease was a specific downregulation of the reporter or if the treatment led to cytotoxicity. Multiplexing both genetic reporter and viability assays in the same well provides additional insight that may suggest a mechanism of action (Figure 6.9). Multiplexing viability assays also can help reveal pipetting errors or differential growth patterns of cells across an assay plate.

To achieve multiplexing, the assay chemistries must be compatible and the signals must be distinguishable. Although it is desirable to measure both viability and apoptosis in the same sample, multiplexing a luminogenic ATP assay and a luminogenic caspase-3 assay is a combination that cannot be achieved because the reagent chemistries are not compatible. The reagent used to measure ATP contains an excess of luciferin—the compound detected in the luminogenic caspase-3 assay. The reagent used to measure caspase-3 activity contains an abundance of ATP—the marker measured in the luminescent viability assay. In cases such as this, multiplexing can be achieved by using appropriate assay reagents or by following a specified sequence of reagent additions. Figure 6.10 demonstrates that measuring viability and apoptosis from the same sample is possible using the GF-AFC protease substrate first in sequence to measure viable cell number followed by addition of the luminogenic caspase-3 reagent to quantify apoptosis (Niles, Moravec, and Riss 2008). The sequence of reagent addition is dictated by the parameter to be measured from a viable cell population. The GF-AFC protease substrate requires live cells, whereas the caspase-3 reagent lyses cells and thus must be added last.

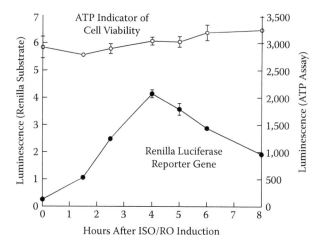

FIGURE 6.9 Sequential homogeneous multiplexing assay to measure *Renilla* luciferase reporter gene activity in live cells, followed by measuring ATP as an indicator of cell viability as a control in the same sample wells. Reporter gene activity in living cells increased over time, but cell numbers (ATP contents) remained constant. Multiplex data confirmed the increase in luminescence was due to specific upregulation of luciferase expression rather than an increase in viable cells. (*Source:* Modified from Riss, T.L. et al. 2005. Selecting cell-based assays for drug discovery screening. *Promega Cell Notes* 13, 16–21.)

Another approach to overcome assay chemistry incompatibility is splitting a sample into two different containers. If the cell population releases a marker into the surrounding environment, a small aliquot of culture medium can be sampled and moved to a different assay plate to segregate assay chemistries. Figure 6.11 shows the measurement of lactate dehydrogenase released from a

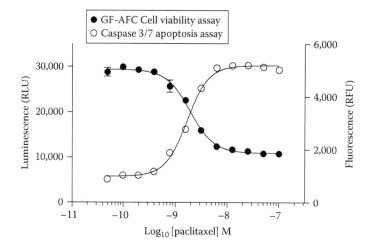

FIGURE 6.10 Multiplexing of fluorescent cell viability assay measuring cleavage of GF-AFC with the luminogenic apoptosis assay measuring caspase-3 cleavage of Z-DEVD-aminoluciferin. The GF-AFC reagent was added to wells containing 10,000 cells per well, incubated 30 min at 37°C, and fluorescence measured as an indicator of cell viability. Luminogenic caspase-3/7 reagent was added, incubated 30 min at room temperature and luminescence measured as an indicator of apoptosis. The inverse relationship confirms loss of viability due to apoptosis. (*Source:* Modified from Promega Corporation Technical Bulletin 371. *CellTiter-Fluor*™ *Cell Viability Assay.*)

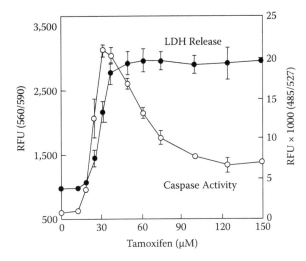

FIGURE 6.11 Multiplex measurement of cytotoxicity and apoptosis markers from the same sample of cells accomplished by separating incompatible assay chemistries into different containers. HepG2 cells (10,000 per well) were plated in a solid white and clear bottom 96-well plate and cultured overnight. Various concentrations of tamoxifen were added to the wells and incubated for 4 hr at 37°C. To assay LDH activity, 50 μL per well of culture supernatant was transferred to a 96-well plate to which 50 μL per well of LDH assay reagent was added. LDH samples were incubated at ambient temperature for 30 min prior to stopping the reaction to measure fluorescence at 560/590 nm. For caspase-3/7 determination, 50 μL per well of *bis*-DEVD-R110 reagent was added to the original culture plate containing cells, and incubated at ambient temperature for 45 min prior to determining fluorescence at 485/527 nm. (*Source:* Modified from *Cell Viability Protocols and Application Guide*. Madison, WI: Promega Corporation.)

population of tamoxifen-treated necrotic cells (Korzeniewski and Callewaert 1983). An aliquot of culture medium was removed to a separate assay plate without disturbing the remaining cells in the original plate that was then treated with a cell lysis reagent containing a fluorogenic peptide substrate to measure caspase-3 activity as a marker of apoptosis. Although this multiplexing approach is not considered homogeneous because of a liquid transfer step, it still presents the advantage of providing two different measurements from the same population of cells.

6.6 SUMMARY AND CONCLUSIONS

A variety of assay technologies have been developed to enable miniaturized screening of viability and apoptosis. Each approach has advantages and disadvantages that must be considered before choosing the most appropriate assay. The ultimate goal is to obtain the highest quality data set in the shortest time by the most cost-effective method possible.

It is important the assay technology does not lead to artifacts caused by interactions of the reagent chemistry with physiological events in the cells. Although the tetrazolium and resazurin reduction assays historically gained popularity because they are homogeneous and do not require the removal of culture medium from the sample, the limited detection sensitivity and the need to incubate viable cells with reagent for 1 to 4 hr to generate an adequate signal are among the reasons the ATP viability assay has become the most popular for HTS. Similar advantages of speed and improved sensitivity to enable miniaturization led to the current popularity of the luminogenic caspase assay approach for detecting apoptosis in HTS.

Understanding the biology and kinetics of the cell death process and realizing that cell culture is an artificial model system can help guide characterization experiments to develop the best assay possible. Probably the most common problem facing design of viability and apoptosis assays for HTS is deciding on the duration of incubation with the test compounds before making a measurement.

Often the results of screening campaigns from pharmaceutical or biotechnology companies are not published to protect their intellectual property positions; however, a growing collection of screening data from NIH-funded programs is now available to the public (including descriptions of methods used) through the PubChem bioassay web site. These postings provide insight into the types of assays used in the academic screening community. The assays posted have undergone sufficient development to verify their readiness for HTS. Additional resources are available to aid in assay development, including an *Assay Guidance Manual* resulting from a collaboration between Eli Lilly & Co. and NIH's Chemical Genomics Center. This resource is currently available on the Internet and continues to be expanded to cover more topics.

The weaknesses of some assay approaches can be compensated for during counter screening or by simultaneously multiplexing with a secondary assay. Options are available for multiplexing viability and apoptosis assays with each other or as internal normalization controls for other cell-based assays. The ability to multiplex assay measurements in entire populations of cells can be an efficient substitute for a high content imaging approach, especially when screening a large collection of compounds or when automated imaging instrumentation is not available.

REFERENCES

Ahmed, S.A., Gogal, R.M., and Walsh, J.E. 1994. A new rapid and simple nonradioactive assay to monitor and determine the proliferation of lymphocytes: an alternative to [³H]thymidine incorporation assays. *J. Immunol. Meth.* 170. 211–224.

Augustine, N.H., Pasi, B.M., and Hill, H.R. 2007. Comparison of ATP production in whole blood and lymphocyte proliferation in response to phytohemagglutinin. *J. Clin. Lab. Anal.* 21, 265–270.

Auld, D.S. et al. 2008. Characterization of chemical libraries for luciferase inhibitory activity. *J. Med. Chem.* 51, 2372–2386.

Barltrop, J. and Owen, T. 1991. 5-(3-carboxymethoxyphenyl)-2-(4,5-dimethylthiazoly)-3-(4-sulfophenyl)tetrazolium, inner salt (MTS) and related analogs of 3-(4,5-dimethylthiazolyl)-2,5-diphenyltetrazolium bromide (MTT) reducing to purple water-soluble formazans as cell viability indicators. *Bioorg. Med. Chem. Lett.* 1, 611–614.

Borenfreund, E. and Puerner, J.A. 1985. Toxicity determined *in vitro* by morphological alterations and neutral red absorption. *Toxicol. Lett.* 24, 119–124.

Bröker, L.E. et al. 2005. Cell death independent of caspases: a review. *Clin. Cancer Res.* 11, 3155–3162.

Bulanova, E.G. et al. 1995. Bioluminescent assay for human lymphocyte blast transformation. *Immunol. Lett.* 46, 153–155.

Cavanaugh, P.F. et al. 1990. A semi-automated neutral red based chemosensitivity assay for drug screening. *Invest. New Drugs* 8, 347–354.

Chakrabarti, R. et al. 2000. Vitamin A as an enzyme that catalyzes the reduction of MTT to formazan by vitamin C. *J. Cellular Biochem.* 80, 133–138.

Chatterjee, R. 2007. Cases of mistaken identity. *Science* 315, 928–930.

Collier, A. and Pritsos, C. 2003. The mitochondrial uncoupler dicumerol disrupts the MTT assay. *Biochem. Pharm.* 66, 281–287.

Cory, A. et al. 1991. Use of an aqueous soluble tetrazolium/formazan assay for cell growth assays in culture. *Cancer Commun* 3, 207–212.

Crouch, S.P. et al. 1993. The use of ATP bioluminescence as a measure of cell proliferation and cytotoxicity. *J. Immunol. Meth.* 160, 81–88.

Denizot, F. and Lang, R. 1986. Rapid colorimetric assay for cell growth and survival: modifications to the tetrazolium dye procedure giving improved sensitivity and reliability. *J. Immunol. Meth.* 89, 271–277.

Digan, M.E. et al. 2005. Evaluation of division-arrested cells for cell-based high-throughput screening profiling. *J. Biomol. Screen.* 10, 615–623.

Ekwall, B. et al. 1998. MEIC evaluation of acute systemic toxicity VI: prediction of human toxicity by rodent LD_{50} values and results from 61 *in vitro* methods. *ATLA* 26, 617–658.

Essodaigui, M., Broxterman. H.J., and Garnier-Suillerot, A. 1998. Kinetic analysis of calcein and calcein-acetoxymethylester efflux mediated by the multidrug resistance protein and P-glycoprotein. *Biochemistry* 37, 2243–2250.

Fan, F. et al. 2005. BacTiter-Glo™ assay for antimicrobial drug discovery and general microbiology. *Promega Notes* 89, 25–27.

Fan, F. and Wood, K. 2007. Bioluminescent assays for high-throughput screening. *Assay Drug Dev. Technol.* 5:127–136.

Gargiulo, R.J. et al. 1981. Analytical fluorogenic substrates for proteolytic enzymes. United States Patent 4275153.

Garner, D.L. et al. 1994. Dual DNA staining assessment of bovine sperm viability using SYBR-14 and propidium iodide. *J. Androl.* 15, 620–629.

Hall, M.P. et al. 1998. Stabilization of firefly luciferase using directed evolution. In *Bioluminescence and Chemiluminescence: Perspectives for the 21st Century*, Roda, A., et al., Eds. Chichester: John Wiley & Sons, 392–395.

Hanna et al. 2001 CellTiter-G10™ luminescent cell viability assay: a sensitive and rapid method for determining cell viability. *Promega Cell Notes* 2, 11–13.

Hansen, M.B., Nielsen, S.E. and Berg, K. 1989. Re-examination and further development of a precise and rapid dye method for measuring cell growth/cell kill. *J. Immunol. Meth.* 119, 203–210.

Inglese, J. 2008. Characterization of chemical libraries for luciferase inhibitory activity. *J. Med. Chem.* 51, 2372–2386.

Inglese, J. et al. 2006. Quantitative high-throughput screening: a titration-based approach that efficiently identifies biological activities in large chemical libraries. *Proc. Natl. Acad. Sci. USA* 103, 11473–11478.

Ishiyama, M. et al. 1993. A new sulfonated tetrazolium salt that produces a highly water-soluble formazan dye. *Chem. Pharm. Bull.* 41, 1118–1122.

Ivnitski-Steele, I. et al. 2008. High-throughput flow cytometry to detect selective inhibitors of ABCB1, ABCC1, and ABCG2 transporters. *Assay Drug. Dev. Technol.* 6, 263–276.

Junker, L.M. and Clardy, J. 2007. High-throughput screens for small-molecule inhibitors of *Pseudomonas aeruginosa* biofilm development. *Antimicrob. Agents Chemother.* 51, 3582–3590.

Karászi, É. et al. 2001. Calcein assay for multidrug resistance reliably predicts therapy response and survival rate in acute myeloid leukaemia. *Br. J. Haematol.* 112, 308–314.

Kashem. M. et al. 2007. Three mechanistically distinct kinase assays compared: measurement of intrinsic ATPase activity identified the most comprehensive set of ITK inhibitors. *J. Biomol. Screen.* 12, 70–83.

Kerr, J.F., Wyllie, A.H., and Currie, A.R. 1972. Apoptosis: a basic biological phenomenon with wide-ranging implications in tissue kinetics. *Br. J. Cancer* 26, 239–257.

Korzeniewski, C. and Callewaert, D.M. 1983. An enzyme-release assay for natural cytotoxicity. *J. Immunol. Meth.* 64, 313–320.

Lockshin, R.A. and Zakeri, Z. 2001. Programmed cell death and apoptosis: origins of the theory. *Nat. Rev. Mol. Cell Biol.* 2, 545–550.

Lundin, A. et al. 1986. Estimation of biomass in growing cell lines by ATP assay. *Meth. Enzymol.* 133, 27–42.

Melnick, J.S. et al. 2006. An efficient rapid system for profiling the cellular activities of molecular libraries. *Proc. Natl. Acad. Sci. USA* 103, 3153–3158.

Monsees, T., Miska, W. and Geiger, R. 1994. Synthesis and characterization of a bioluminogenic substrate for α-chymotrypsin. *Anal. Biochem.* 221, 329–334.

Mosmann, T. 1983. Rapid colorimetric assay for cellular growth and survival: application to proliferation and cytotoxicity assays. *J. Immunol. Meth.* 65, 55–63.

Niles, A., Moravec, R., and Hesselberth, E. 2007. A homogeneous assay to measure live and dead cells in the same sample by detecting different protease markers. *Anal. Biochem.* 366, 197–206.

Niles, A.L. et al. 2007b. Using protease biomarkers to measure validity and cytotoxicity. *Promega Cell Notes* 19, 16–20.

Niles, A., Moravec, R., and Riss, T.L. 2008. Multiplex caspase activity and cytotoxicity assays. *Methods Mol. Biol.* 414, 151–162.

O'Brien, M. et al. 2005. Homogeneous, bioluminescent protease assays: caspase-3 as a model. *J. Biomol. Screen.* 10, 137–148.

O'Brien, P. and Haskins, J.R. 2006. *In vitro* cytotoxicity assessment. In *Methods in Molecular Biology 356: High Content Screening* Taylor, D.L. et al. Totowa, NJ: Humana, 415–425.

Pagliacci, M. et al. 1993. Genistein inhibits tumour cell growth *in vitro* but enhances mitochondrial reduction of tetrazolium salts: a further pitfall in the use of the MTT assay for evaluating cell growth and survival. *Eur. J. Cancer* 29, 1573–1577.

Paull, K.D. et al. 1988. The synthesis of XTT: a new tetrazolium reagent that is bioreducible to a water-soluble formazan. *J. Heterocyclic Chem.* 25, 911–914.

Riss, T.L. et al., 2005. Selecting cell-based assays for drug discovery screening. *Promega Cell Notes* 13, 16–21.

Riss, T.L. and Moravec, R.A. 2004. Use of multiple assay endpoints to investigate the effects on incubation time, dose of toxin, and plating density in cell-based cytotoxicity assays. *Assay Drug Dev. Technol.* 2, 51–62.

Rossi, C. et al. 2007. Identifying druglike inhibitors of myelin-reactive T cells by phenotypic high-throughput screening of a small-molecule library. *J. Biomol. Screen.* 12, 481–489.

Rubinstein, L.V. et al. 1990. Comparison of *in vitro* anticancer drug screening data generated with a tetrazolium assay versus a protein assay against a diverse panel of human tumor cell lines. *J. Natl. Cancer Inst.* 82, 1113–1117.

Severson, W.E. et al. 2007. Development and validation of a high-throughput screen for inhibitors of SARS CoV and its application in screening of a 100,0000-compound library. *J. Biomol. Screen.* 12, 33–40.

Shoemaker, R.H. 2006. The NCI60 human tumour cell line anticancer drug screen. *Nat. Rev. Cancer* 6, 813–823.

Shum, D. et al. 2008. A high density assay format for the detection of novel cytotoxic agents in large chemical libraries. *J. Enz. Inhib. Med. Chem.* 23, 931–945.

Simenov, A. et al. 2008. Fluorescence spectroscopic profiling of compound libraries. *J. Med. Chem.* 51, 2363–2371.

Skehan P. 1990. New colorimetric cytotoxicity assay for anticancer-drug screening. *J. Natl. Cancer Inst.* 82, 1107–1112.

Skehan, P. et al. 1989. Evaluation of colorimetric protein and biomass stains for assaying drug effects upon human tumor cell lines. *Proc. Amer. Assoc. Cancer Res.* 30, 612–616.

Squatrito, R., Connor, J., and Buller, R. 1995. Comparison of a novel redox dye cell growth assay to the ATP bioluminescence assay. *Gynecol. Oncol.* 58, 101–105.

Tada, H. et al. 1986. An improved colorimetric assay for interleukin 2. *J. Immunol. Meth.* 93, 157–165.

Tominaga, H. et al. 1999. A water-soluble tetrazolium salt useful for colorimetric cell viability assay. *Anal. Commun.* 36, 47–50.

Ulukaya, E., Colakogullari, M., and Wood, E.J. 2004. Interference by anti-cancer chemotherapeutic agents in the MTT-tumor chemosensitivity assay. *Chemotherapy* 50, 43–50.

Vukicevic, S. et al. 1992. Identification of multiple active growth factors in basement membrane Matrigel suggests caution in interpretation of cellular activity related to extracellular activity related to extracellular matrix components. *Exp. Cell Res.* 202, 1–8.

Wesierska-Gadek, J. et al. 2005. A new multiplex assay allowing simultaneous detection of the inhibition of cell proliferation and induction of cell death. *J. Cell. Biochem.* 96, 1–7.

White, A.G. et al. 1989. Lymphocyte activation: changes in intracellular adenosine triphosphate and deoxyribonucleic acid synthesis. *Immunol Lett.* 22, 47–50.

Wigglesworth, K.J. et al. 2008. Use of cryopreserved cells for enabling greater flexibility in compound profiling. *J. Biomol. Screen.* 13, 354–362.

Wyllie, A., Kerr, J., and Currie, A. 1980. Cell death: the significance of apoptosis. *Int. Rev. Cytol.* 68, 251–306.

Xia, M. et al. 2008. Compound cytotoxicity profiling using quantitative high-throughput screening. *Envir. Health Perspect.* 116, 284–291.

Zaman, G.J.R. et al. 2007. Cryopreserved cells facilitate cell-based drug discovery. *Drug Disc. Today* 12, 521–526.

7 Assay Development for Antimicrobial Drug Discovery

H. Howard Xu

CONTENTS

7.1 INTRODUCTION

The discovery of antimicrobial agents is one of the most significant advances in the history of medicine. From Ehrlich's salvarsan to Fleming's penicillin and other classes of antibiotics and antifungals, these drugs have, for nearly a century, saved the lives of millions of people suffering from life-threatening infections and improved the lives of many more infected by less dangerous microbes. However, for each new drug introduced for clinical use, resistance has sooner or later emerged (Powers, 2004; Gulshan and Moye-Rowley, 2007).

In recent decades, misuse of antimicrobial agents in medicine and agriculture has accelerated the emergence and spread of multidrug-resistant pathogenic microorganisms (Prasad and Kapoor, 2005; Amyes, 2007; Nordmann et al., 2007; Nicasio, Kuti, and Nicolau, 2008; Vergidis and Falagas, 2008). In addition, the problems of antibiotic spectrum, toxicity, and allergic reactions to antibiotics continually plague applications of existing antimicrobials. The need to discover and develop new antimicrobial therapeutics will continue for the foreseeable future. This chapter will focus on recent assay development approaches for discovery of antibacterial and antifungal drugs. Since viruses are non-cellular microorganisms and the approaches to discovery of antiviral drugs are quite distinct from those for antibacterials and antifungals, antiviral drugs will not be addressed. However, excellent reviews covering antiviral drug discovery are available (Westby et al., 2005; Tanikawa, 2006; De Clercq, 2007; von Itzstein, 2007).

Traditionally, the most productive paradigm for antimicrobial lead discovery has been the use of cell-based growth inhibition as a screen for inhibitory molecules, followed by optimization of candidates via medicinal chemistry to improve potency and selectivity of the initial active compounds. Most antimicrobials resulting from such lead optimization programs were initially identified as active ingredients of natural product extracts. However, this approach has generated diminishing returns in the past few decades. Fewer new chemical entities exerting bioactivity have been discovered; most of the major active components of natural product extracts have been identified and even continually rediscovered during repeated screens (von Nussbaum et al., 2006).

Since 1995, when the genome of the first cellular organism was completely sequenced (Fraser et al., 1995), microbial genomics has advanced at an astonishing rate. Now, hundreds of microbial genomes have been completely sequenced (Fraser et al., 1995; Goffeau et al., 1996; Blattner et al., 1997; Kunst et al., 1997; Tomb et al., 1997; Kuroda et al., 2001; Parkhill et al., 2001; Tettelin et al., 2001; Read et al., 2003; Ward and Fraser, 2005; Fedorova et al., 2008), paving the way for the establishment and advancement of a new discipline: microbial functional genomics—the study of the structures and functions of genes and their encoded products on a genome-wide scale. The study of essential genes has become a prime area of investigation in microbial functional genomics. Since antimicrobial drugs in current use target only a small percent of the entire complement of essential gene products, which by virtue of their essentiality are potential drug targets, the identification and subsequent studies of under- or unexplored essential genes and their products can be expected to accelerate the discovery of novel antimicrobial agents capable of combating multi-drug resistant pathogens (Haney et al., 2002).

The number of essential gene products for a given bacterium is estimated to be between 200 and 300, but currently used antibiotics target only about 20, or no more than 10%, of these essential proteins (Gil et al., 2004). The potential value of the untapped drug targets has fostered various novel approaches for essential gene identification in a number of bacterial species (Pucci, 2007). Comparative genomics combined with conditional expression has been used to identify essential genes in *Escherichia coli* (Arigoni et al., 1998; Freiberg et al., 2001). Additionally, promoter replacement techniques have been developed to determine *Staphylococcus aureus* essential genes (Jana et al., 2000; Zhang et al., 2000; Fan et al., 2001). Genome-scale transposon-based insertion mutagenesis has been employed to identify essential genes in *Mycoplasma genitalium* (Hutchison et al., 1999), *Haemophilus influenzae* (Reich, Chovan, and Hessier, 1999; Akerley et al., 2002), and *E. coli* (Gerdes et al., 2003). Moreover, gene disruption methodologies have been used to identify essential genes in *Streptococcus pneumoniae* (Thanassi et al., 2002) and *Bacillus subtilis* (Kobayashi et al., 2003). An antisense RNA expression approach has identified a comprehensive set of essential genes

TABLE 7.1
Number of Essential Genes Determined for Various Bacterial Species

Species	Number of Essential Genes	Method	References
Bacillus subtilis	271	Gene disruption	Kobayashi et al., 2003
Mycoplasma genitalium	265–350	Transposon mutagenesis	Hutchison et al., 1999
Streptococcus pneumoniae	113	Gene disruption	Thanassi et al., 2002
Haemophilus influenzae	478	Transposon mutagenesis	Akerley et al., 2002
Escherichia coli	620	Transposon mutagenesis	Gerdes et al., 2003
E. coli (PEC database)	302	Various methods	PEC
Staphylococcus aureus	150	Antisense RNA expression	Ji et al., 2001
Staphylococcus aureus	168	Antisense RNA expression	Forsyth et al., 2002
Salmonella enterica	490	Transposon site hybridization	Knuth et al., 2004
Francisella novicida	312	Transposon mutagenesis	Gallagher et al., 2007
Acinetobacter baylyi ADP1	499	Transposon mutagenesis	de Berardinis et al., 2008
Typical bacterial species	206	Theoretical analyses	Gil et al., 2004

in *S. aureus* (Ji et al., 2001; Forsyth et al., 2002). Finally, combined analysis of several computational models and experimental strategies led Gil and colleagues (2004) to propose that a minimal bacterial essential set contains 206 genes.

Different species of bacteria may require distinct sets of essential genes for their growth and survival. In addition, various methodologies for determining the essentiality of genes have used different approaches experimentally and may have individual limitations. It is not surprising that these methods produced very different estimates of the numbers of essential genes for respective bacteria (Table 7.1). Similarly, essential genes from a number of fungal organisms have been identified via gene deletion resulting in haplo-insufficiency (Giaever et al., 1999, 2002), gene replacement and conditional expression (Roemer et al., 2003), antisense RNA expression (De Backer et al., 2001), and conditional promoter replacement (Hu et al., 2007). The availability of these sets of essential genes served as the foundation for novel biochemical and cell-based approaches to the discovery of antimicrobial drugs.

7.2 ANTIBACTERIAL DRUG DISCOVERY

The approaches used to discover antibacterial drugs include cell-free target-based biochemical screens (Section 7.2.1), cell-based assays combined with underexpression (cell sensitization, Section 7.2.2) or overexpression (cell resistance, Section 7.2.3), and other cell-based assays followed by sophisticated molecular biology and genetic analyses (Section 7.2.4). These approaches are discussed in detail in the following sections.

7.2.1 TARGET-BASED BIOCHEMICAL ASSAYS

FtsZ, a tubulin-like GTPase cell division protein encoded by the *ftsZ* gene, plays an essential role in bacterial cell division. During cell division, FtsZ forms polymers in the presence of GTP that recruit other division proteins to make the cell division apparatus (Bramhill, 1997; Wang et al., 2003). Scientists at Merck & Co. developed a fluorescent FtsZ polymerization assay (Trusca and Bramhill, 2002) and screened over 100,000 extracts of microbial fermentation broths and plants (Wang et al., 2003).

The assay takes advantage of the fact that fluorescently tagged FtsZ protein can form polymers of large molecular weight that are retained on the 0.2 μm filters of 96-well filter plates. Retained polymers can be quantified by measuring the fluorescence on the filter wells with a plate reader using 485 nm excitation and 535 nm emission filters (Trusca and Bramhill, 2002; Wang et al., 2003). Active extracts were identified by reduction in retained fluorescence. Bioassay-guided purification of the activity in one of the most promising extracts led to the isolation of viriditoxin (Wang et al., 2003), a compound initially reported as a toxic metabolite from *Aspergillus viridinutans*. Viriditoxin exhibited inhibitory activity of polymerization with an IC_{50} (concentration of an inhibitor required to decrease enzyme activity by 50% compared to control) of 8.2 μg/mL. At 50 μg/mL, viriditoxin inhibited cell division by promoting formation of filamentous *E. coli* cells. It exhibited moderate antimicrobial activities against several Gram-positive bacterial pathogens with single-digit MICs (minimum inhibitory concentrations) (Wang et al., 2003).

Recently, Haydon and colleagues (2008) described a new class of small molecule inhibitors of FtsZ, exemplified by PC190723 (Figure 7.1A). PC190723 possesses potent and selective *in vitro* bactericidal activity against staphylococci, including methicillin-resistant and multidrug-resistant *S. aureus*. This compound was also efficacious in an *in vivo* model of infection, curing mice infected with lethal doses of *S. aureus* (Figure 7.1B) (Haydon et al., 2008).

Peptide deformylase (PDF), which is not found in mammalian cells, is an essential bacterial metalloenzyme that has received much attention as one of the targets for developing novel antibiotics (Clements et al., 2001; Hackbarth et al., 2002). Bacterial protein synthesis, unlike cytosolic protein synthesis in mammalian cells, is initiated by *N*-formylmethionine (Adams and Capecchi, 1966).

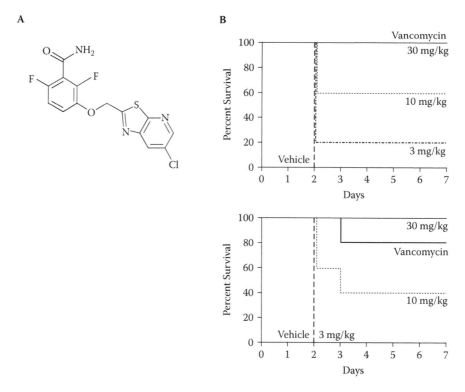

FIGURE 7.1 Characterization of cell division inhibitor PC190723. (A) Chemical structure of PC190723. (B) *In vivo* efficacy of PC190723 in a murine model of infection. Mice were injected intraperitoneally (IP) with a lethal inoculum of *S. aureus* ATCC 19636 at time 0. One hour after infection the animals received 3 mg/kg (uneven dashed line), 10 mg/kg (dotted line), or 30 mg/kg (dark line) of PC190723; negative control (vehicle only, dashed black line); or 3 mg/kg of the vancomycin control antibiotic (thick dark line) by subcutaneous (SC, top) or intravenous (IV, bottom) administration. Mortality was recorded daily for 7 days. (*Source:* From Haydon, D.J. et al. 2008. *Science* 321, 1673, 2008. Reprinted with permission from AAAS.)

The *N*-formylmethionine of a nascent protein synthesized in bacteria is removed by the sequential activities of PDF and a methionine aminopeptidase to generate the mature protein. The gene encoding PDF was cloned and overexpressed in *E. coli* by Meinnel and coworkers (1993). The PDF enzyme has an unusual metal ion (Fe^{2+}) as its catalyst. However, the ferrous ion in this enzyme is unstable and can be quickly and irreversibly oxidized to ferric ion, rapidly inactivating the enzyme. PDF-based assay development therefore depended on the ability of nickel ion to replace ferrous ion *in vitro*, increasing the stability of the enzyme and maintaining its enzymatic activity (Groche et al., 1998; Clements et al., 2001; Hackbarth et al., 2002).

Specifically, PDF assays can be performed as described by Clements and coworkers (2001) with the following conditions in a final volume of 100 μL: 8 ng of PDF with nickel ion (PDF·Ni), an appropriate concentration of compound to be tested, 80 mM HEPES (pH 7.4), 0.7 M KCl, 0.035% Brij, 1 mM $NiCl_2$, and 4 mM f-Met-Ala-Ser. The reaction is carried out at 37°C for 30 min. The free amino group of the product is detected using fluorescamine by the addition of 50 μL of 0.2 M sodium borate (pH 9.5) followed by 50 μL of fluorescamine (0.2 mg/mL in dry dioxane). Fluorescence can be quantified with a plate reader using an excitation wavelength of 390 nm and an emission wavelength of 495 nm. Vehicle controls plus or minus enzyme provide the 0 and 100% inhibition values, respectively. Data can be analyzed by conversion of fluorescence units to percent inhibition. Once the hit compounds are identified, secondary assays using pure compounds can be performed using multiple concentrations of an inhibitor to obtain IC_{50} values. Using this assay, Clements and coworkers (2001) screened a library of compounds featuring metal chelating groups and identified several compounds as inhibitors of PDF, including the natural hydroxamic acid antibiotic actinonin (previously identified as a PDF inhibitor) (Chen et al., 2000) and a related N-formyl-hydroxylamine derivative (BB-3497) (Clements et al., 2001), with IC_{50}s of 10 and 7 nM, respectively.

The last step in the fatty acid biosynthetic pathway is catalyzed by enoyl-acyl carrier protein (ACP) reductase, which is responsible for reduction of the double bond in the enoyl-ACP derivative (Heath and Rock, 1995; Payne et al., 2002). While *fabI* genes encode enoyl-ACP reductases (FabI enzymes) in *S. aureus* and *E. coli*, an alternative enoyl-ACP reductase, FabK, replaces the function of FabI in a number of bacterial species such as *Streptococcus pneumoniae* (Heath and Rock, 2000). More interestingly, a number of bacterial species (such as *Enterococcus faecalis* and *Pseudomonas aeruginosa*) possess both the FabI and FabK enzymes (Heath and Rock, 2000). To discover FabI-specific antibacterial inhibitors, Payne and colleagues at GlaxoSmithKline (GSK) developed assays for various versions of enoyl-ACP reductases (Payne et al., 2002; Seefeld et al., 2003) based on the following reaction scheme:

Assays were performed in 96-well microtiter plates, with wells containing components specific for each enzyme. *S. aureus* FabI assays contained 100 mM sodium ADA [ADA: N-(2-acetamido)-2-iminodiacetic acid, pH 6.5], 4% glycerol, 25 µM crotonoyl-ACP, 50 µM NADPH, and 20 nM of *S. aureus* FabI. Assays for *S. pneumoniae* FabK contained MDTG buffer [100 mM 2-(N-morpholino) ethanesulfonic acid (MES), 51 mM diethanolamine, 51 mM triethanolamine (pH 6.5), 4% glycerol], 25 µM crotonoyl-ACP, 50 µM NADPH, and 1.5 nM of *S. pneumoniae* FabK. Appropriate concentrations of compounds were screened. The enzyme activity was monitored using changes in consumption of NAD(P)H via absorbance at 340 nm by a plate reader (Payne et al., 2002). HTS screening of over 300,000 compounds identified a benzodiazepine derivative as an inhibitor of *S. aureus* FabI with moderate *in vitro* activity but no antibacterial activity (Miller et al., 2002; Payne et al., 2002). Subsequent lead optimization in conjunction with x-ray crystal structure-based design led to aminopyridine derivatives with enhanced *in vitro* activities against both FabI and FabK enzymes and improved antibacterial potency and spectrum of activity (Payne et al., 2002; Seefeld et al., 2003).

Bacterial aminoacyl tRNA synthetases have been recognized as a group of important drug targets for the discovery of antibacterial agents (Payne et al., 2007). With mupirocin (targeting isoleucyl tRNA synthetase) already on the market as a topical antibiotic, it should be possible to discover and develop into drugs antibacterial compounds that target other aminoacyl tRNA synthetases. Scientists at GSK performed high-throughput biochemical screens (Jarvest et al., 2002) using *S. aureus* methionyl tRNA synthetase (MRS) in a full aminoacylation assay (Pope et al., 1998). Aminoacylation assays (200 µL) were performed using an appropriate concentration of MRS, radio-labeled methionine and appropriate tRNA (tRNA^met) at 22°C in a buffered reaction containing 50 mM Tris-HCl (pH 7.9), 10 mM $MgCl_2$, 50 mM KCl, and 2 mM dithiothreitol. After specific intervals, 50-µL aliquots were quenched using 100 µL of 7% trichloroacetic acid and incubated on ice for 10 min. Trichloroacetic acid-precipitable material was harvested using Millipore Multiscreen 0.45-µm polyvinylidene difluoride 96-well plates and counted by liquid scintillation. The initial HTS hit had an IC_{50} of 350 nM and exhibited no Gram-positive antibacterial activity (Jarvest et al., 2002). Optimization generated derivatives with nanomolar IC_{50} values toward MRS and potent antibacterial activity against staphylococcal and enterococcal pathogens including strains resistant to clinical antibiotics (Jarvest et al., 2002, 2003). One of these optimized lead compounds (SB-425076) also demonstrated *in vivo* efficacy in an *S. aureus* abscess infection model (Jarvest et al., 2002).

While target-based biochemical screens have identified interesting specific inhibitors with *in vitro* and/or *in vivo* activities, none of the inhibitor series has reached clinical stages of development.

7.2.2 CELL-BASED ASSAYS INVOLVING UNDEREXPRESSION OF ESSENTIAL GENES

7.2.2.1 Antisense RNA Expression Approach

An RNA that interferes with the activity of a messenger RNA is defined as an antisense RNA. The principle of identification of essential genes using antisense RNA expression is that cell growth inhibition occurs when the inducible expression of the antisense RNA molecule (usually expressed from a plasmid-based antisense DNA fragment) corresponding to an essential gene is expressed at a level sufficiently high to inhibit the normal function of the cognate mRNA (Ji et al., 2001; Forsyth et al., 2002; Forsyth and Wang, 2008).

One of the advantages of essential gene identification using this approach is that the cell clones thus obtained can be easily developed into cell-based assays to screen for inhibitors that target the proteins encoded by the essential genes from which the antisense RNAs are derived (Forsyth et al., 2002; Forsyth and Wang, 2008). Specifically, the promoter controlling the expression of antisense RNA can be easily modulated by the inducer concentration to cause a range of growth defects from no growth to no observable effect. Consequently, a concentration of the inducer can be determined experimentally which has slight effect on cell growth but specifically sensitizes the cell toward inhibitors that target the very same protein target whose cellular function is impacted by the modulated antisense RNA expression (Forsyth et al., 2002).

For example, the expression of antisense RNA corresponding to *fab* (later named *fabF*) mRNA in *S. aureus* was induced by an appropriate concentration of xylose (the inducer) to a level that sensitized cells to specific inhibitors of β-ketoacyl carrier protein synthase II, encoded by *fabF*, compared to non-induced cells (Figure 7.2) (Forsyth et al., 2002). The assay developed using the *fabF* antisense RNA was subsequently optimized and adapted as a two-plate assay by scientists at Merck & Co. to screen natural product extracts (Wang et al., 2006, 2007; Young et al., 2006). Specifically, *S. aureus* cells carrying plasmid S1-1941 bearing a DNA fragment in an antisense orientation to *fabF* gene (FabF AS-RNA strain) or vector control were inoculated in Miller's LB broth containing 34 μg/mL of chloramphenicol and incubated overnight at 37°C with shaking at 220 rpm. Each culture was diluted to a final OD_{600} of 0.003 into a flask containing Miller's LB medium plus 1.2% agar (autoclaved and cooled to 48°C), 0.2% glucose, 15 μg/mL chloramphenicol and 50 mM of xylose (Young et al., 2006). Two assay plates, one seeded with the FabF AS-RNA strain (AS plate) and the other seeded with the control strain (control plate), were prepared by pouring 100 mL of each of the above mixtures into a 20 cm by 20 cm bioassay dish. Immediately, well casters were placed into the agar and the agar was allowed to solidify for 30 min at room temperature (Young et al., 2006). The samples (20 μL) were applied to wells on both the control and AS plates and were incubated for 18 hr at 37°C.

Those samples inhibitory to the bacterial cells produced zones of inhibition surrounding the wells. Any differences in sizes of the inhibition zones between the AS plate and the control plate for a given sample were measured in millimeters. A sample producing a larger zone of inhibition on the AS plate than that on the control plate was considered an active sample (Young et al., 2006). Several classes of FabF/FabH (β-ketoacyl-acyl carrier protein synthase III) inhibitors were identified by scientists at Merck, including phomallenic acids (Young et al., 2006), platensimycin (Wang et al., 2006, 2007), and platencin (Wang et al., 2007) with moderate to potent antibacterial activity (structures shown in Figure 7.3).

The major advantage of the antisense RNA modulated cell-based assays is that the target specifically sensitized cells would respond to lower concentrations of target-specific inhibitors in libraries of pure compounds or natural product extracts that could not be as easily identified if a traditional cell-based growth inhibition screen were performed. In contrast to target-based biochemical assays (as described in Section 7.2.1), the hit compounds discovered in antisense

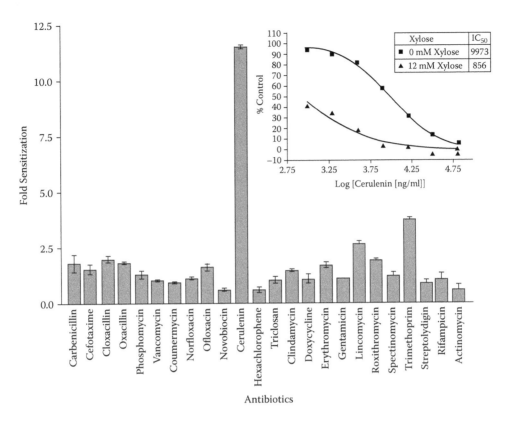

FIGURE 7.2 Induction of antisense to *fab* mRNA only sensitizes cells to the specific inhibitor cerulenin. (*Source:* From Forsyth, R.A. et al. 2002. *Mol. Microbiol.* 43, 1387. Reprinted with permission from Blackwell Publishing Ltd.)

RNA-modulated cell-based assays can penetrate bacterial cell envelopes and possess antibacterial activity—desirable properties unusual for the majority of initial hits from biochemical assays.

7.2.2.2 Promoter Replacement Approach

Another approach for target-specific cell sensitization is via promoter replacement developed by DeVito and colleagues (2002). The strategy involves replacing the natural essential gene on the chromosome with a plasmid-borne copy under the control of an inducible promoter, allowing the investigators to modulate how much of the encoded product is produced in the derivative cell clone.

Specifically, these investigators constructed an array of *E. coli* strains in which an essential gene under the control of a regulatable promoter was introduced into the cell harbored on a plasmid and subsequently the native copy of the gene on the chromosome was removed via recombination (DeVito et al., 2002). This resulted in a bacterial cell in which the intracellular level of one essential target protein could be adjusted with a specific inducer. After modulating target concentration in the cell using inducer to an appropriate level that specifically sensitized the cell to the inhibitors of the target protein (enzyme), this whole cell assay was used to screen a chemical library using growth inhibition as an end point.

Using this strategy, it is possible to find leads that are both active cell growth inhibitors and are also specific to a molecular target known beforehand (DeVito et al., 2002). DeVito and coworkers screened the reference strains (parent strains used for genetic manipulations) and the underexpression strains (such as ↓*murA* and ↓*metG*) against the same compound library. Hit compounds

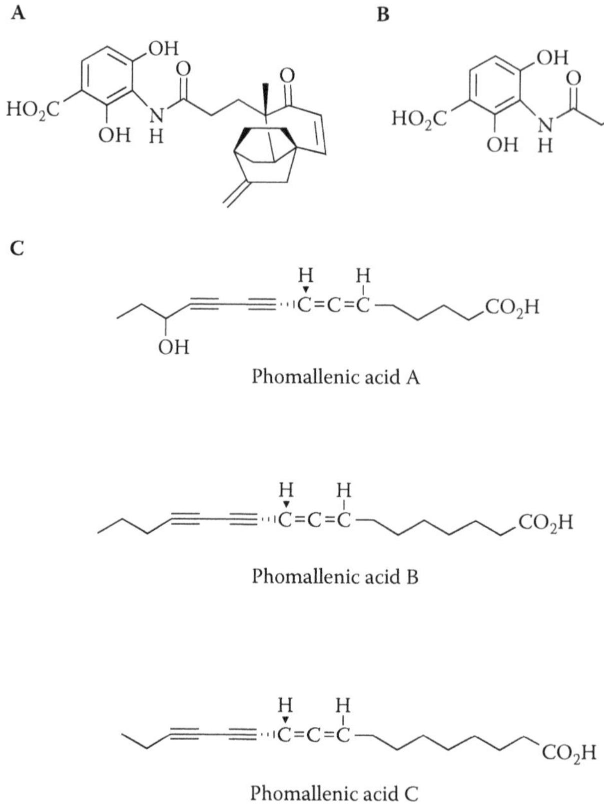

FIGURE 7.3   (A) Structure of platencin. (*Source:* From Wang, J. et al. 2007. *Proc. Natl. Acad. Sci. USA* 104, 7612. Reprinted with permission from National Academy of Sciences of the USA.) (B) Structure of platensimycin. (*Source:* From Wang, J. et al. 2006. *Nature* 441, 358. Reprinted with permission from Macmillan Publishers.) (C) Structures of phomallenic acids. (*Source:* From Young, K. et al. 2006. *Antimicrob. Agents Chemother.* 50, 519. Reprinted with permission from American Society for Microbiology.)

significantly inhibiting the growth of each strain were compared. Compounds that were hits in underexpression strains but not in reference strains were potentially target-specific inhibitors. Hits from the primary screens were retested for antibacterial activity toward the engineered strains under various concentrations of inducer. If upregulation of the target (by increasing inducer concentration) rendered the bacterial cell less sensitive to the compound (see Section 7.2.3), the hit would be confirmed as target-specific. Indeed, five compounds identified using the ↓*murA* strain were confirmed as MurA-specific antibacterial inhibitors via upregulation of target concentration as well as the MurA enzyme assays (DeVito et al., 2002).

7.2.3   Cell-Based Assays Involving Overexpression of Essential Genes

The principle of assays employing overexpression of essential genes (hence the corresponding essential proteins) lies in the concept that elevated levels of a drug target protein confer drug resistance because higher concentrations of a specific inhibitor are required to bind to the excess target to inhibit cell growth (Chopra, 1998). Li and colleagues (2004) developed a multicopy suppression assay to determine cellular targets and resistance mechanisms of novel antibacterial compounds. First, a screen of over 8000 small molecules led to the identification of 49 growth inhibition lead compounds. Then, the investigators grew a pool of *E. coli* MC1061 (hyperpermeable mutant) clones containing random *E. coli* MG1655 genomic DNA fragments in the presence of inhibitory concentrations of the lead compounds to select for multicopy suppressors (Li et al., 2004).

Of the 49 leads tested, suppressor clones could be isolated for 33 of the compounds, while no resistant clones were found for the remainder (Li et al., 2004). Following plasmid isolation and DNA sequencing of the inserts, the majority of suppressors were found to encode multidrug efflux pump AcrB, indicative of the role of efflux pumps in resistance to the lead compounds. Two lead compounds that led to the identification of suppressor clones containing the *folA* gene (encoding the dihydrofolate reductase enzyme) were found to be competitive inhibitors of dihydrofolate reductase *in vitro* and to target this enzyme *in vivo* (Li et al., 2004).

Similarly, the author's group described a strategy of identifying cellular targets of antibacterial inhibitors using a collection of engineered *E. coli* clones each overexpressing an essential gene (Xu, Real, and Bailey, 2006). The idea was to screen diverse compound libraries or natural product extracts to identify antibacterial inhibitors or extracts and then determine the cellular targets by using mixed pools of clones or individual arrays of clones of *E. coli* cells, each overexpressing one essential gene. Proof of concept results indicated that the known targets (MurA and FabI) of two antibacterial compounds (phosphomycin and triclosan) could be "identified" using either a mixed culture of eight clones or individual clones. For example, in the individual clone assay, eight overexpressing *E. coli* clones were arrayed in a 96-well microtiter plate containing LB medium with ampicillin for maintenance of plasmids in the presence of inhibitory concentrations of either triclosan (a known specific inhibitor of FabI, enol-acyl carrier protein reductase) or phosphomycin (a known specific inhibitor of MurA, UDP-*N*-acetylglucosamine enolpyruvyl transferase), with or without the inducer IPTG (isopropyl β-D-thiogalactopyranoside). Kinetic monitoring of cell growth within each well of the microplate revealed robust growth in wells containing both the inhibitor and the cell clone overexpressing its specific target protein but not irrelevant cell clones, demonstrating a simplistic and specific assay for identification of targets of inhibitors (Xu, Real, and Bailey, 2006). More recent results demonstrate the utility of this approach for identifying targets of three antibacterial inhibitors from 48 *E. coli* overexpression clones (unpublished data).

7.2.4 OTHER ASSAYS FOR DISCOVERY OF ANTIBACTERIAL AGENTS

Whole cell growth inhibition screens combined with subsequent target identification using molecular methods have proven viable approaches to the discovery of novel antibacterial inhibitors. Andries and colleagues (2005) at Johnson & Johnson employed whole cell assays to discover a series of antimycobacterial diarylquinolines (DARQs). Chemical optimization of a lead compound led to DARQ derivatives exhibiting potent *in vitro* activities against several mycobacteria including *Mycobacterium tuberculosis* (Andries et al., 2005; Ji et al., 2006), with MICs below 0.5 µg/mL. Antimycobacterial efficacy *in vivo* was confirmed for three of the derivatives.

The most active compound of the class, R207910, is a pure enantiomer with two chiral centers (Figure 7.4). Structurally and mechanistically, DARQs are different from both quinoline and fluoroquinolone classes. One of the major structural differences between DARQs and quinolines and quinolones is the specificity of the functionalized lateral (3′) chain possessed by the DARQ class (Andries et al., 2005). Sequence analyses of nearly completed genomes of mycobacterial strains selected for resistance to R207910 indicated that this drug candidate targets ATP synthase (*atpE*), a target distinct from those of quinolones and quinolines. The cellular target of R207910 was further confirmed with complementation studies in which wild-type *M. smegmatis* transformed with a construct expressing the mutant F0 subunit was rendered resistant to R207910 (Andries et al., 2005).

Sometimes a novel antibacterial inhibitor may be discovered as a by-product of searching for something very different. The discovery of plectasin is a good example. Plectasin is a peptide antibiotic discovered from studies of the *Pseudoplectania nigrella* fungus during a screening process

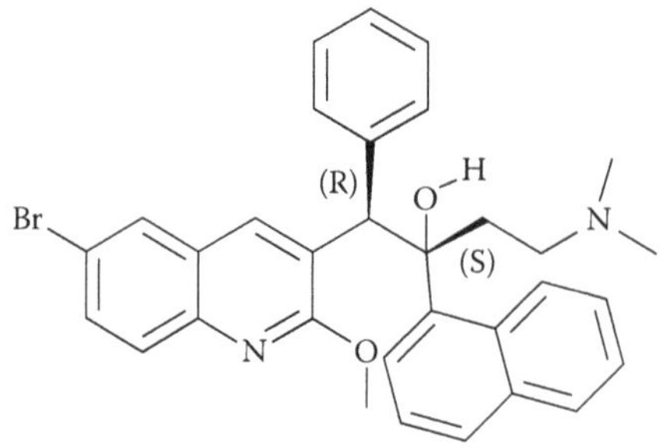

FIGURE 7.4 Absolute configuration of R207910 [1-(6-bromo-2-methoxy-quinolin-3-yl)-4-dimethylamino-2-naphthalen-1-yl-1-phenyl-butan-2-ol]. (*Source:* Andries, K. et al. 2005. *Science*, 307, 223. Reprinted with permission from AAAS.)

for novel bacterial and fungal enzymes of industrial value (such as in detergents or animal feeds) by scientists at Novozymes (Mygind et al., 2005). A selection procedure was established to screen for secreted proteins from bacterial and fungal sources (Becker et al., 2004). mRNAs were extracted from mycelia of *P. nigrella* and converted into cDNAs that were cloned into plasmids. The recombinant plasmids were subjected to insertional mutagenesis by a transposon (TnSig) encoding a β-lactamase gene that lacked a signal peptide (Becker et al., 2004; Mygind et al., 2005). Fusion of this transposon in the correct reading frame to a protein or peptide carrying a signal peptide resulted in ampicillin resistance of the corresponding clone.

A cDNA fragment discovered using this assay contains an open reading frame (ORF) encoding a peptide of 95 amino acids. This peptide consists of a signal peptide, a pro-piece and a 40-residue C terminal domain (residues 56 to 95) and exhibited 50% to 55% sequence identity to several defensins of invertebrates. This peptide can mature into plectasin which corresponds to the C terminal portion of the original peptide (residues 56 to 95) with a molecular weight of 4,398.80 Da (Mygind et al., 2005). Plectasin is the first defensin to be isolated from a fungus. *In vitro*, the recombinant peptide was especially active against *S. pneumoniae*, including strains resistant to conventional antibiotics. Plectasin showed extremely low toxicity in mice and was as efficacious as vancomycin and penicillin in treating experimental peritonitis and pneumonia in mice caused by *S. pneumoniae* (Mygind et al., 2005).

7.3   ANTIFUNGAL DRUG DISCOVERY

Antifungal drug discovery is surprisingly similar to the discovery of antibacterial drugs. Facilitated by the complete sequencing of fungal organisms (including pathogens such as *Candida albicans* and *Aspergillus fumigatus*), target-based biochemical screens (Section 7.3.1) and cell-based assays combined with target identification (Sections 7.3.2, 7.3.3 and 7.3.4) have sprung up over the past few years.

7.3.1   TARGET-BASED BIOCHEMICAL ASSAYS

Protein *N*-myristoyl transferases catalyze the *N*-myristoylation step in which the 14-carbon fatty acid myristate is added to a terminal glycine residue of a protein during translation. Proteins carrying the terminal lipid moiety are involved in many essential eukaryotic cellular processes (Farazi, Waksman, and Gordon, 2001). Specifically, myristoyl-CoA: protein *N*-myristoyl transferease (Nmt1p) was determined to be essential for both *Cryptococcus neoformans* (Lodge et al., 1994) and *Candida albicans* (Weinberg et al., 1995), rendering it an attractive target for the development of antifungal drugs.

Pennise and coworkers (2002) described a continuous fluorometric assay for this enzyme for the purpose of high-throughput screens and enzyme kinetics studies. They designed an acceptor

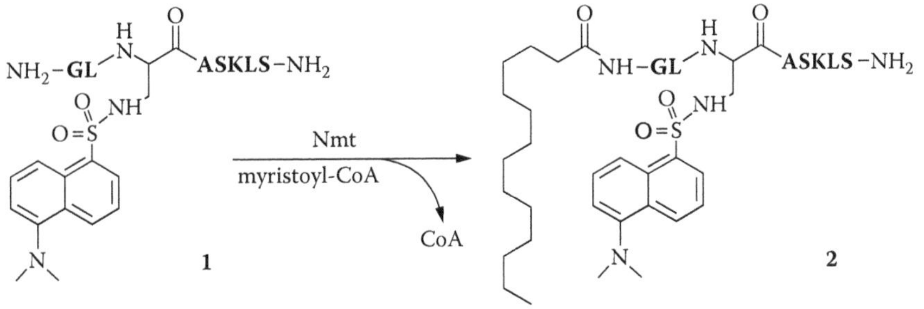

FIGURE 7.5 Reaction catalyzed by myristoyl-CoA protein *N*-myristoyltransferase in eukaryotic cells. (*Source:* Pennise, C.R. et al. 2002. *Anal. Biochem.* 300, 275. Reprinted with permission from Elsevier.)

peptide substrate [octapeptide GL-Dap(*N*-dansyl)-ASKLS-NH$_2$] with a dansyl moiety (an environmentally sensitive fluorescence probe) positioned proximally to the reaction center (Figure 7.5). Incubation of structure **1** with the enzyme (Nmt) and myristoyl-CoA resulted in a time-dependent increase in fluorescence at 500 nm with excitation at 340 nm (Figure 7.5). Covalent attachment of the lipid moiety to the free amine of the terminal glycine residue of stucture **1** positioned the nonpolar, hydrophobic group near the dansyl moiety, changing the local chemical environment of the reporter group and causing a dramatic shift in its fluorescence properties. To maintain the stability of fluorescence signal during assays, a nonionic detergent, 0.5 mM *n*-dodecyl-β-D-maltoside, was added to the assay mixture (Pennise et al., 2002).

Novel derivatives of benzofuran inhibitors have been synthesized to enhance inhibitory activity against *C. albicans* Nmt enzyme, with some exhibiting *in vivo* antifungal activity in mouse models of *C. albicans* infection (Masubuchi et al., 2001, 2003; Kawasaki et al., 2003).

7.3.2 CELL-BASED ASSAYS INVOLVING UNDEREXPRESSION OF ESSENTIAL GENES

Haplo-insufficiency is a situation in which the total level of a gene product (a particular protein) produced by a cell is about half of the normal level, which is not sufficient to permit the cell to function normally. Giaever and coworkers (1999) developed and validated a genomic scale cell-based assay based on induced haplo-insufficiency (named haplo-insufficiency profiling [HIP]) to identify targets of known drugs by profiling drug sensitivities of strains. They constructed a collection of 233 heterozygous *Saccharomyces cerevisiae* strains, each carrying a deletion in one copy of the two gene alleles, including several genes encoding known drug targets.

During the construction of deletion strains, unique oligonucleotide tags (molecular barcodes) were included into the strains, permitting subsequent profiling of relative strain abundance (thus target identification) in a fitness test of pooled strains in the presence of sublethal concentrations of drugs. When grown individually, heterozygous deletion strains exhibited reduced growth rates compared to the wild-type strain in the presence of sublethal concentrations of drugs, indicative of enhanced sensitivity of the heterozygous deletion strains (Giaever et al., 1999).

When mixed cultures of heterozygous deletion strains were grown in the presence of a drug, strains that grew more slowly with respect to the wild type represented those most sensitive to the drug and indicated a possible drug target. For example, a pool of 12 heterozygous strains with equal cell numbers was grown in the presence of 0.5 µg/mL of tunicamycin, a well characterized glycosylation inhibitor (Kuo and Lampen, 1974). Aliquots of cells were taken from the pool over time and genomic DNA was isolated and used as template for PCR amplification of fluorescently labeled tags (Giaver et al., 1999). Subsequent oligo microarray analyses indicated that the hybridization signal on the oligonucleotide array representing the alg7/ALG7 strain (in which one of the ALG7 genes was deleted; ALG7 encodes Asn-linked glycosyl transferase, a known target of tunicamycin) was diminished at 22 hr and undetectable by 48 hr (Giaever et al., 1999).

Subsequently, Giaever and coworkers (2004) expanded the heterozygous deletion strains to cover the entire genome (~6000 genes) of *S. cerevisiae* and employed HIP to identify functional interactions of 10 diverse small molecules with cellular proteins. These compounds include anticancer and antifungal agents, statins, alverine citrate, and dyclonine. Using these assays, they not only identified previously known targets but also revealed novel cellular interactions. Furthermore, a chemical core structure shared among three therapeutically distinct compounds was found to inhibit the Erg24 heterozygous deletion strain, demonstrating that cells may respond similarly to compounds of related structure (Giaever et al., 2004).

Researchers at Elitra Pharmaceuticals and their collaborators developed a technology to construct conditional mutant strains of *C. albicans*, a human fungal pathogen, for purposes of essential gene identification and antifungal drug discovery (Roemer et al., 2003). This technology was designated gene replacement and conditional expression (GRACE) and involves two successive steps (Figure 7.6). First, a precise gene replacement of one allele of the diploid pair is made in the parent strain, generating a heterozygous mutation for the gene of interest (Figure 7.6). Second, the native promoter of the remaining gene of interest is replaced with the tightly regulatable tetracycline (Tet) promoter (Figure 7.6). Consequently, one gene copy is deleted and the other is placed under the control of the Tet promoter. The resulting GRACE strains for all genes can be evaluated for growth defects as the result of transcriptional repression of the remaining gene copies. A total of 567 essential genes in *C. albicans* were identified (Roemer et al., 2003).

For drug screening purposes, the titratable repression of the Tet promoter by addition of tetracycline allows control over the cellular concentration of a specific target protein. Similar to antisense RNA-based cell sensitization described in Section 7.2.2.1, this approach provides a MOA (mechanism of action)-based assay because cells with reduced target protein levels become hypersensitive

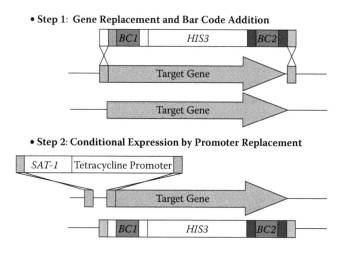

FIGURE 7.6 The GRACE method of target validation. Step 1: heterozygote strains were constructed by transforming the wild-type *Candida albicans* starting strain, CaSS1, using a PCR-generated disruption cassette containing HIS3 selectable marker flanked with appropriate homologous sequence to precisely replace one allele of the target gene. Two distinct bar codes were introduced into the disruption cassette during PCR amplification. Two primer pairs that anneal to the common arms (white and dark bars, respectively) flanking each tag enable simple PCR amplification of the strain-identifying bar codes. Thus, all heterozygote strains are uniquely tagged with distinct strain identifying bar codes. Step 2: bar-coded heterozygous strains were transformed using a PCR-generated tetracycline promoter replacement cassette containing the SAT-1-dominant selectable marker engineered for expression in *C. albicans*. Homologous flanking sequence was added during PCR amplification to precisely replace the endogenous promoter of the remaining wild-type allele with the Tet promoter replacement cassette after transformation (*Source:* Roemer, T. et al. 2003. *Mol. Microbiol.* 50, 167. Reprinted with permission from Blackwell Publishing Ltd.)

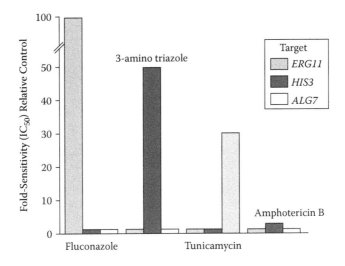

FIGURE 7.7 *Candida albicans*-sensitized whole cell assays. GRACE strains conditionally regulating the ERG11, HIS3, and ALG7 known drug targets were constructed and IC_{50} values determined against a matrix of antifungal compounds including their cognate inhibitors, fluconazole, 3-amino triazole, and tunicamycin, respectively. All strains were assayed in a suitable tetracycline concentration to underexpress the drug target to the extent that growth rate was reduced ~90%. Sensitized cells displayed a range of 30- to 100-fold lower IC_{50} values specifically detected between the drug target and its known inhibitor. IC_{50} determination to amphotericin B, whose mechanism of action is distinct from fluconazole, 3-amino triazole, and tunicamycin and involves disrupting plasma membranes, revealed no elevated drug sensitivity among any of the three sensitized *C. albicans* strains. (*Source:* Roemer, T. et al. 2003. *Mol. Microbiol.* 50, 167. Reprinted with permission from Blackwell Publishing Ltd.)

to compounds that inhibit the target. To illustrate this strategy, three GRACE strains with repression (via appropriate tetracycline concentrations) in *Erg11*, *His3*, and *Alg7* genes were individually exposed to a series of concentrations of three drugs: fluconazole, 3-amino triazole, and tunicamycin, respectively, and IC_{50} values were obtained (Roemer et al., 2003). In all these strains, the sensitizing effect was specific only to the drug that targeted the gene product whose expression was repressed (Figure 7.7). For example, reducing the level of His3 hypersensitized cells to 3-amino triazole 30-fold, while resulting in no increased sensitivity to amphotericin B, tunicamycin, or fluconazole, which inhibit targets other than His3 (Figure 7.7).

Applying the same HIP strategy developed in *S. cerevisiae* (Giaever et al., 1999, 2004), scientists at Merck constructed nearly 2900 heterozygous deletion strains of *C. albicans* (Xu et al., 2007). These strains have been used in a cell-based assay called the *C. albicans* fitness test (CaFT), to determine the MOA of inhibitory compounds and natural products (Jiang et al., 2008) in *C. albicans*, and to validate targets in both *C. albicans* and *A. fumigatus* (Rodriguez-Suarez et al., 2007).

Xu and colleagues (2007) screened this collection of heterozygous deletion strains using 35 known or novel compounds and determined their mechanisms of action. These compounds included FDA-approved antifungal drugs such as fluconazole, voriconazole, caspofungin, 5-fluorocytosine, and amphotericin B, as well as additional compounds targeting ergosteril, fatty acid and sphingolipid biosynthesis, microtubules, actin, secretion, rRNA processing, translation, glycosylation and protein folding mechanisms (Xu et al., 2007).

Jiang and coworkers (2008) applied CaFT to natural product extracts that were previously screened and found to possess intrinsic activities against both *C. albicans* and *A. fumigatus*. The fitness test profile against one extract (ECC577, derived from a culture broth of *Fusarium larvarum*) indicated that most of the strains hypersensitive to this extract were heterozygous deletion strains involving genes encoding subunits of the eukaryotic mRNA C/P (cleave and polyadenylation)

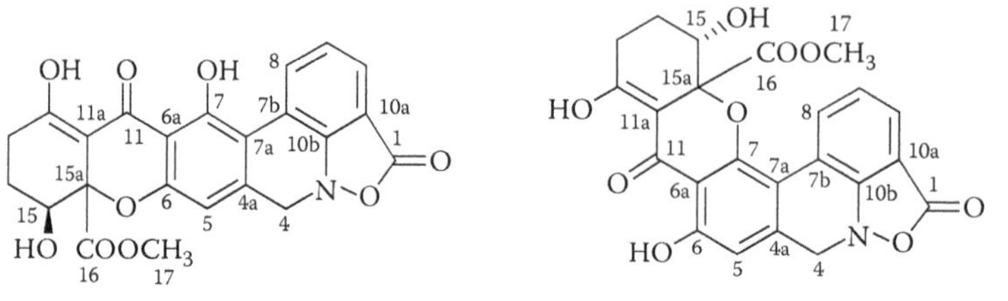

1 **2**

Parnafungin A1 (**1a**, *syn* –$COOCH_3$) Parnafungin B1 (**2a**, *syn* –$COOCH_3$)
Parnafungin A2 (**1b**, *anti* –$COOCH_3$) Parnafungin B2 (**2b**, *anti* –$COOCH_3$)

FIGURE 7.8 Structures of parnafungins. The major diastereomers, parnafungins A1 and B1, display the 15-hydroxyl and adjacent methyl carboxylate in a syn relative configuration. The minor diastereomers, parnafungins A2 and B2, display anti configuration. (Source: Parish, C.A. et al. 2008. J. Am. Chem. Soc. 130, 7060. Copyright 2008, American Chemical Society.)

complex. One of these genes is homologous to the bovine polyadenylate-binding protein 2 (PABP2) that stimulates poly(A) polymerase (PAP) activity by enhancing its binding affinity for the mRNA substrate. Subsequent bioactivity-guided purification and structure elucidation led to the discovery of the active component in ECC577, parnafungin (Figure 7.8), which produced the same CaFT profile as the original extract (Jiang et al., 2008; Parish et al., 2008). *In vitro* biochemical assays and gene sequence analysis of parnafungin-resistant *C. albicans* mutants further confirmed that parnafungin is a specific inhibitor of PAP. Parnafungin exhibited potent and broad spectrum activity against multiple fungal pathogens and demonstrated efficacy in a murine model of disseminated candidiasis (Jiang et al., 2008).

7.3.3 Cell-Based Assays Involving Overexpression of Essential Genes

Overexpression of fungal essential genes has also been employed to determine the targets of antifungal inhibitors (Launhardt, Hinnen, and Munder, 1998; Tsukahara et al., 2003). To validate the concept that overexpressed genomic DNA fragments are capable of curing an induced phenotype, Launhardt and coworkers (1998) grew *S. cerevisiae* cells carrying a library of *S. cerevisiae* genomic DNA fragments on a multicopy plasmid at a lethal concentration of ketoconazole, an antifungal drug. Eighteen colonies emerged on plates with ketoconazole. After further characterization, 13 of the 18 ketoconazole-resistant colonies indeed contained plasmid inserts that could transfer the ketoconazole resistance phenotype to wild-type cells (Launhardt, Hinnen, and Munder, 1998). Among the 13 clones, 4 contained insert DNAs encoding full length or partial length lanosterol C-14 demethylase (product of ERG11 gene), confirming the known target of triazole inhibitors (Launhardt, Hinnen, and Munder, 1998).

Tsukahara and colleagues (2003) screened for small molecular compounds that inhibit cell wall assembly of GPI (glycosylphosphatidylinositol)-anchored mannoproteins. A potent inhibitor, 1-[4-butylbenzyl]isoquinoline (BIQ), was obtained, which inhibits the vegetative growth of both *S. cerevisiae* and *C. albicans* and reduces adherence to a rat intestine epithelial cell monolayer (Tsukahara et al., 2003). Next, these investigators grew *S. cerevisiae* cells harboring a genomic library of the same species constructed in a multicopy vector on medium supplemented with an inhibitory concentration of BIQ. Twenty-seven BIQ-resistant colonies were obtained and were all found to contain an ORF, YJL091c. Subcloning results confirmed that YJL091c conferred the BIQ-resistant phenotype and suppressed other phenotypes induced by the inhibitor, thus suggesting that the product of this gene is the target of BIQ (Tsukahara et al., 2003).

7.3.4 OTHER ASSAYS FOR DISCOVERY OF ANTIFUNGAL AGENTS

With the increase in resistance to the limited number of existing antifungal drugs, combination therapy with more than one agent may prove a more effective treatment approach. Zhang and colleagues (2007) developed a cell-based high-throughput synergy screening (HTSS) assay to identify microbial natural products as combination antifungal agents. A microbial natural product library with ~20,000 extracts was screened for hits that synergized the effect of a low dosage of ketoconazole that alone showed little detectable fungicidal activity (Zhang et al., 2007).

Of the 12 hit extracts exhibiting broad spectrum antifungal activities, 7 showed little cytotoxicity against human hepatoma cells. Six compounds were isolated and purified from these extracts and identified as beauvericin, berberine, cyclosporine A, geldanamycin, lovastatin, and radicicol (Zhang et al., 2007). Beauvericin, previously discovered to possess insecticidal properties but no antifungal activity, was found to be the most potent compound producing a synergistic effect with ketoconazole against diverse fungal pathogens. In an immunocompromised mouse model, combinations of beauvericin (0.5 mg/kg) and ketoconazole (0.5 mg/kg) exhibited *in vivo* efficacy by prolonging survival of a host infected with *Candida parapsilosis* and reducing fungal colony counts in animal organs (Zhang et al., 2007).

Another excellent example of a cell-based screening approach for antifungal drug discovery was recently illustrated by Gassner and coworkers (2007). They developed a cell-based high-throughput yeast halo assay to identify biologically active compounds from diverse libraries (NCI) and marine natural product extracts (Figure 7.9). Compounds were not confined to discrete wells, but were pin-transferred

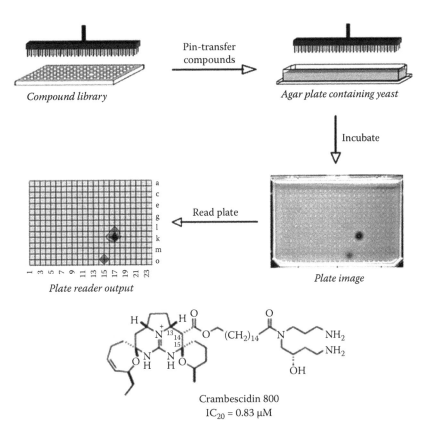

FIGURE 7.9 (See color insert following page 114.) High-throughput yeast halo assay strategy (top) and structure of a potent antifungal agent (crambescidin 800) identified (bottom) (Source: Gassner, N.C. et al. 2007. *J. Nat. Prod.* 70, 383. Copyright 2007 American Chemical Society.)

directly onto agar plates seeded with yeast cells and allowed to diffuse freely from the site of addition (Figure 7.9). Because the diffusion of the sample provided a concentration gradient and a corresponding inhibitory halo, this screen is quantitative and allows inhibitory potencies to be estimated. A total of 46 active compounds were identified from the 3104 NCI compound libraries. Fractionation and purification of one highly active marine natural product extract resulted in the identification of the active ingredient as crambescidin 800, a new potent antifungal agent (Gassner et al., 2007).

7.4   CONCLUSIONS

The advances of laboratory robotics and high-throughput screening coincided with the completion of the first microbial genome in the mid 1990s (Fraser et al., 1995). The pharmaceutical and biotechnology companies devoted a great deal of resources and effort to exploit newly discovered drug targets by rapidly screening, in high-throughput mode, libraries of hundreds of thousands of compounds against multitudes of target enzymes (target-based biochemical screens) (Payne et al., 2007). However, this strategy presented a major drawback: most inhibitors thus obtained did not possess antimicrobial activities and subsequent structural modifications to improve cellular activity, if they worked at all, were painstakingly slow and costly. Although some hits eventually became lead series via lead optimization, this initial strategy proved to be largely fruitless since no lead compounds from these efforts have advanced to clinical stages.

Recent developments in novel cell-based antimicrobial drug discovery strategies described in this chapter provide a glimpse of hope for accelerating the pace of antimicrobial drug discovery and development. The common theme among these approaches rests on the integration of cell-based, traditional growth inhibition screening with the concurrent or follow-on identification of cellular targets of the antimicrobial inhibitors discovered. Modulation of cellular target levels (overexpression or underexpression) on a genome-wide scale appears to be the most effective and promising strategy to identify novel classes of antimicrobial agents with perhaps entirely new mechanisms of actions. Although no hits from such latest endeavors have moved into clinical trials, with sufficient time (five to ten years), effort, and resources (that are becoming more scarce), these and other innovative approaches will bear fruit.

REFERENCES

Adams, J.M. and M.R. Capecchi. 1966. *N*-formylmethionyl-sRNA as the initiator of protein synthesis. *Proc. Natl. Acad. Sci. USA* 55, 147–155.
Akerley, B.J. et al. 2002. A genome-scale analysis for identification of genes required for growth or survival of *Haemophilus influenzae*. *Proc. Natl. Acad. Sci. USA* 99, 966–971.
Amyes, S.G. 2007. Enterococci and streptococci. *Int. J. Antimicrob. Agents* 29, S43–S52.
Andries, K. et al. 2005. A diarylquinoline drug active on the ATP synthase of *Mycobacterium tuberculosis*. *Science* 307, 223–227.
Arigoni, F. et al. 1998. A genome-based approach for the identification of essential bacterial genes. *Nat. Biotechnol.* 16, 851–856.
Becker, F. et al. 2004. Development of *in vitro* transposon assisted signal sequence trapping and its use in screening *Bacillus halodurans* C125 and *Sulfolobus solfataricus* P2 gene libraries. *J. Microbiol. Meth.* 57, 123–133.
Blattner, F.R. et al. 1997. The complete genome sequence of *Escherichia coli* K-12. *Science* 277, 1453–1474.
Bramhill, D. 1997. Bacterial cell division. *Annu. Rev. Cell. Dev. Biol.* 13, 395–424.
Chen, D.Z. et al. 2000. Actinonin, a naturally occurring antibacterial agent, is a potent deformylase inhibitor. *Biochemistry* 39, 1256–1262.
Chopra, I. 1998. Overexpression of target genes as a mechanism of antibiotic resistance in bacteria. *J. Antimicrob. Chemother.* 41, 584–588.
Clements, J.M. et al. 2001. Antibiotic activity and characterization of BB-3497, a novel peptide deformylase inhibitor. *Antimicrob. Agents Chemother.* 45, 563–570.

De Backer, M.D. et al. 2001. An antisense-based functional genomics approach for identification of genes critical for growth of *Candida albicans*. *Nat. Biotechnol.* 19, 235–241.

de Berardinis, V. et al. 2008. A complete collection of single-gene deletion mutants of *Acinetobacter baylyi* ADP1. *Mol. Syst. Biol.* 4, 174–182.

De Clercq, E. 2007. The design of drugs for HIV and HCV. *Nat. Rev. Drug Discov.* 6, 1001–1018.

DeVito, J.A. et al. 2002. An array of target-specific screening strains for antibacterial discovery. *Nat. Biotechnol.* 20, 4878–4883.

Fan, F. et al. 2001. Regulated ectopic expression and allelic-replacement mutagenesis as a method for gene essentiality testing in *Staphylococcus aureus*. *Plasmid* 46, 71–75.

Farazi, T.A., G. Waksman, and J.I. Gordon. 2001. The biology and enzymology of protein *N*-myristoylation. *J. Biol. Chem.* 276, 39501–39504.

Fedorova, N.D. et al. 2008. Genomic islands in the pathogenic filamentous fungus *Aspergillus fumigatus*. *PLoS Genet.* 4, e1000046.

Forsyth, A. and L. Wang. 2008. Techniques for the isolation and use of conditionally expressed antisense RNA to achieve essential gene knockdowns in *Staphylococcus aureus*. *Meth. Mol. Biol.* 416, 307–321.

Forsyth, R.A. et al. 2002. A genome-wide strategy for the identification of essential genes in *Staphylococcus aureus*. *Mol. Microbiol.* 43, 1387–1400.

Fraser, C.M. et al. 1995. The minimal gene complement of *Mycoplasma genitalium*. *Science* 270, 397–403.

Freiberg, C. et al. 2001. Identification of novel essential *Escherichia coli* genes conserved among pathogenic bacteria. *J. Mol. Microbiol. Biotechnol.* 3, 483–489.

Gallagher, L.A. et al. 2007. A comprehensive transposon mutant library of *Francisella novicida*, a bioweapon surrogate. *Proc. Natl. Acad. Sci. USA* 104, 1009–1014.

Gassner, N.C. et al. 2007. Accelerating the discovery of biologically active small molecules using a high-throughput yeast halo assay. *J. Nat. Prod.* 70, 383–390.

Gerdes, S.Y. et al. 2003. Experimental determination and system level analysis of essential genes in *Escherichia coli* MG1655. *J. Bacteriol.* 185, 5673–5684.

Giaever, G. et al. 1999. Genomic profiling of drug sensitivities via induced haploinsufficiency. *Nat. Genet.* 21, 278–283.

Giaever, G. et al. 2002. Functional profiling of the *Saccharomyces cerevisiae* genome. *Nature* 418, 387–391.

Giaever, G. et al. 2004. Chemogenomic profiling: identifying the functional interactions of small molecules in yeast. *Proc. Natl. Acad. Sci. USA* 101, 793–798.

Gil, R. et al. 2004. Determination of the core of a minimal bacterial gene set. *Microbiol. Mol. Biol. Rev.* 68, 518–537.

Goffeau, A. et al. 1996. Life with 6000 genes. *Science* 274, 546, 63–67.

Groche, D. et al. 1998. Isolation and crystallization of functionally competent *Escherichia coli* peptide deformylase forms containing either iron or nickel in the active site. *Biochem. Biophys. Res. Commun.* 246, 342–346.

Gulshan, K. and W.S. Moye-Rowley. 2007. Multidrug resistance in fungi. *Eukaryot. Cell* 6, 1933–1942.

Hackbarth, C.J. et al. 2002. *N*-alkyl urea hydroxamic acids as a new class of peptide deformylase inhibitors with antibacterial activity. *Antimicrob. Agents Chemother.* 46, 2752–2764.

Haney, S.A. et al. Projan. 2002. Genomics in anti-infective drug discovery: getting to endgame. *Curr. Pharm. Des.* 8, 1099–1118.

Haydon, D.J. et al. 2008. An inhibitor of FtsZ with potent and selective anti-staphylococcal activity. *Science* 321, 1673–1675.

Heath, R.J. and C.O. Rock. 1995. Enoyl-acyl carrier protein reductase (*fabI*) plays a determinant role in completing cycles of fatty acid elongation in *Escherichia coli*. *J. Biol. Chem.* 270, 26538–26542.

Heath, R.J. and C.O. Rock. 2000. A triclosan-resistant bacterial enzyme. *Nature* 406, 145–146.

Hu, W. et al. 2007. Essential gene identification and drug target prioritization in *Aspergillus fumigatus*. *PLoS Pathog.* 3, e24.

Hutchison, C.A. et al. 1999. Global transposon mutagenesis and a minimal *Mycoplasma* genome. *Science* 286, 2165–2169.

Jana, M. et al. 2000. A method for demonstrating gene essentiality in *Staphylococcus aureus*. *Plasmid* 44, 100–104.

Jarvest, R.L. et al. 2002. Nanomolar inhibitors of *Staphylococcus aureus* methionyl tRNA synthetase with potent antibacterial activity against gram-positive pathogens. *J. Med. Chem.* 45, 1959–1962.

Jarvest, R.L. et al. 2003. Optimisation of aryl substitution leading to potent methionyl tRNA synthetase inhibitors with excellent gram-positive antibacterial activity. *Bioorg. Med. Chem. Lett.* 13, 665–668.

Ji, B. et al. 2006. Bactericidal activities of R207910 and other newer antimicrobial agents against *Mycobacterium leprae* in mice. *Antimicrob. Agents Chemother.* 50, 1558–1560.

Ji, Y. et al. 2001. Identification of critical staphylococcal genes using conditional phenotypes generated by antisense RNA. *Science* 293, 2266–2269.

Jiang, B. et al. 2008. PAP inhibitor with *in vivo* efficacy identified by *Candida albicans* genetic profiling of natural products. *Chem. Biol.* 15, 363–374.

Kawasaki, K. et al. 2003. Design and synthesis of novel benzofurans as a new class of antifungal agents targeting fungal *N*-myristoyltransferase 3. *Bioorg. Med. Chem. Lett.* 13, 87–91.

Knuth, K. et al. 2004. Large-scale identification of essential Salmonella genes by trapping lethal insertions. *Mol. Microbiol.* 51, 1729–1744.

Kobayashi, K. et al. 2003. Essential *Bacillus subtilis* genes. *Proc. Natl. Acad. Sci. USA* 100, 4678–4683.

Kunst, F. et al. 1997. The complete genome sequence of the gram-positive bacterium *Bacillus subtilis*. *Nature* 390, 249–256.

Kuo, S.C. and J.O. Lampen. 1974. Tunicamycin: an inhibitor of yeast glycoprotein synthesis. *Biochem. Biophys. Res. Commun.* 58, 287–295.

Kuroda, M. et al. 2001. Whole genome sequencing of meticillin-resistant *Staphylococcus aureus*. *Lancet* 357, 1225–1240.

Launhardt, H., A. Hinnen, and T. Munder. 1998. Drug-induced phenotypes provide a tool for the functional analysis of yeast genes. *Yeast* 14, 935–942.

Li, X. et al. 2004. Multicopy suppressors for novel antibacterial compounds reveal targets and drug efflux susceptibility. *Chem. Biol.* 11, 1423–1430.

Lodge, J.K. et al. 1994. Targeted gene replacement demonstrates that myristoyl-CoA: protein *N*-myristoyltransferase is essential for viability of *Cryptococcus neoformans*. *Proc. Natl. Acad. Sci. USA* 91, 12008–12012.

Masubuchi, M. et al. 2001. Design and synthesis of novel benzofurans as a new class of antifungal agents targeting fungal *N*-myristoyltransferase 1. *Bioorg. Med. Chem. Lett.* 11, 1833–1837.

Masubuchi, M. et al. 2003. Synthesis and biological activities of benzofuran antifungal agents targeting fungal *N*-myristoyltransferase. *Bioorg. Med. Chem.* 11, 4463–4478.

Meinnel, T., Y. Mechulam, and S. Blanquet. 1993. Methionine as translation start signal: a review of the enzymes of the pathway in *Escherichia coli*. *Biochimie* 75, 1061–1075.

Miller, W.H. et al. 2002. Discovery of aminopyridine-based inhibitors of bacterial enoyl-ACP reductase (FabI). *J. Med. Chem.* 45, 3246–3256.

Mygind, P.H. et al. 2005. Plectasin is a peptide antibiotic with therapeutic potential from a saprophytic fungus. *Nature* 437, 975–980.

Nicasio, A.M., J.L. Kuti, and D.P. Nicolau. 2008. The current state of multidrug-resistant Gram-negative bacilli in North America. *Pharmacotherapy* 28, 235–249.

Nordmann, P. et al. 2007. Superbugs in the coming new decade; multidrug resistance and prospects for treatment of *Staphylococcus aureus*, *Enterococcus* spp. and *Pseudomonas aeruginosa* in 2010. *Curr. Opin. Microbiol.* 10, 436–440.

Parish, C.A. et al. 2008. Isolation and structure elucidation of parnafungins, antifungal natural products that inhibit mRNA polyadenylation. *J. Am. Chem. Soc.* 130, 7060–7066.

Parkhill, J. et al. 2001. Genome sequence of *Yersinia pestis*, the causative agent of plague. *Nature* 413, 523–527.

Payne, D.J. et al. 2002. Discovery of a novel and potent class of FabI-directed antibacterial agents. *Antimicrob. Agents Chemother.* 46, 3118–3124.

Payne, D.J. et al. 2007. Drugs for bad bugs: confronting the challenges of antibacterial discovery. *Nat. Rev. Drug Discov.* 6, 29–40.

PEC Database. Profiling of *E. coli* chromosome. http://www.shigen.nig.ac.jp/ecoli/pec/index.jsp.

Pennise, C.R. et al. 2002. A continuous fluorometric assay of myristoyl-coenzyme A: protein *N*-myristoyltransferase. *Anal. Biochem.* 300, 275–277.

Pope, A.J. et al. 1998. Characterization of isoleucyl-tRNA synthetase from *Staphylococcus aureus* II: mechanism of inhibition by reaction intermediate and pseudomonic acid analogues studied using transient and steady-state kinetics. *J. Biol. Chem.* 273, 31691–31701.

Powers, J.H. 2004. Antimicrobial drug development: the past, the present, and the future. *Clin. Microbiol. Infect.* 10, 23–31.

Prasad, R. and K. Kapoor. 2005. Multidrug resistance in yeast Candida. *Int. Rev. Cytol.* 242, 215–248.

Pucci, M.J. 2007. Novel genetic techniques and approaches in the microbial genomics era: identification and/or validation of targets for the discovery of new antibacterial agents. *Drugs R&D* 8, 201–212.

Read, T.D. et al. 2003. The genome sequence of *Bacillus anthracis* Ames and comparison to closely related bacteria. *Nature* 423, 81–86.

Reich, K.A., L. Chovan, and P. Hessler. 1999. Genome scanning in *Haemophilus influenzae* for identification of essential genes. *J. Bacteriol.* 181, 4961–4968.

Rodriguez-Suarez, R. et al. 2007. Mechanism-of-action determination of GMP synthase inhibitors and target validation in *Candida albicans* and *Aspergillus fumigatus*. *Chem. Biol.* 14, 1163–1175.

Roemer, T. et al. 2003. Large-scale essential gene identification in *Candida albicans* and applications to anti-fungal drug discovery. *Mol. Microbiol.* 50, 167–181.

Seefeld, M.A. et al. 2003. Indole naphthyridinones as inhibitors of bacterial enoyl-ACP reductases FabI and FabK. *J. Med. Chem.* 46, 1627–1635.

Tanikawa, K. 2006. Recent advances in antiviral agents: antiviral drug discovery for hepatitis viruses. *Curr. Pharm. Des.* 12, 1371–1377.

Tettelin, H. et al. 2001. Complete genome sequence of a virulent isolate of *Streptococcus pneumoniae*. *Science* 293, 498–506.

Thanassi, J.A. et al. 2002. Identification of 113 conserved essential genes using a high-throughput gene disruption system in *Streptococcus pneumoniae*. *Nucleic Acids Res.* 30, 3152–3162.

Tomb, J.F. et al. 1997. The complete genome sequence of the gastric pathogen *Helicobacter pylori*. *Nature* 388, 539–547.

Trusca, D. and D. Bramhill. 2002. Fluorescent assay for polymerization of purified bacterial FtsZ cell-division protein. *Anal. Biochem.* 307, 322–329.

Tsukahara, K. et al. 2003. Medicinal genetics approach towards identifying the molecular target of a novel inhibitor of fungal cell wall assembly. *Mol. Microbiol.* 48, 1029–1042.

Vergidis, P.I. and M.E. Falagas. 2008. Multidrug-resistant Gram-negative bacterial infections: the emerging threat and potential novel treatment options. *Curr. Opin. Investig. Drugs* 9, 76–83.

von Itzstein, M. 2007. The war against influenza: discovery and development of sialidase inhibitors. *Nat. Rev. Drug Discov.* 6, 967–974.

von Nussbaum, F. et al. 2006. Antibacterial natural products in medicinal chemistry: exodus or revival? *Angew. Chem. Int. Ed.* 45, 5072–5129.

Wang, J. et al. 2003. Discovery of a small molecule that inhibits cell division by blocking FtsZ, a novel therapeutic target of antibiotics. *J. Biol. Chem.* 278, 44424–44428.

Wang, J. et al. 2006. Platensimycin is a selective FabF inhibitor with potent antibiotic properties. *Nature* 441, 358–361.

Wang, J. et al. 2007. Discovery of platencin, a dual FabF and FabH inhibitor with *in vivo* antibiotic properties. *Proc. Natl. Acad. Sci. US A* 104, 7612–7616.

Ward, N. and C.M. Fraser. 2005. How genomics has affected the concept of microbiology. *Curr. Opin. Microbiol.* 8, 564–571.

Weinberg, R.A. et al. 1995. Genetic studies reveal that myristoylCoA: protein *N*-myristoyl transferase is an essential enzyme in *Candida albicans*. *Mol. Microbiol.* 16, 241–250.

Westby, M. et al. 2005. Cell-based and biochemical screening approaches for the discovery of novel HIV-1 inhibitors. *Antiviral Res.* 67, 21–40.

Xu, D. et al. 2007. Genome-wide fitness test and mechanism-of-action studies of inhibitory compounds in *Candida albicans*. *PLoS Pathog.* 3, e92.

Xu, H.H., L. Real, and M.W. Bailey. 2006. An array of *Escherichia coli* clones over-expressing essential proteins: a new strategy of identifying cellular targets of potent antibacterial compounds. *Biochem. Biophys. Res. Commun.* 349, 1250–1257.

Young, K. et al. 2006. Discovery of FabH/FabF inhibitors from natural products. *Antimicrob. Agents Chemother.* 50, 519–526.

Zhang, L. et al. 2000. Regulated gene expression in *Staphylococcus aureus* for identifying conditional lethal phenotypes and antibiotic mode of action. *Gene* 255, 297–305.

Zhang, L. et al. 2007. High-throughput synergy screening identifies microbial metabolites as combination agents for the treatment of fungal infections. *Proc. Natl. Acad. Sci. USA* 104, 4606–4611.

8 Image-Based High Content Screening

Yan Feng and Christopher J. Wilson

CONTENTS

8.1 INTRODUCTION

Fluorescence microscopy has long been a powerful tool for cell biology as well as drug discovery research. It can measure both the contents and locations of multiple biomolecules or probes in cells simultaneously. Thus it provides unparalleled levels of detailed information about cellular responses to drug treatment or other perturbations. It has been particularly useful for rapidly validating drug leads in multiple cellular systems relevant to their efficacy and toxicity profiles.

Although widely adopted, fluorescence microscopy had been a mostly manual and non-quantitative tool until about 10 years ago.

The rapid development of digital imaging technology, especially the wide availability of highly sensitive, low noise, charge-coupled display (CCD) cameras led to increasing demands to automate fluorescence microscopy. Cellomics' ArrayScan, introduced in 1997, was the first dedicated automated microscope for drug screening purposes. It enabled automated collection of images from multiwell plates in multiple fluorescence channels as well as automated image analysis for several specific applications. Now, more than 10 vendors currently provide high content screening (HCS) instruments covering a wide range of technical capabilities.

Automated microscopy-based screening is also known as HCS because it collects many cellular parameters simultaneously. In the past decade, HCS assays have been developed for drug discovery to determine hit validation, toxicity, mechanism of action (MOA) prediction, and primary high-throughput screening (HTS) (Giuliano, Hashins, and Taylor, 2003; Taylor, 2007). Technically, HCS assays still require significant investments in equipment, IT infrastructure, and expertise. Other assay platforms that can address the same questions should not be overlooked. Ultimately the usefulness of HCS raises certain questions. Why choose to develop HCS assays when other assay formats may be available? What advantages does HCS bring that other assays lack? How can we fully utilize the high information content from the images effectively? This chapter will discuss some basic steps for setting up HCS assays and highlight some of the challenges faced in setting up the technology platform in drug discovery settings.

8.2 GENERAL CONSIDERATIONS FOR PLANNING HCS SCREENING

Several practical factors and constraints play important roles in the decision to develop an image-based HCS assay. These can be broadly grouped into the following areas:

- Sample characteristics (typically mammalian cells)
- Scale, automation, image acquisition, and processing
- Assay development cycle time
- Assay timing
- Fixed versus live cell imaging
- Multiplexing
- Computational requirements
- Data storage requirements

We will discuss these factors in some detail and attempt to highlight both the attractive features and issues associated with HCS.

HCS assays typically utilize adherent mammalian cell lines or primary cells and cell samples present a major consideration for the platform. Cells are required to adhere to the bottom surfaces of multiwell plates to resist the many washing steps required to remove fixative and stain with antibodies or dyes. It is also desirable to use cells that disperse well (do not clump) and grow evenly across plates at 50% to 70% confluence. Cells that have these characteristics enable much simpler image processing for identifying nuclei and other features because they minimize overlap between neighboring cells and assignment of cellular features to nuclei is simplified. U2-OS, A549, and HeLa are examples of "well behaved" cell lines that exhibit desirable characteristics for imaging and are commonly found in cell biology laboratories.

Growth of these lines is relatively straightforward, so they can be scaled up for HTS campaigns (where greater than one billion cells may be needed) and genetic manipulation is possible—all are desirable characteristics as well. Other primary cells like fibroblasts from humans and mice and mixed populations of primary neuronal cells can also be effective because they adhere and spread well. HEK293, a common screening cell line, is a good example of cells that are difficult to work with because of adherence problems. The problems can be overcome using various types of plate coatings and plate washer tricks, but in general are best avoided if possible.

HCS assays have also been described using whole zebrafish embryos (Peterson et al., 2000) and other multicellular models such as multinucleated myotubes from differentiated C2C12 mouse cells. These assays clearly proved interesting and novel ways to measure biological activity and despite probable scalability issues, they represent a potential new wave of image-based assays using more complex cellular models.

Scale-up is another key issue that thwarts widespread adoption of HCS as a platform of first choice. The two key constraints driving this issue are the wash steps required for plate processing (non-homogeneous) and the long read times for plates. This is not to say that large-scale screens (more than one million wells) have not been carried out for specific HCS assays, but other homogeneous and faster methods that can measure similar biological readouts in cells would likely win out over HCS assays.

HCS assays have routinely been formatted in 96- and 384-well plates and reports of 1536-well assays have emerged at conferences. Equipment suited to manipulate these plates and manage the associated liquid handling steps in bulk is absolutely required to achieve any scale over a few 96-well plates at a time. Bulk liquid dispensers such as the WellMate and liquid transfer devices like the Beckman FX are commonly found in most assay development and screening labs and are essential for HCS. Plate washing is usually needed for HCS assays and presents an important step to optimize. As discussed earlier, cell adherence must be maintained and adequate cell washing at many steps in the staining process is critical (see Section 8.5). The BioTek ELx405 series is a good example of a robust washer commonly used for HCS assays. When using a confocal-based system, it is technically feasible to run green fluorescent protein (GFP) translocation or other dye staining assays without a wash (although image quality would be expected to improve with washing). This is practically not the case with wide field imaging systems because the background fluorescence from media components overwhelms the GFP signal and significantly reduces contrast. The confocal system achieves this by removing the out-of-focus light above the cell layer (medium).

Image acquisition time is another bottleneck for scaling up HCS assays. The many factors that affect imaging time include autofocus speed, number of fields per well, number of channels (or wavelengths) per well, exposure time for each wavelength, filter switching, and IT hardware (databases, file share arrangements, and networks). As an example, a 384-well GFP assay with Draq5 nuclear stain, using a 300-ms exposure in the GFP channel and a 100-ms exp in the Draq5 channel and capturing one field per well, takes approximately 20 min on a GE INCell® Analyzer 1000. Increasing the exposure and the number of fields will directly increase read time. When using a higher magnification objective such as a 20×, gathering enough cells for statistical comparisons of wells becomes an important consideration and commonly at least four fields per well are captured. This can increase read time for a 384-well plate to over an hour.

Informal discussions within the HCS community indicate that the 15 to 30 min per 384-well plate at ~0.75 µm resolution capturing ~200 cells is typical for most imaging platforms. For all systems, it is customary to integrate a plate delivery robot so that imaging of stacks of plates can occur without need for manually feeding the instrument and running over nights and weekends.

Although read times can present significant bottlenecks in large-scale screening campaigns (more than one million compounds), smaller-scale drug discovery efforts and genomic target finding screens (siRNA or cDNA libraries) are amenable to HCS assays. The testing of 100 compounds in eight-point dose response formats in triplicate yields 2400 data points or roughly nine 384-well plates. This number of plates, a typical request in a drug discovery effort, is manageable for most molecular cell biologists in teams that have access to the appropriate automation equipment. Similarly, a typical genomic library screen usually consists of 5,000 to 20,000 data points and can be readily scaled, usually over several weeks.

As discussed in Section 8.4, each of several types of HCS assays requires a different development time that will influence the decision to pursue the assay. Antibody-based assays can be developed relatively quickly if good antibodies and biological controls are available and represent a major advantage of using an HCS platform. Because staining procedures are fairly generic, it is usually straightforward to test several antibodies and antibody dilutions against a set of cells and/or conditions in one plate. If an adequate antibody is identified, assay development can quickly move to more advanced development stages focusing on Z′ improvement and optimization of other parameters. Conversely, if no antibodies are available, the development process effectively stops until reagents become available. GFP-based assays are somewhat different in that they usually require some molecular and cellular biology steps to create the fluorescent protein fusion and the cell line. This process is highly variable and an estimation of timing requires a case-by-case evaluation. Several GFP translocation assays can be purchased from BioImage (now part of ThermoFisher). Other staining and morphological assays such as Mitotracker staining for mitochondrial health typically resemble antibody staining assays in their development cycle timing: if reagents are available, assay development moves quickly. For all assay development projects, additional time is required to develop imaging algorithms to identify and measure cellular features. Again, this step may be straightforward and quick as with nuclear translocation assays; in other cases such as myotube differentiation assays, image analysis may be difficult and require significantly greater investments of resources.

Timing of HCS assays requires some forethought. For compound-based assay development, cells are typically plated and allowed to adhere and recover overnight. The test compound and stimuli (such as growth factors) are added to the wells the following day. The cells are then incubated for minutes to hours or days, depending on the biology and the kinetics of the response. Fixative can be added directly to cells and media, presumably inactivating most cellular reactions within seconds to minutes.

The timing of the steps is critical and must be centered on a good understanding or prediction of the biology. Fast reactions, such as Akt redistribution to the peripheral membrane after growth factor addition, occur quickly and must be monitored minute by minute. Accumulation of protein such as soluble β-catenin occurs on the order of hours because it requires de novo translation of new protein. Finally, changes to cellular physiology such as epithelial–mesenchymal transition (EMT) require days. The timing of each assay includes technical details that must be addressed. For example, shorter assays typically require advanced automation and scheduling to ensure the accurate timing of liquid transfers across plate handling steps. Assays that require days involve less timing control, but evaporation becomes an issue that can result in "edge effects" if not dealt with (typically by increasing volumes within wells). For all assays, it is important to pick the right time for the right biology to ensure that the desired phenomenon is measured.

One advantage of microscopy and GFP technologies is the ability to use populations of live cells rather then fixed cells. This ability may be important for several reasons including the ability to take kinetic measurements of cellular processes like calcium flux or movement of cells in a cell migration assay. In addition, excluding fixatives removes the possibility of artificial re-localization of proteins induced by the fixation agent. Several instruments include chambers with controllable heating, humidity, and CO_2 for long term imaging over hours and days (Table 8.1). A significant challenge for live cell imaging beyond assay development stages is contending with automation at scale.

TABLE 8.1
Partial List of High Content Imaging Instruments

Instrument	Manufacturer	Technology	Live Cell	Liquid Handling	Data Management	Information
ArrayScan VTi	Thermo Fisher	Wide field with Apotome™	Available	Available	STORE™	http://www.cellomics.com
ImageXpress Micro	Molecular Devices	Wide field	Available	Available	MDCStore™	http://www.moleculardevices.com
ImageXpress Ultra	Molecular Devices	Laser scanning confocal	No	No	MDCStore™	http://www.moleculardevices.com
INCell Analyzer 1000	GE Healthcare	Wide field with structured light	Yes	Yes	In Cell Miner™	http://www.gelifesciences.com
INCell Analyzer 3000	GE Healthcare	Laser scanning confocal	Yes	Yes	In Cell Miner™	http://www.gelifesciences.com
Opera	Perkin Elmer	Laser confocal with Nipkow disk	Available	Available	File directory	http://las.perkinelmer.com
CellWoRx	Applied Precision	Wide field with oblique illumination	No	No	STORE™	http://www.api.com
Pathway 435	BD Biosciences	Wide field with Nipkow disk	No	No	File directory	http://www.atto.com
iCyte	Compucyte	Wide field laser scanning	No	No	File directory	http://www.compucyte.com
MIAS-2	Maia Scientific	Wide field	No	No	File directory	http://www.maia-scientific.com
Explorer ᶜX3	Acumen	Wide field laser scanning	No	No	File directory	http://www.ttplabtech.com

As discussed previously, read time is often a bottleneck in the HCS process and will dictate upstream liquid handling processes unless a pause step (as with fixed cells) is included. Assays with many 384-well plates in live cell mode become difficult, even with a fully integrated robotic system, because the cell growth, cell plating, compound addition, and other steps require precise coordination with imaging that is often not worth the cost.

Many fluorescent dyes and proteins now available enable multiple detection channels and the ability to "multiplex" related assays. HCS assays typically use at least two channels: one for a DNA stain and another for the fluorophore of interest. In general, the maximum number of channels utilized at one time ranges from two to five. Instrument hardware and driver software determine the number of channels and fluorophores to be acquired. Some factors to consider here include illumination source (arc lamp or laser), filter and mirror requirements, number of cameras or PMT detectors, camera sensitivity, and desired detection wavelength range. Other considerations for multiplexing include read time, resolution, and assay time (for live cell imaging).

The final consideration focuses on the computational and data storage needs for HCS assays. HCS assays generate large amounts of primary image data. For example, a medium-scale screen of 25,000 compounds in duplicate can easily be achieved in one week with a single HCS instrument and relatively limited liquid handling automation. With four individual channels and four image fields per well collected, the experiment could generate ~600 GB of raw image data, and a full HTS screen at 1 M compounds will generate over 25 TB of raw image data. Some image analysis routines also generate large amounts of data, especially when single cell analysis data is stored. An estimated 2.5 billion data points are generated in a typical medium sized experiment; a full million compound screen will produce over 100 billion data points (Figure 8.1).

Based on the output data scale, setting up an HCS platform requires significant communication with information technology and data storage resources specialists. After images and data are analyzed it is often possible to retain the data on storage media with slower access speeds since most images are rarely retrieved for inspection and generally may not have to be immediately accessible. In addition, the deletion of images or raw cellular data should be considered. In most assays, they represent intermediate steps to the final well-based data. Obviously customary business practices and intellectual property requirements must be considered when deleting data to ensure compliance with institutional policies.

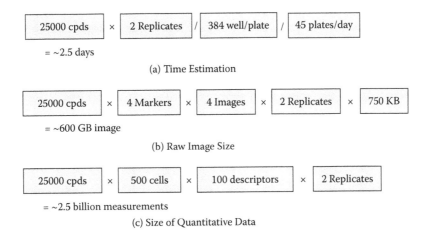

FIGURE 8.1 HCS Data Size. (a) Typical medium sized HCS screen of 25,000 compounds in duplicate took one week to complete, assuming imaging takes one hour per 384-well plate. (b) 600 GB of raw image data were generated, assuming four image fields for each treatment and four marker channels were taken with a 2×2 binned 1 M pixel CCD camera and 12-bit digitizing format. (c) 2.5 billion measurements were generated, assuming 500 cells were quantified with 100 descriptors each in every treatment condition.

8.3   TYPES OF HCS ASSAYS

HCS assays can be roughly divided into four main formats: (1) intensity-based assays, (2) translocation assays, (3) cytometry assays, and (4) morphology assays (Figure 8.2). Intensity-based assays score for the change of intensity of a particular biomolecule in cells, for example, upregulation of lineage-specific markers or modification of regulatory proteins. Typically, average cellular or nuclear intensity per cell is treated as assay output. This is the simplest format of high content assay, taking advantage of a sensitive CCD camera and high light throughput of most HCS imagers.

In principle, intensity measurement is no different from the well-based immunoblot or other homogeneous well-based assay format such as an in-cell Western. However, because individual nuclei are identified and used as the mask, precise normalization to cell number can be achieved. In addition, microscopy with high quality cameras offers much higher signal-to-noise ratios, thus making many more assays possible. Finally, if a threshold can be set and cells classified as responders or non-responders, an intensity-based assay can be turned into a cytometry assay.

Translocation assays score for movement of a particular protein from one cellular compartment to another. Immunofluorescence using specific antibodies or expression of GFP fusion proteins is the most common method for detection. Ratios or differences of concentrations in different cellular compartments can be used as quantitative scores for such events. Nuclear–cytoplasmic translocation of transcription factors or signal transduction molecules such as NFκB (Ding et al., 1998), NFAT (Venkatesh et al., 2004), FOXO (Kau et al., 2003), p38MK2, and HDAC4/5 are assays that are of interest historically. Because of its simple format and robust quantitation methods, nuclear–cytoplasmic translocation can be adopted in engineered biosensors to detect binding or cleavage events in cells, such as p53/HDM2 interaction (Giuliano and Taylor, 1998) or caspase-3 activation (Knauer et al., 2005).

Cytometry assays treat each cell as a single entity and score for population changes of cells with certain phenotypic characteristics. Fluorescence-activated cell sorting (FACS) is the most common format for cytometry-based assays. FACS requires large numbers (>10^5) of cells and does

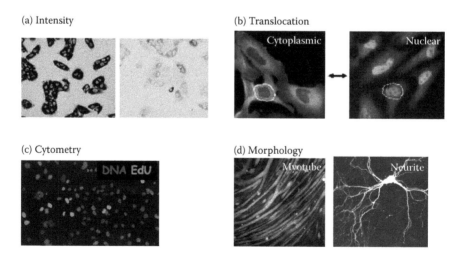

FIGURE 8.2   (See color insert following page 114.) HCS assay formats. (a) Intensity measurement of phospho-S6. (b) Cytoplasmic–nuclear translocation of GFP-Foxo3a. Ratio of GFP-Foxo3a intensity in the nuclear mask region (red circle) and in the cytoplasmic mask region (yellow ring) was used as quantitative measurement. (c) Cytometry assay measuring percentage of S-phase cells. S-phase cells were labeled with ethynyl-dU incorporation and azido-rhodamine label (red) and all cells were labeled with Hoechst 33342 dye (blue). (d) Morphology assays on myotube formation and neurite formation. Myotube length and width and neurite length and branches were among the parameters of interest.

not provide enough resolution to score intracellular or morphological changes in cells. HCS-based cytometry requires far fewer cells ($\sim 10^2$) and can be performed at high spatial resolution, thus providing higher throughput and data not available from conventional FACS. The cytometry-based cell cycle assay has been adopted widely in oncology research in which interference with cell cycles and cell growth is one of the main goals of therapeutic intervention. Cell cycle stages are determined by DNA content and elevations of other specific cell cycle markers such as phospho-histone H3 for mitotic cells and cleaved PARP for apoptotic cells (Wilson et al., 2006).

Morphology-based assays are generally more challenging for image analysis and thus prove more difficult to develop into fully automated high-throughput formats. A few exceptions such as neurite outgrowth and myotube differentiation that have no other surrogate assay formats have been studied more extensively (Liu et al., 2007).

8.4   EXAMPLE PROTOCOL

Sample preparation is the first critical step for obtaining high quality image data. Here we provide a typical protocol used in our laboratory as the starting point for HCS assay development. In practice, we have found more than 90% of assays can be covered by this simple procedure.

8.4.1   CELL CULTURE

Cells are plated at $\sim 10^5$ cells/mL, 30 µL for 384-well and 100 µL for 96-well plates, respectively. We use a Multidrop (ThermoFisher), Microfill (BioTek), or WellMate (Matrix) to assist automation of cell plating. Cells are then grown at 37°C in 5% CO_2 overnight to allow attachment, then treated with a test agent (compound, siRNA, cDNA, etc.) for minutes, hours or day, depending on the assay.

8.4.2   STAINING

We use an ELX405 plate washer (BioTek) running at lowest dispensing and aspiration speed to assist all fixation and staining processes. A concentrated fixative—either formaldehyde in phosphate buffered saline (PBS) or Mirsky's fixative (National Diagnostics)—is added directly to the wells. After 15 to 60 min, the wells are washed once with PBS.

For cells expressing GFP fusion protein, Hoechst 33342 or Draq5 nuclear stain is added to the fixative to ready the cells for image acquisition. For immunofluorescence staining, the cells are incubated with PBS-TB (PBS with 0.2% Triton X-100, 0.1% bovine serum albumin [BSA]) for 10 min to permeabilize the cell membranes. Primary antibody is then added at ~ 0.5 to 5 µg/mL in PBS-TB and incubated at room temperature for 1 hr or at 4°C overnight. The wells are then washed two times with PBS-TB. Fluorescently labeled secondary antibody is added at 2 µg/mL together with 10 µg/mL Hoechst 33342 nuclear stain in PBS-TB and incubated at room temperature for 1 to 2 hr. The cells are then washed with PBS-TB and once with PBS before image acquisition.

8.5   IMAGE ACQUISITION HARDWARE

There are many aspects to consider when selecting an imager: imaging capability, image quality, speed, price, automation compatibility, image analysis, data storage infrastructure, etc. In general, imaging capability and image quality are the most important. For example, if you want to image small structures such as dendritic spines, a high magnification confocal instrument is the best choice. If you want only to count intensity changes in responding cells, a low magnification scanner may offer you much higher speed and reduce the IT burden associated with image data storage. Many of the current vendors provide specific analysis packages.

Although commercial and proprietary image analysis packages exist, it can still be a very involved process to develop your own analysis algorithms. Determining whether a vendor has the appropriate analysis algorithm for your main application is crucial in selecting a vendor. In many cases, testing your own sample application on several available imagers is the only way to determine which one yields the best results.

Currently, more than ten commercial vendors provide automated fluorescence imaging hardware, including confocal and non-confocal imagers. In principle, a confocal imager can provide images with reduced out-of-focus light. However, in practice, most of the applications use low magnification objectives (4, 10, or 20×) with a flat monolayer of cells, and thus a wide field microscope can provide high quality images adequate for image analysis. All the HCS imagers can focus on each field and acquire fluorescence images at a variety of magnifications and wavelengths in fully automated mode. Robotic plate handlers can usually be integrated to the imager for large screening efforts. Several instruments also provide live imaging (CO_2 and humidity) and liquid handling capabilities. Table 8.1 is a partial list summarizing their main features. Technical details can easily be obtained by contacting individual vendors.

8.6   IMAGE ANALYSIS AND VISUALIZATION

Many image analysis software packages are available for high content screening, but most users use the packages that come with their particular instruments. This is largely because image files and associated metadata are often written in formats that are not easily processed by other instrument vendors or third party analysis software packages (non-standard text formats for metadata or proprietary, non-TIFF image formats). In addition several vendors (Cellomics and MDS) have database applications that require additional software to retrieve images and metadata. In many cases, vendor analysis software is adequate for most intensity changes or translocation assays and can achieve "wellular" discrimination of positive and negative controls with respectable assay quality metrics (plate Z' scores >0.5).

In addition to vendor software, attempts have been made to cope with the complexities of specific vendor outputs and develop generic image and data analysis software that can be used with all acquisition platforms. These include the free and open source image and data analysis software programs known as Cell Profiler and Cell Profiler Analyst from the Broad/MIT and Whitehead Institutes (http://www.cellprofiler.org/index.htm) (Carpenter et al., 2006) and Zebrafish image analysis software (ZFIQ) from Methodist Hospital in Houston (http://www.cbi-platform.net/download. htm) (Liu et al., 2008). The Cell Profiler source boasts the largest academic community focused on high content screening image and data analysis.

The commercial Pipeline Pilot package of SciTegic-Accelrys (http://accelrys.com/products/scitegic/) includes advanced component collections capable of image analysis. Because Pipeline Pilot was created for customers that handled large data sets, the high content screening protocols are straightforward to develop and have very useful and robust web reporting features—often better than vendor software. More advanced image analysis software can be found in the Definiens package (http://www.definiens.com/) that uses a novel over-segmentation and reconstruction approach to identify cellular regions of interest (Baatz et al., 2006). Finally another example of a free and robust, but rudimentary, software package is Image J from the National Institutes of Health (http://rsb.info.nih.gov/ij). This package is extremely usefully for viewing and quickly manipulating images for display and analysis. The software has a recordable macro language for analysis of images in batches and also has methods for Java programmers to write plug-ins that extend the functionality of the software.

Typical HCS image analysis revolves around first identifying nuclei using an image from DNA stained (Hoechst or Draq5) cells, identifying other regions of the cells, and associating those regions back to discrete nuclei. Generally, image analysis is composed of three steps: (1) region finding or segmentation, (2) region qualification or filtering, and (3) region measurement. Most vendors have developed adaptive thresholding methods for identifying nuclei in images where the background is uneven (due to illumination or staining artifacts). These methods typically rely on users to input the approximate sizes of the nuclei and some form of sensitivity.

Problematic segmentation of nuclei occurs in several situations such as dividing cells, clumped and apoptotic nuclei, and regions where the cells are out of focus. Often these can be eliminated using size filters (region qualification). However, small numbers of problematic nuclei (below 10% of all nuclei) usually cause little or no interference with subsequent well summary measurements. The most common method for measuring the cytoplasm is to expand the nuclei several pixels and create a collar region (also called annulus region), excluding the original nucleus. Pixels in the collar and nucleus can be measured in both the DNA stained channels and the other stain or fluorescent channels. Typically the assumption is made that images from different channels will align perfectly with the base nuclei channel.

Measurements are gathered to summarize pixel values within a region and the most frequently used parameters are the mean, median, and coefficient of variance (CV). Other methods for cytoplasm identification include watershed, region growing, and others. It is worth noting that these other methods usually require a consistently stained cytoplasm in one channel. Assays in which the intensity of the cytoplasm stain is expected to decrease should not be used for region identification as the region area will be directly impacted by intensity and will produced biased results (weaker staining shows little cytoplasm, even though the cytoplasm viewed by phase imaging has not shrunk). In these cases, the collar method is typically less biased. For example, nuclear to cytoplasmic shuttling of NFκB subunit p65 can be appropriately measured using nuclei finding and a simple collar method with most cell types (Ding et al., 1998). It is also common to find pits, vesicles, or granules in cells. These methods are useful for β-arrestin and GPCR translocation assays. They usually rely on granule counting or ratios of granule intensity to cytoplasm intensity, and are typically quite robust (Barak et al., 1997).

For most assay applications, cellular measurements are statistically reduced to measurements that reflect the population of cells in a well. The most common summary or "wellular" statistic is the average or median of the number of cells within a well. Other less common metrics are based on variation within a population or comparisons of treatment and control groups using the t-test or other non-parametric test such as the Kolmogorov–Smirnov (Perlman et al., 2004). Because these wellular measurements end up representing activity of a test agent, it is usually advantageous to use wellular summary metrics that are meaningful (e.g., average or median). The more complex statistical summaries, although occasionally more robust, often suffer from downstream interpretability problems with non-HCS scientists. It is often useful to use three to four or more wellular measures to analyze data. For example cell count, nuclear area, and nuclear intensity measurements are often indicative of proliferation or cytotoxicity effects and can be used in addition to the assay measurement for the biology of interest. Once wellular measurements are chosen for an assay they can be treated like other plate reader outputs and analyzed by appropriate screening and assay methodologies (Z' calculations, signal-to-noise ratio, hit calling, and IC_{50} curve fitting).

8.7 MULTIPARAMETER DATA ANALYSIS

If only a single wellular readout is needed, simple statistics used in other HTS works can be readily applied (Zhang, Chung, and Oldenburg, 1999). In most cases, well median or the average of one or several independent parameters is sufficient for a given image-based assay. However, in some cases, multiple individual cellular parameters and their correlations contain crucial data and must be extracted into wellular readouts in a way that cannot be achieved by simple statistics that treat each parameter independently. We will cite two specific examples to illustrate this challenge.

In a cell cycle assay, the percentage of cells in each cell cycle stage is the most crucial measurement. Cells can be classified into G1or G2 phase by their DNA content, the M phase (mitotic or interphase) by the phospho-histone H3 mitotic marker, and the S phase by a pulsed label with dU to detect DNA synthesis. Each cell has four intensity parameters and has to be classified into the G1, G2, S, or M phase based on these parameters. Because we are potentially dealing with billions of cells, cell cycle classification must be automated.

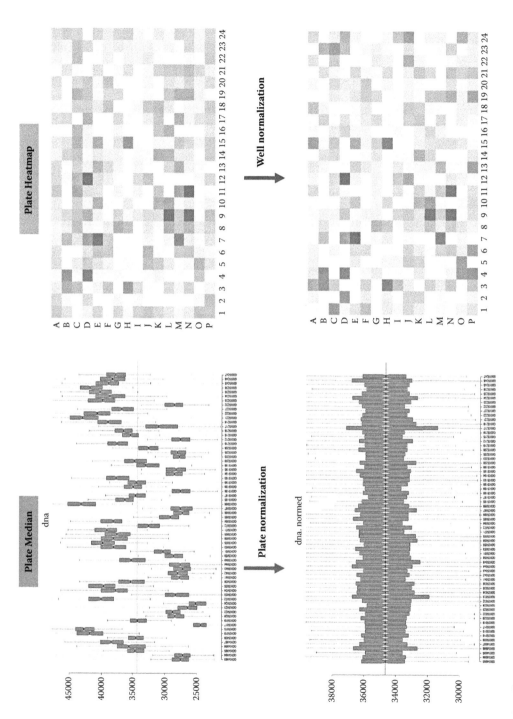

FIGURE 8.3 (See color insert following page 114.) Pre-processing data normalization. Normalization reduces plate-to-plate and well-to-well variations, allowing uniform analysis of entire HCS dataset in other modules.

Before we deal with classification, systematic errors such as plate-to-plate and well-to-well variations must be handled; otherwise the errors may thwart any automated classification effort in general. A simple median polish normalization procedure applied to each of the four parameters can readily take care of systematic errors (Tukey, 1977). The effects of normalization on plate-to-plate variation and plate edge effect, two typical systematic errors in high-throughput screening, are shown in Figure 8.3. Then an automated classification procedure can be applied on a plate-by-plate basis using a decision tree-based algorithm (Figure 8.4). Alternatively, a supervised method such as neural network or support vector can be applied. This approach requires a small, often manually curated training set (Tao, Hoyt, and Feng, 2007).

A typical HCS experiment may generate several gigabytes of numbers extracted from the images describing the amounts and locations of biomolecules on a cell-to-cell basis. Many of these numbers have no obvious biological meaning. For example, while the amount of DNA per nucleus has obvious significance, the importance of other nuclear measures (DNA texture, nuclear ellipticity, etc.) is much less clear. This often leads biologists to ignore non-obvious measurements, even though they may report useful data about compound activities. One standard method used in other fields to analyze large, multidimensional datasets is factor analysis. In mathematical terms, the so-called common factor model—a set of measured random variables—is a linear function of common factors and unique factors. In HCS, the common factors reflect the set of

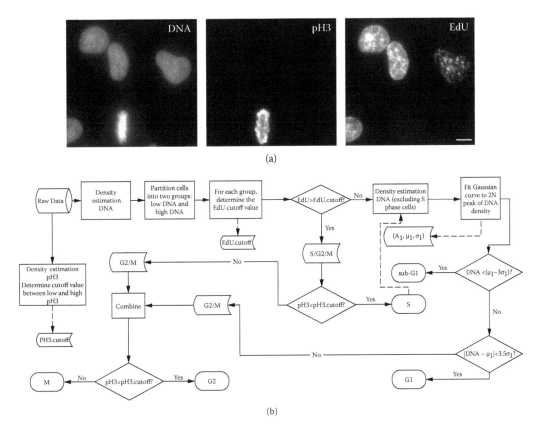

FIGURE 8.4 (See color insert following page 114.) Decision tree cell cycle phase classification. (a) Cells were stained to show their levels of DNA, phospho-H3, and EdU in each nucleus. (b) Every cell was classified as G1 (2N DNA), G2 (4N DNA), M (pH3 positive) or S (EdU positive) phases using the automated four-parameter decision tree model. (c) The results of the classification were shown in two scatter plots where cells in different phases were labeled with different colors, G1 (orange), G2 (green), M (red), and S (blue), respectively.

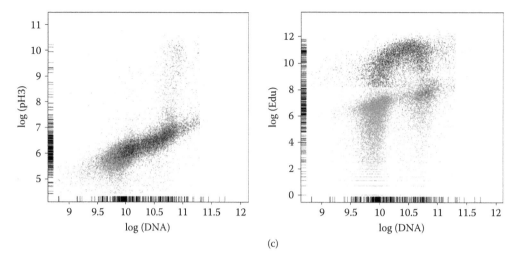

(c)

FIGURE 8.4   (*Continued*).

major phenotypic attributes measured in the assay. The model allows a large data reduction but retains most of the information content and quantifies phenotypes using data-derived factors that are biologically interpretable in many cases. For this reason, factor analysis is highly appropriate for high content imaging because it seeks to identify underlying processes (Young et al., 2008). Fitting to the factor model can be accomplished using the score procedure included in SAS® (Hatcher, 1994).

8.8   CONCLUSION AND FUTURE PERSPECTIVES

In this chapter we outlined some of the finer technical details and considerations for setting up HCS or image-based assays. We also discussed more advanced data processing techniques such as cellular classification to improve assay quality. HCS assays continue to be useful across a broad range of biological applications and our prediction is that this trend will continue. Improvements to imaging hardware (mainly speed of acquisition) and data analysis software may result in significant benefits, including use of HCS as a platform of choice for high throughout screening.

ACKNOWLEDGMENTS

We thank Charles Tao, Jonathan Hoyt, and Daniel Young for experimental and data analysis support.

REFERENCES

Baatz, M. et al. 2006. Object-oriented image analysis for high content screening: detailed quantification of cells and sub cellular structures with the Cellenger software. *Cytometry A* 69, 652–658.

Barak, L.S. et al. 1997. A β-arrestin/green fluorescent protein biosensor for detecting G protein-coupled receptor activation. *J. Biol. Chem.* 272, 27497–27500.

Carpenter, A.E. et al. 2006. CellProfiler: image analysis software for identifying and quantifying cell phenotypes. *Genome Biol.* 7, R100.

Ding, G.J. et al. 1998. Characterization and quantitation of NF-κB nuclear translocation induced by interleukin-1 and tumor necrosis factor-α: development and use of a high capacity fluorescence cytometric system. *J. Biol. Chem.* 273, 28897–28905.

Giuliano, K.A., Haskins, J.R., and Taylor, D.L. 2003. Advances in high content screening for drug discovery. *Assay Drug Dev. Technol.* 1, 565–577.

Giuliano, K.A., and Taylor, D.L. 1998. Fluorescent-protein biosensors: new tools for drug discovery. *Trends Biotechnol.* 16, 135–140.

Hatcher, L.A. 1994. *Step-by-Step Approach to Using SAS for Factor Analysis.* SAS Institute.

Kau, T.R. et al. 2003. A chemical genetic screen identifies inhibitors of regulated nuclear export of a Forkhead transcription factor in PTEN-deficient tumor cells. *Cancer Cell* 4, 463–476.

Knauer, S.K. et al. 2005. Translocation biosensors to study signal-specific nucleo-cytoplasmic transport, protease activity and protein-protein interactions. *Traffic* 6, 594–606.

Liu, D. et al. 2007. Screening of immunophilin ligands by quantitative analysis of neurofilament expression and neurite outgrowth in cultured neurons and cells. *J. Neurosci. Meth.* 163, 310–320.

Liu, T. et al. 2008. ZFIQ: a software package for zebrafish biology. *Bioinformatics* 24, 438–439.

Perlman, Z.E. et al. 2004. Multidimensional drug profiling by automated microscopy. *Science* 306, 1194–1198.

Peterson, R.T. et al. 2000. Small molecule developmental screens reveal the logic and timing of vertebrate development. *Proc. Natl. Acad. Sci. USA* 97, 12965–12969.

Tao, C.Y., Hoyt, J., and Feng, Y. 2007. A support vector machine classifier for recognizing mitotic subphases using high-content screening data. *J. Biomol. Screen.* 12, 490–496.

Taylor, D.L. 2007. Past, present, and future of high content screening and the field of cellomics. *Meth. Mol. Biol.* 356, 3–18.

Tukey, J.W. 1977. *Exploratory Data Analysis*: Springfield, MA: Addison-Wesley.

Venkatesh, N. et al. 2004. Chemical genetics to identify NFAT inhibitors: potential of targeting calcium mobilization in immunosuppression. *Proc. Natl. Acad. Sci. USA* 101, 8969–8974.

Wilson, C.J. et al. 2006. Identification of a small molecule that induces mitotic arrest using a simplified high-content screening assay and data analysis method. *J. Biomol. Screen.* 11, 21–28.

Young, D.W. et al. 2008. Integrating high-content screening and ligand-target prediction to identify mechanism of action. *Nat. Chem. Biol.* 4, 59–68.

Zhang, J.H., Chung, T.D., and Oldenburg, K.R. 1999. A simple statistical parameter for use in evaluation and validation of high-throughput screening assays. *J. Biomol. Screen.* 4, 67–73.

9 Application of RNA Interference in Drug Discovery

Natalie M. Wolters and Jeffrey P. MacKeigan

CONTENTS

9.1 INTRODUCTION

Andrew Fire and Craig Mello proved that double-stranded RNA (dsRNA) conferred specific gene silencing in *Caenorhabditis elegans* (Fire et al. 1998). The incredible significance of this discovery did not disappear within the scientific community, and both were awarded the 2006 Nobel Prize in Medicine and Physiology. The loss of function phenotype resulting from RNAi was first observed in the petunia; and the mechanism elucidated by Fire and Mello after *C. elegans* were injected with long strands of dsRNA that amazingly displayed potency several orders of magnitude greater than the antisense single-stranded molecules (Fire et al. 1998).

Although methods for silencing gene expression in worms and invertebrates were well established, researchers had few options for targeting specific genes within mammalian genomes. Prior to the advent of RNAi, an effective but costly method was to generate knockout mice that were used to produce knockout mouse embryonic fibroblasts (MEFs).

Also available at the time were single-stranded antisense molecules. Even in its earliest stages, RNAi proved to be more potent and robust than the then current antisense technologies. RNAi-mediated knockdown decreased protein levels more potently and at concentrations several orders of magnitude lower than antisense (Bass 2001; Elbashir et al. 2001). In addition, RNAi was surprisingly robust, as injection of dsRNA into the tail provided gene silencing throughout the entire organism, with knockdown persisting through to the progeny. Further studies in *C. elegans* determined that knockdown could be induced by a wide variety of mechanisms, including bathing the animals in a solution containing dsRNA or directly feeding worms dsRNA (Timmons and Fire 1998). The approaches were rapidly extended to *Drosophila* embryos (Kennerdell and Carthew 1998) and invertebrate cultured cell lines (Clemens et al. 2000), allowing both model organisms to be exploited for genome-wide RNAi screens for phenotypic changes and even drug discovery efforts.

9.2 RNAi FOR DRUG DISCOVERY

9.2.1 LIBRARIES FOR RNAi-BASED SCREENS

The identification of the subset of druggable genes to be examined is an important factor in the implementation of a successful RNAi screen. Initially, genome-wide RNAi screens carried out in both *C. elegans* and *Drosophila* (Kamath et al. 2003), were made possible by the experimental parameters associated with these screens, such as the ease of delivery and administration of large strands of dsRNA. More recently, genome-wide RNAi screens of mammalian cells have become quite common. Although the genome does not exclude any gene from study, early studies focused on a particular gene subset that contained the most relevant target genes that were also readily druggable (Aza-Blanc et al. 2003; MacKeigan, Murphy, and Blenis 2005).

RNAi libraries are usually organized into gene families based on gene ontology. These categories span a wide variety of genes implicated in numerous processes, including kinases (MacKeigan, Murphy, and Blenis 2005; Paddison et al. 2004); phosphatases (MacKeigan, Murphy, and Blenis 2005; Paddison et al. 2004; Moffat et al. 2006); tumor suppressors (Moffat et al. 2006); DNA modifying enzymes (Moffat et al. 2006); cell cycle components (Berns et al. 2004); and a variety of signaling molecules (Berns et al. 2004). The number and nature of the genes selected will reflect the molecular pathways of each RNAi screen and should be chosen with the intention to maximize potential drug targets, while minimizing irrelevant targets.

Remember to factor in multiple sequences targeting a single gene, as replication of the phenotype with multiple siRNAs is one of the most crucial factors in generating high quality RNAi screen results. When searching for novel targets within a biochemical pathway or cellular process, one should keep the number of gene families inclusive and expansive, and not reduce the novelty of the drug targets. However, when looking to answer questions within the context of a specific molecular mechanism, the paring down of an RNAi library to only relevant genes is both practical and resourceful. For instance, if a researcher wanted to identify a novel regulator of a phospho-protein like mTOR, probing a hand-picked set of human kinases and phosphatases within the relevant cell type would retain all the pertinent genes to determine a direct regulator of the phosphorylation event.

RNAi libraries have typically been screened in one of two ways: (1) a gene-by-gene approach or (2) a gene pooling approach. Gene-by-gene screens target a single gene per well with multiple (four or more) RNAi reagents targeting the same gene in the same well of a 384-well plate (Figure 9.1). siRNAs and shRNA screens have utilized gene-by-gene screening. In particular for siRNA screens, forward and reverse transfection methods yielded significant results. Forward transfection is the more traditional method and involves first plating cells, and then adding the siRNAs in complex with the transfection reagent. More recently groups have experimented with reverse transfecting cells that may in certain situations provide higher transfection efficiencies (Amarzguioui 2004; Ovcharenko et al. 2005). The reverse transfection first prepares spotted siRNA and lipid reagent into naïve 96- or 384-well plates and then cells are dispensed onto the mixture bypassing the extra day in culture. Viral shRNA particles use exclusively forward infection methods as integration into the host cell genome is maximized in cells already proliferating in culture.

In contrast to gene-by-gene screening, gene pooling with multiple shRNA viral particles may target hundreds of genes in a single 96- or 384-well plate. The advantage of gene pooling is that it reduces the number of wells to be screened, thereby reducing screening costs. In fact, gene pooling has been used for entire shRNA libraries (Brummelkamp et al. 2006) and selected for a particular phenotype. To identify the individual shRNA that generated the phenotype of interest, PCR or oligonucleotide barcodes are used to identify the enriched or depleted shRNA (Berns et al. 2004; Paddison et al. 2004). Each specific shRNA has a unique barcode of DNA sequence and the difference in the ratios of fluorochromes predicts the shRNAs with altered frequencies in the pooled shRNA setting (Figure 9.1).

A recent successful application of the shRNA barcode approach for drug resistance genes was completed. The authors used RNAi to identify for genes that conferred resistance to trastuzumab

RNAi screening protocol

1. Infect shRNA virus or transfect siRNA into cells

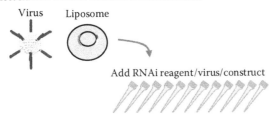

Add RNAi reagent/virus/construct

2. Allow cells to reach desired knockdown (48 h–120+h)

Argonaute 2

Dicer

3. Sensitize cells with agonist/antagonist

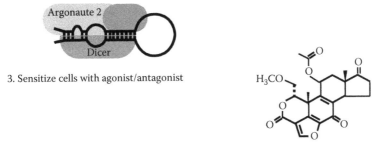

4. Read absorbance or acquire high content images from assay

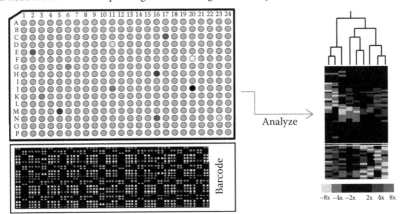

Barcode

Analyze

−8x −4x −2x 2x 4x 8x

FIGURE 9.1 (See color insert following page 114.) RNAi screening protocol. Custom libraries or genome-wide RNAi libraries are designed, constructed, and synthesized with up to five duplexes for each gene. In gene-by-gene screening RNAi duplexes for each target are combined and arrayed in a 96- or 384-well format. Viral shRNAs constructs or liposome mediated siRNA oligonucleotides are forward (or reverse) transfected onto disease-relevant cell types. On the basis of validation data, one should achieve greater than 95% infection or transfection efficiency and, on average, greater than 80% knockdown for each mRNA target species tested. Infection or transfection proceeds for 48+ hr to allow for sufficient knockdown before cells are sensitized with agonist or antagonist compounds before reading the assay. RNAi barcode pooling approaches also involve determining the precise barcode that generated the desired phenotypic results.

after screening 8000 different genes (Berns et al. 2007). They identified the lipid phosphatase PTEN, and as expected its loss hyperactivated Akt and downstream signaling. Importantly, loss of PTEN accounts for a large percentage of resistant breast cancer cases (Saal et al. 2005). As a follow-up to the pooled shRNA screen, patient biopsies from HER-2 overexpressing breast cancer patients who also received trastuzumab monotherapy were analyzed for PTEN expression. As expected, tumors expressing low levels of PTEN responded poorly to therapy, and were predicted to be resistant to trastuzumab. Using an RNAi-based approach, this screen uncovered the importance of PTEN (and subsequent high PI3K/Akt signaling) expression status for predicting patient outcome. These RNAi approaches highlight the ability of RNAi to identify biomarkers that may eventually determine efficacies of different drug combinations for specific cancer subtypes.

9.2.2 Experimental Design

One of the most significant parameters within an RNAi experiment, whether targeting an individual gene or an entire genome, is the type of RNAi (siRNA, shRNA, or miRNA) to carry out the gene-by-gene or pooled knockdown. Although each agent is capable of mediating knockdown, the decision whether to use siRNA-, shRNA-, or miRNA-based gene targeting is extremely important. To make an informed decision on which experimental approach to employ, researchers must consider the length of the proposed experiment (2 to 10 days versus 5 to 15+ days), the cell type to be used (amenable to lipid transfection versus infection). As siRNAs have been the most extensively characterized tools of RNAi, we recommend their use in a highly transfectable cell type. Chemically synthesized siRNAs are readily available from several companies, and many maintain databases of sequences selected for potency or pre-validated knockdown.

The delivery method chosen for the RNAi agent in any experiment will depend heavily on the cell type used and the nature of the assay. Groups have found success with lipid-based transfections (Aza-Blanc et al. 2003; Li et al. 2006; MacKeigan, Murphy, and Blenis 2005; Paddison et al. 2004; Pelkmans et al. 2005), electroporation (Brummelkamp et al. 2003), and viral transduction (Berns et al. 2004; Kolfschoten et al. 2005; Moffat et al. 2006; Ngo et al. 2006; Westbrook et al. 2005) of both short interfering and short hairpin RNAs. Lipid-based transfection reagents are the most common means of delivery and have achieved high success rates, although extensive optimization is needed for each individual cell line. When choosing a reagent for a lipid-based approach, the lipid must be able to efficiently knock down gene expression at relatively dilute concentrations to avoid lipid-mediated toxicity. Prior to initiating RNAi screens for drug targets one must titrate a handful of lipids at various time points and concentrations to optimize conditions for each cell type.

In addition to the temporal advantages associated with the use of siRNAs, the method has been very well characterized, allowing efficient and optimal experimental design. The specific amount of siRNA delivered to a cell can be tightly controlled. Assuming a uniform transfection efficiency, siRNAs can be titrated down into the 1 to 25 nanomolar range and still allow substantial knockdown of the gene of interest. This is an especially important consideration for limiting potential off-target effects. Establishing the lowest effective siRNA concentration that generates the specific phenotype (such as apoptosis, autophagy, or pathway modulation) also generates effective knockdown and minimizes potential off-target effects caused by an overabundance of siRNA (Elis et al. 2008).

For experiments requiring the use of cultured cells that are especially difficult to transfect, viral plasmids containing shRNA motifs have been engineered with sequences spanning the entire genome. Retroviral- (Berns et al. 2004; Kolfschoten et al. 2005; Ngo et al. 2006; Westbrook et al. 2005); adenoviral- (Subramanian and Chinnadurai 2006; Cao et al. 2005); and lentiviral- (Moffat et al. 2006; Golding et al. 2006) based systems have been utilized by numerous groups with great success. As lentiviruses and adenoviruses are capable of infecting both dividing and non-dividing cells, transduction of shRNA viral particles may be the only feasible way to achieve knockdown in extremely-slow growing and non-dividing cell types.

Viral vectors allow the generation of a stable knockdown, as their shRNA-containing genomes integrate into the host cell and thus replicate with each cell division. This is an important distinguishing characteristic of a viral-based approach, and is necessary for experiments that require more than the 2- to 10-day window of knockdown. Researchers must pay particular attention to validating experiments and eliminating the possibilities of off-target effects. The determination of the lowest effective dose using a viral shRNA transduction has yet to be perfected. The titration of viral-based shRNAs is somewhat difficult compared to evaluating siRNA delivery and one must establish the multiplicity of infection (MOI) of each viral vector as a means to normalize the amount of each virus. This has proven problematic in initial attempts to generate viral shRNA libraries. Although this step cannot eliminate all off-target effects, it can get rid of the potential over-infection of any particular viral shRNA or miRNA construct.

Although siRNA- and shRNA-mediated gene knockdowns are the most thoroughly characterized, Hannon and colleagues (2008) generated RNAi reagents with endogenous microRNA-like properties. The goal of miRNA knockdown libraries is to increase their potency and/or specificity. Using miRNA precursors as the backbones for the delivery of hairpins, these shRNA-mirs contain unique hairpin loops of complementary sense and antisense strands. However, they are flanked by the stem sequences found within miRNAs, providing extensive secondary structures beyond the early designed shRNA molecules.

The shRNA-mirs have been shown to efficiently and specifically target and inhibit gene expression in both transient (Zeng, Wagner, and Cullen 2002) and stable (Dickins et al. 2005) settings. In both cases, shRNA-mirs produce a more potent gene silencing effect than traditionally designed shRNAs (Boden et al. 2004; Silva et al. 2005). Notably, when shRNA-mir cassettes integrate into the genome as a low number or even single copy, effective gene silencing is still detected (Dickins et al. 2005). This efficiency is especially important for lowering the potential for off-target effects and mediating sufficient knockdown to observe phenotypes. Since viral vectors are available, shRNA-mirs may provide excellent options for those who require long-term knockdown or delivery into a difficult-to-transfect cell type.

9.2.3 Optimization of RNAi Target Sequence

Whether utilizing a siRNA-, shRNA-, or miRNA-based approach to modulate gene expression, optimization of the target sequence is paramount. During any attempt to identify drug targets, one must ensure the strongest knockdown and simultaneously limit the potential off-target effects. Designing potent siRNAs requires a set of rules that can be applied to efficient and specific siRNA design (Figure 9.2). The first consideration in the design of any siRNA sequence should be the nature of the target sequence.

The most straightforward approach to determine the efficacy of knockdown is to directly test four to five siRNAs in your disease-relevant cell type that expresses your target gene, using quantitative RT-PCR with specific primers and, of course, Western blotting when antibodies exist. This should produce at least two siRNAs with the most potent knockdowns—usually exceeding 80%, with 70% a historical benchmark. However, determining individual siRNA potency in a high-throughput drug discovery platform is not practical. As a result, the ability to predict siRNA efficacy before actually testing the siRNA proves to be an important tool for high-throughput and genome-wide RNAi screens. With this in mind, individual laboratories and consortiums have devised algorithms to predict both the efficacies and specificities of RNAi sequences (Boese et al. 2005; Huesken et al. 2005; Yuan et al. 2004).

Properties such as thermodynamic values, sequence asymmetry, and polymorphisms that contribute to RNA duplex stability are taken into account by these databases (Pei and Tuschl 2006). In addition, artificial neural networks have been utilized to train algorithms based on the analysis of randomly selected siRNAs (Huesken et al. 2005). These programs siphon significant trends from large sets of RNA sequences whose efficacies are known and validated. Certain base pair (bp) positions have a tendency to possess distinct nucleotides (Figure 9.2). In effective nucleotides, position 1 is preferentially an adenosine (A) or uracil (U), and many strands are enriched with these nucleotides along the first 6 to 7 bps of sequence (Pei and Tuschl 2006). The conserved RISC cleavage site at nucleotide position 10 favors an adenosine, which may be important, while other nucleotides are

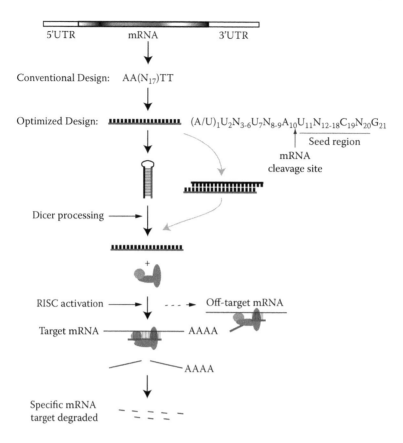

FIGURE 9.2 Critical for efficient gene silencing is the design of the siRNA sequence. The siRNA librar-ies are designed with informatics algorithms against all known human genes. Algorithms take into account siRNA sequence base composition (G + C/A + T content), secondary structure of target mRNA, and positional effects within the mRNA, and choose the best sequence motifs to ensure that the siRNA sequence targets a single gene. This allows efficient Dicer processing, RISC activation, and specific target mRNA degradation.

overrepresented at distinct sites in potent sequences at positions 7, 11, 19, and 21. Although the selec-tion bias for these primary sequences is largely unknown, these base pair positions may confer ther-modynamic stability advantages for effective loading and targeting of the complementary mRNA.

The efficacy of an siRNA sequence highly depends on where it hybridizes along the mRNA tar-get. The secondary structure of the mRNA may occlude the siRNA molecule from annealing and tar-geting for the mRNA for degradation. To date no known algorithm exists that can take into account each mRNA's distinct secondary structure. Until such advanced algorithms are amenable to siRNA target design, the field commonly uses four distinct sequences that span the length of the coding sequence and sometimes 3′ UTR of each gene to maximize a couple siRNAs that are accessible to the siRNA-loaded RISC. During high-throughput screening, this can be accomplished by using four distinct siRNA target sequences against a single gene into one well (gene-by-gene). Although proper controls and procedures must be carried out to appropriately interpret the data, this method greatly reduces material requirements while ensuring knockdowns of most if not all gene targets.

9.2.4 OFF-TARGET EFFECTS

Many off-target effects associated with RNAi are caused not by homology to the coding sequences of other genes, but rather by short stretches of 6 to 7 bps within the 5′ end of the guide strand

(Jackson et al. 2003, 2006). These seed regions align perfectly with regions in the 3′ UTRs of other mRNA molecules (Birmingham et al. 2006). Seed regions were originally found within microR-NAs, and the siRNAs that impart off-target effects do so in a manner analogous to that of endogenous microRNAs.

Analysis of 1000 siRNA sequences known to non-specifically degrade mRNA showed a bias for complementation to 3′ UTRs within positions 2 to 7 of their guide strands (Jackson et al. 2006). Regardless, experimental design must take care to recapitulate phenotypes observed within an RNAi experiment through repetition (i.e., more than two potent siRNA sequences yield the same phenotype) or through rescue (i.e., the phenotype can be restored by re-expression of the gene of interest).

9.2.5 SYNTHETIC LETHAL SCREENING

Model organisms have been used successfully for synthetic lethal screens in attempts to identify suitable targets for drug discovery (Hartwell et al. 1997). Initially, synthetic lethal mutants were used in yeasts as genetic tools. However, the utility of RNAi allows synthetic lethal screening to be realized in mammalian cells and systems with simple cell viability or apoptosis as an endpoint. Synthetic lethal screens have become valuable for identifying high priority targets in different mutational backgrounds. One must follow a reductionist approach and adopt a mechanistic mind to truly understand the biology behind each RNAi-generated synthetic lethal screen, as most RNAi screen hits are against genes of unknown function.

After the selection of the specific techniques to be employed, one must determine how to quantify the genes associated with the phenotype of interest. The design of an assay that acts as a reporter or a molecular sensor can prove to be the most challenging step in designing a successful RNAi screen. The assay must be both highly reproducible and sensitive, without noise, and exhibiting minimal standard error. Traditional assay development techniques must be taken into account for all RNAi screens as they are for conventional compound-based screens.

The most straightforward screens utilized well established experimental methods whose relative ease allowed the rapid identification of a manageable number of genes. A successful assay will be: (1) robust, to diminish background and assay noise; (2) reproducible, to identify potential hits accurately; (3) sensitive, to detect genuine changes in the RNAi phenotype without masking by assay noise; and (4) manageable and expandable, to be amenable to a large-scale RNAi screen. The most robust methods are likely to utilize fluorescence. Fluorescent assays include ELISA antibody-based, FACS-based, and traditional GFP-based molecular sensor assays. The construction of GFP-based reporter genes is very amenable to RNAi screens for drug discovery and has proven to be a reproducible means to observe RNAi-mediated phenotypes.

The absolute number of RNAi screen hits will depend on assay quality, execution, and the statistical stringency used to distinguish a weakly positive or strong phenotype. Additional secondary and tertiary cell-based assays must be used as methods of validation and confirmation of the associated phenotype. When screening for novel modulators of a particular process, such as novel modulators of PI3K/Akt or mTOR/S6K signaling, it is always necessary to validate the pathway activation by confirming that known components produce the expected result. For example, when looking for novel negative modulators of the PI3K/AKT pathway, a crucial step in validation would be the demonstration that the kinase PDK1 or phosphatase PTEN behaves as one would expect and can serve as a proper control in each plate of the high-throughput RNAi assay.

Overall, two methods should be utilized to follow-up any hits within an RNAi screen to validate that the phenotype of interest is a product of specific gene knockdown. First, multiple unique siRNAs targeting the same gene, assuming sufficient knockdown, should exhibit the same exact phenotype. The strong correlation between knockdown and phenotype with multiple siRNAs is important to show that the molecular effect is on target. A second approach is to "rescue" the phenotype with exogenously expressed target genes. It is typical to utilize siRNAs that target the 3′ UTR of the gene of interest. These 3′ UTR siRNAs knock down endogenous transcripts and not the exogenously introduced cDNA plasmids that lack both 5′ and 3′ UTRs. Introduction of siRNAs that target only the 3′ UTR coupled with exogenous gene expression should reverse the phenotype.

A second more specific approach is to not alter the siRNA primary sequence, and use site-directed mutagenesis to generate three to five "wobble" mutations in the target gene sequence. Exogenous expression of the wobble-mutated cDNA will be resistant to RNAi silencing. As with any other drug discovery approach, the validation of genes within a screen requires the demonstration that the phenotypes observed were neither false positives, false negatives, nor off-target effects. RNAi-based screens represent large investments and solid follow-up on each gene is a crucial step in the success of any RNAi-based approach.

9.3 ENDOGENOUS NON-CODING RNAS (MIRNAS)

The discovery of an endogenous class of RNAi molecules or microRNAs is perhaps one of the most intriguing advances. These small temporal RNAs were initially described during *C. elegans* development (Lee, Feinbaum, and Ambros 1993; Wightman, Ha, and Ruvkun 1993) and quickly extended to mammalian systems (Lau et al. 2001). These classes of non-coding RNA molecules are prevalent throughout the human genome and miRNA orthologs are also commonly found within lower organisms. miRNAs are critical regulatory molecules for controlling mRNA and protein levels.

Currently, the Sanger Institute curates approximately 6400 hairpin precursor miRNAs that result in 6200 mature miRNA products found in viruses, plants, flies, birds, fish, rodents, and primates (Griffiths-Jones 2004; Griffiths-Jones et al. 2008). As of April 2008, the human miRNAome contained 678 unique miRNAs and each sequence is postulated to repress the expression of 50 to 200 mRNA transcripts. This is a major explosion in the identification of human miRNAs. In April 2007, Tuschl and colleagues reported 340 distinct mature miRNAs from humans (Landgraf et al. 2007). As a result, in only one year, the number of known miRNAs increased 100%, as reflected in the number of miRNA citations (Figure 9.3). The expanding list of miRNAs and the ability to regulate

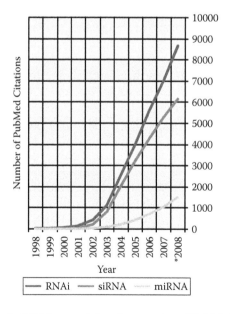

FIGURE 9.3 Major explosion in RNAi research based on number of PubMed citations since the 1998 landmark RNAi studies by Fire and Mello. The graph reflects the emerging study of miRNAs and the importance of targeting miRNA or the pathways they modulate for drug discovery.

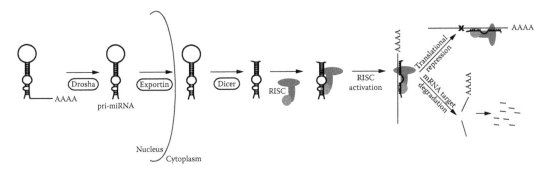

FIGURE 9.4   Mature microRNAs (miRNAs) are single-stranded RNA molecules of 17 to 24 nucleotides that are encoded in the genome. Each miRNA is formed from Drosha processing to generate a pri-miRNA that is exported from the nucleus by exportin, and the stem loop is removed by Dicer. miRNAs are predicted to regulate multiple genes, opening a complex regulatory circuit that results from a single miRNA. The ability of miRNAs to regulate both translational repression and mRNA target degradation situates them as potential diagnostic markers or even therapeutic targets in human disease.

hundreds of targets highlights the enormous regulatory network of endogenous non-coding RNAs and the importance of targeting miRNAs and the pathways they modulate for drug discovery.

The processing of endogenous miRNAs differs from processing siRNA or shRNA sequences (Figure 9.4). siRNAs are delivered exogenously to cells, while shRNA molecules are produced ectopically. In contrast, miRNAs are transcribed by class II RNA polymerases into primary miR-NAs (pri-miRNAs) with significant secondary structures recognized by the Drosha complex that cleaves the miRNA into its second immature form, a 70-nucleotide hairpin containing a 2-bp overhang on its 3′ end. This "pri-miRNA" is then exported from the nucleus to the cytoplasm, where it is processed by Dicer into its fully active miRNA form. The processing of siRNAs, shRNAs, and miRNAs converges at the point of cleavage by Dicer, after which mature siRNAs and miRNAs associate with distinct multiprotein complexes to create an RNA-induced silencing complex (RISC). The proteins present in the RISC vary among species, but a core of proteins including Dicer and the Argonaute (Ago) protein families constitute the silencing complex.

9.4   SUMMARY AND CONCLUSIONS

Completion of the sequencing of the human genome and the discovery of RNAi has dramatically impacted our ability to functionalize the genome and discover novel drug targets. RNAi is a valuable tool for drug discovery, giving industry and academia the ability to functionalize genetic events for drug discovery. RNAi screens produce genome-wide loss-of-function data sets that can situate previously uncharacterized genes into the context of a specific molecular pathway, and ultimately within the context of drug discovery as we attempt to map genes to human disease.

Although most screens carried out thus far have focused on the etiology of cancer, the door for increased understanding of other diseases is visible. RNAi-based screens have already identified novel targets for other diseases, including Alzheimer's (Yang et al. 2006), metabolic syndrome (Tang et al. 2006), and cancer (Bartz and Jackson 2005). This information will allow researchers to better understand the molecular bases of these diseases, which, in time, will help identify new and improved therapeutic targets.

Model organisms have been central to the identification of genes associated with human disease. However, the complexities associated with genetic mammalian cells are sometimes intractable in lower organisms. One solution to this is RNAi screening in human disease relevant cell types in an attempt to gain insight into the intricacies of mammalian cell biology. The first screen carried out in a mammalian cell probed a library of siRNAs against selected targets in the human genome to identify regulators of TRAIL-induced apoptosis (Aza-Blanc et al. 2003). Within two years, RNAi-based screens identified multiple tumor suppressors (Kolfschoten et al. 2005; Westbrook et al. 2005) along with kinases and phosphatases implicated in cell survival and chemoresistance (MacKeigan, Murphy, and Blenis 2005). These studies show the vast potential from large-scale RNAi screens in target identification and the development of new therapeutic targets.

REFERENCES

Amarzguioui, M. 2004. Improved siRNA-mediated silencing in refractory adherent cell lines by detachment and transfection in suspension. *Biotechniques* 36, 766–770.

Aza-Blanc, P. et al. 2003. Identification of modulators of TRAIL-induced apoptosis via RNAi-based phenotypic screening. *Mol. Cell* 12, 627–637.

Bartz, S. and A.L. Jackson. 2005. How will RNAi facilitate drug development? *Sci. STKE* 295, pe39.

Bass, B.L. 2001. RNA interference: the short answer. *Nature* 411, 428–429.

Berns, K. et al. 2004. A large-scale RNAi screen in human cells identifies new components of the p53 pathway. *Nature* 428, 431–437.

Berns, K. et al. 2007. A functional genetic approach identifies the PI3K pathway as a major determinant of trastuzumab resistance in breast cancer. *Cancer Cell* 12, 395–402.

Birmingham, A. et al. 2006. 3′ UTR seed matches, but not overall identity, are associated with RNAi off-targets. *Nat. Meth.* 3, 199–204.

Boden, D. et al. 2004. Enhanced gene silencing of HIV-1-specific siRNA using microRNA designed hairpins. *Nucleic Acids Res.* 32, 1154–1158.

Boese, Q. et al. 2005. Mechanistic insights aid computational short interfering RNA design. *Meth. Enzymol.* 392. 73–96.

Brummelkamp, T.R. et al. 2003. Loss of the cylindromatosis tumour suppressor inhibits apoptosis by activating NF-κB. *Nature* 424, 797–801.

Brummelkamp, T.R. et al. 2006. An shRNA barcode screen provides insight into cancer cell vulnerability to MDM2 inhibitors. *Nat. Chem. Biol.* 2, 202–206.

Cao, H.B. et al. 2005. Down-regulation of IL-8 expression in human airway epithelial cells through helper-dependent adenoviral-mediated RNA interference. *Cell Res.* 15, 111–119.

Clemens, J.C. et al. 2000. Use of double-stranded RNA interference in *Drosophila* cell lines to dissect signal transduction pathways. *Proc. Natl. Acad. Sci. USA* 97, 6499–6503.

Dickins, R.A. et al. 2005. Probing tumor phenotypes using stable and regulated synthetic microRNA precursors. *Nat. Genet.* 37, 1289–1295.

Elbashir, S.M. et al. 2001. Duplexes of 21-nucleotide RNAs mediate RNA interference in cultured mammalian cells. *Nature* 411, 494–498.

Elis, W. et al. 2008. Down-regulation of class II phosphoinositide 3-kinase-α expression below a critical threshold induces apoptotic cell death. *Mol. Cancer Res.* 6, 614–623.

Fire, A. et al. 1998. Potent and specific genetic interference by double-stranded RNA in *Caenorhabditis elegans*. *Nature* 391, 806–811.

Golding, M.C. et al. 2006. Suppression of prion protein in livestock by RNA interference. *Proc. Natl. Acad. Sci. USA* 103, 5285–5290.

Griffiths-Jones, S. 2004. The microRNA Registry. *Nucleic Acids Res.* 32, D109–D11.

Griffiths-Jones, S. et al. 2008. miRBase: tools for microRNA genomics. *Nucleic Acids Res.* 36, D154–158.

Hannon, G.J. et al. 2008. Topoisomerase levels determine chemotherapy response *in vitro* and *in vivo*. http://www.pnas.org/content/105/26/9053.full.pdf. Accessed Oct. 2009.

Hartwell, L.H. et al. 1997. Integrating genetic approaches into the discovery of anticancer drugs. *Science* 278, 1064–1068.

Huesken, D. et al. 2005. Design of a genome-wide siRNA library using an artificial neural network. *Nat. Biotechnol.* 23, 995–1001.

Jackson, A.L. et al. 2003. Expression profiling reveals off-target gene regulation by RNAi. *Nat. Biotechnol.* 21, 635–637.

Jackson, A.L. et al. 2006. Widespread siRNA "off-target" transcript silencing mediated by seed region sequence complementarity. *RNA* 12, 1179–1187.

Kamath, R.S. et al. 2003. Systematic functional analysis of the *Caenorhabditis elegans* genome using RNAi. *Nature* 421, 231–237.

Kennerdell, J.R. and R.W. Carthew. 1998. Use of dsRNA-mediated genetic interference to demonstrate that frizzled and frizzled 2 act in the wingless pathway. *Cell* 95, 1017–1026.

Kolfschoten, I.G. et al. 2005. A genetic screen identifies PITX1 as a suppressor of RAS activity and tumorigenicity. *Cell* 121, 849–858.

Landgraf, P. et al. 2007. A mammalian microRNA expression atlas based on small RNA library sequencing. *Cell* 129, 1401–1414.

Lau, N.C. et al. 2001. An abundant class of tiny RNAs with probable regulatory roles in *Caenorhabditis elegans*. *Science* 294, 858–862.

Lee, R.C., R.L. Feinbaum, and V. Ambros. 1993. The *C. elegans* heterochronic gene lin-4 encodes small RNAs with antisense complementarity to lin-14. *Cell* 75, 843–854.

Li, S. et al. 2006. RNAi screen in mouse astrocytes identifies phosphatases that regulate NF-κB signaling. *Mol. Cell* 24, 497–509.

MacKeigan, J.P., L.O. Murphy, and J. Blenis. 2005. Sensitized RNAi screen of human kinases and phosphatases identifies new regulators of apoptosis and chemoresistance. *Nat. Cell Biol.* 7, 591–600.

Moffat, J. et al. 2006. A lentiviral RNAi library for human and mouse genes applied to an arrayed viral high-content screen. *Cell* 124, 1283–1298.

Ngo, V.N. et al. 2006. A loss-of-function RNA interference screen for molecular targets in cancer. *Nature* 441, 106–110.

Ovcharenko, D. et al. 2005. High-throughput RNAi screening *in vitro*: from cell lines to primary cells. *RNA* 11, 985–993.

Paddison, P.J. et al. 2004. A resource for large-scale RNA-interference-based screens in mammals. *Nature* 428, 427–431.

Pei, Y. and T. Tuschl. 2006. On the art of identifying effective and specific siRNAs. *Nat. Meth.* 3, 670–676.

Pelkmans, L. et al. 2005. Genome-wide analysis of human kinases in clathrin- and caveolae/raft-mediated endocytosis. *Nature* 436, 78–86.

Saal, L.H. et al. 2005. PIK3CA mutations correlate with hormone receptors, node metastasis, and ERBB2, and are mutually exclusive with PTEN loss in human breast carcinoma. *Cancer Res.* 65, 2554–2559.

Silva, J.M. et al. 2005. Second-generation shRNA libraries covering the mouse and human genomes. *Nat. Genet.* 37, 1281–1288.

Subramanian, T. and G. Chinnadurai. 2006. Temperature-sensitive replication-competent adenovirus shRNA vectors to study cellular genes in virus-induced apoptosis. *Meth. Mol. Med.* 130, 125–34.

Tang, X. et al. 2006. An RNA interference-based screen identifies MAP4K4/NIK as a negative regulator of PPARγ, adipogenesis, and insulin-responsive hexose transport. *Proc. Natl. Acad. Sci. USA* 103, 2087–2092.

Timmons, L. and A. Fire. 1998. Specific interference by ingested dsRNA. *Nature* 395, 854.

Westbrook, T.F. et al. 2005. A genetic screen for candidate tumor suppressors identifies REST. *Cell* 121, 837–848.

Wightman, B., I. Ha, and G. Ruvkun. 1993. Post-transcriptional regulation of the heterochronic gene lin-14 by lin-4 mediates temporal pattern formation in *C. elegans*. *Cell* 75, 855–862.

Yang, J.P. et al. 2006. A novel RNAi library based on partially randomized consensus sequences of nuclear receptors: identifying the receptors involved in amyloid-β degradation. *Genomics* 88, 282–292.

Yuan, B. et al. 2004. siRNA Selection Server: an automated siRNA oligonucleotide prediction server. *Nucleic Acids Res.* 32, W130–W134.

Zeng, Y., E.J. Wagner, and B.R. Cullen. 2002. Both natural and designed micro RNAs can inhibit the expression of cognate mRNAs when expressed in human cells. *Mol. Cell* 9, 1327–1333.

10 Assay Development Using Primary and Primary-Like Cells

Sabrina Corazza and Erik J. Wade

CONTENTS

10.1 INTRODUCTION

The process of identifying new chemical entities (NCEs) often involves the use of cell-based or biochemical assays by which the effects of new compounds on a specific target of interest are monitored. Transformed cell lines have been the traditional mainstays of assay development since the beginning of high-throughput screening (HTS) in the early 1980s.

The use of recombinantly expressed targets is a well established and widely used system in almost every step of the pipeline for new candidate discovery in the pharmaceutical industry. A variety of cell lines such as Chinese hamster ovary (CHO) and human embryonic kidney (HEK293) have been used for recombinant protein expression and assay design. These cell lines exhibit relatively short doubling times in culture, adhere to the cell culture plates without the requirement for specific or expensive factors, and can be expanded across multiple passages in culture while maintaining appropriate phenotype. Therefore, the use of recombinantly expressed targets cloned into cell lines provides screening laboratories with virtually unlimited material of consistent quality. More recently, these and other cells have been also adapted to automated cultivation by cell culture robots such as the CellMate ™ and the SelectT ™ (Automation Partnership) or are used as stock aliquots from large batches of quality controlled frozen cells (Zhu, 2007), further improving the consistency of the cell supply step.

Despite the clear advantages of using recombinantly expressed targets, certain disadvantages must be taken into consideration. Transformed cells have been subjected to clonal selection that may perturb their biology in unknown ways. In addition, they provide no guarantee that a cell line will contain required interaction partners or accessory proteins for a target protein. The expression level of the target gene is often forced to non-physiological levels that can alter the pharmacology of the target or its interaction with partners or accessory proteins, thus affecting the behavior of the target with respect to well known modulating molecules. The overexpression of a target may also distort the behavior of an assay toward weak modulators or those interacting through sites other than those used by well known target modulating molecules. It is difficult or impossible to rule out subtle effects that may negatively impact the quality and reliability of an assay during validation.

By the end of 2003, the high-throughput screening (HTS) approach generated 74 leads in clinical development and only two marketed drugs (Fox, 2004). The attrition rate of drugs identified by HTS is at least as high as that for earlier approaches and is clearly influenced by factors discussed in several papers reviewing the role of HTS in drug discovery (Kola and Landis, 2004; Macarron, 2006). The high attrition rate suggests a need to continue to improve screening assays to better reflect conditions experienced *in vivo*, at least to the degree possible for an *in vitro* system.

The key goal of any HTS is to discover as many meaningful hits as possible from a compound collection while achieving an acceptable false positive rate. This requires careful consideration of factors such as the nature of the target, an appropriate substance library, and an assay predictive of all possible interactions of an unknown compound with the target of interest.

Cell-based assays more closely resembling the *in vivo* physiological expression and functionality of the target are therefore highly sought after for both the initial phases of HTS and also the prediction and evaluation of efficacies, side effects, and toxicities of compounds in humans. The most common reason for failure of NCEs during drug discovery and development arises from inadequate metabolic and pharmacokinetic parameters (Schafer, 2008). New approaches have been developed to improve the early determination and prediction of drug metabolism of NCEs. Most are *in vitro* methods that require recombinant cell lines and, more recently, primary or primary-like cells.

Primary human cells offer the best prospects for developing more meaningful assays. In cases where primary human cells are not available, primary mammalian cells can be used if their functions are known to be preserved in evolution. Primary cells derived from *in vitro*-differentiated human or murine stem cells may also be suitable, particularly where larger amounts of material are required than can be provided by explants of primary cells. All these cell types theoretically represent more physiological systems for target gene expression and biology, even though their use presents drawbacks that complicate their implementation. An overview of the advantages, difficulties, applications, and methods for assay development using primary and stem cells derived primary cells is discussed in the next sections. In conclusion, new cell models help increase the number of new safe and efficacious compounds identified and thereby help reduce the escalating costs of drug discovery and development.

10.2   PRIMARY CELLS

Primary cells can be defined as cells taken directly from an organism and adapted to survive and eventually grow in culture, at least for a limited period. In fact, with the exception of some cells derived from tumors, primary cells have limited lifespans in culture. Cells can be isolated from tissues for ex vivo culture in several ways. They can be released from soft tissues by enzymatic digestion with enzymes such as collagenase, trypsin, or pronase that break down extracellular matrices. Alternatively, in the explant culture method, pieces of tissue are placed in growth media and the cells that grow are available for culture.

One possibility to overcome the limited lifespans of primary cells is by converting them from explants to cell lines by repeated passages. These cells typically continue to proliferate for a number of generations but soon the culture goes through the process of senescence and stops dividing while generally retaining viability. A small fraction of the cells continue to proliferate but they may have undergone genetic changes that lead to cell transformation and are not necessarily any closer to the parental primary cells than existing cell lines. In addition, the efficiency of this procedure

TABLE 10.1
Primary Cells with Well Known Isolation Protocols

Tissue	Cell Type	Primary	Immortalized	Species
Gut	L cells	Yes	Yes	Human
Adipose tissue	Adipocytes	Yes	No	Human, rat
Skeletal muscle	Muscle cells	Yes	No	Human
CNS	Hypothalamous	No	Yes	Human
Liver	Hepatocytes	Yes	No	Human, rat
Blood	PBL, PMN, T cells	Yes	No	Human

Note: For more methods and bibliographic references see: http://www.tissuedissociation.com.

varies among species and it works poorly with the human cells that are of the greatest interest to the pharmaceutical industry.

Methods for the isolation and cultivation of primary cells from humans and other organisms have been refined into a number of well validated protocols for tissues (see Table 10.1). These primary cells offer good starting points for developing cell-based assays and running more physiological screening campaigns, though clearly a need continues to exist for additional primary cells to allow broader coverage of human tissues and disease models, especially for difficult-to-obtain types such as neurons and cardiomyocytes.

The successful isolation of primary cells is dependent on several factors, some of which are not subject to optimization, such as species, type of tissue, age and sex of donor, and presence of genetic modifications (e.g., knockout animals). Other factors such as the dissociation medium, enzymes and concentrations, temperature, and incubation times can be optimized to ensure the quality and consistency of a primary or cell line preparation. The identification and availability of key growth factors is an important determinant of which primary cells can be maintained in culture. For example lactoferrin is a pleiotropic factor with potent antimicrobial and immunomodulatory activities. Recently it has been shown that lactoferrin at physiological concentrations can also promote bone growth, potently stimulating the proliferation and differentiation of primary osteoblasts (Naot, 2005).

A primary cell culture often consists of mixed populations of cell types: some may continue to proliferate, while others survive without proliferating or exhibit reduced growth rates, leading to unpredictable shifts in the proportions of cells in a population. This possibility should be considered when using primary cells, and adequate controls of the cell population composition should always be used within a single cell preparation and in batch-to-batch preparations.

Fluorescence-activated cell sorting (FACS) is a widely used method to assess the composition of a freshly isolated cell population and for purifying mixed cells populations after isolation from tissues (Kamihira, 2007; Herzenberg, 2002). Cells are first treated with antibodies recognizing specific marker genes uniquely expressed by the cell population of interest and analyzed by FACS to measure the composition of the population. In Figure 10.1, for example, primary mouse macrophages were analyzed by FACS after treatment with rat anti-CD204 and anti-F4/80 (pale green) antibodies. As control populations, untreated macrophages and the same cells treated with only secondary anti-rat FITC antibody were analyzed. The positive cell population is clearly shifted to the right. FACS can be used also to sort and separate desired cell populations from other cells—a useful methodology but a stressful process for the cells. As a result, FACS is not suitable for all cell types, particularly those whose differentiation is affected by stress.

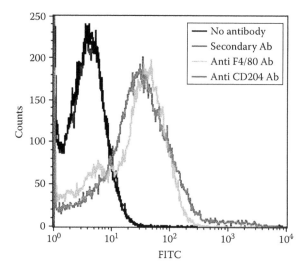

FIGURE 10.1   (See color insert following page 114.) Primary mouse macrophages were analyzed by FACS after treatment with two different rat antibodies: anti-CD204 (dark green) and anti-F4/80 (pale green). As controls, untreated macrophages (black line) and the same cells treated with only secondary anti-rat FITC antibody (red line) were analyzed. The positive cell population clearly shifted to the right.

The use of freshly isolated primary cells for HTS is tied to the need for sufficient cells of consistent quality, with a defined growth rate, and free of contaminants such as mycoplasma, bacteria, fungi yeast, and viruses. The yield of a preparation is clearly a critical factor as is batch-to-batch variability, as discussed below. The quality of any single cell preparation must be verified; morphological observation and phenotypic profiling are the most common methods. In addition, PCR profile analysis of key genes and immunofluorescence to verify the presence of specific and distinctive marker genes are often used to verify cell population compositions in preparations (Fico et al., 2008). For specific use and application of primary cells, for example, for screening on a specific target gene, it may be useful to run functional tests on the cell population to assess the functionality of the key target genes to be used later for screening.

10.3   PRIMARY-LIKE CELLS

Primary-like cells can be defined as cell lines obtained by differentiation of embryonic stem (ES) cells toward specific cell lineages using specifically adapted protocols. The cells obtained closely resemble primary cells directly isolated from fresh tissues, with the advantage of a potentially unlimited quantity of material available. For primary cells, the major challenges are availability of the desired cell type and effective isolation of a sufficient number to make their use practical. ES cells, in contrast, can self-renew and can reproduce in culture almost indefinitely under the correct conditions. This unique feature clearly represents a remarkable advantage because it provides starting material for the production of primary-like cells in large quantities. The major challenges are identifying easy and reproducible differentiation protocols and ensuring maintenance of the purity of the differentiated cell population produced. We will use "primary cells" in the remainder of this chapter to collectively refer to primary and primary-like cells except where their differences require explanation.

10.4   USES OF PRIMARY CELLS

The three major areas of interest in which primary cells offer significant advantages in the drug discovery process are cell-based assays and screening; absorption, distribution, metabolism, excretion,

TABLE 10.2
Advantages and Disadvantages in Using Primary Cells versus Other Cell Types

Features	Primary Cells	ES Differentiated Cells	Recombinant Systems
Primary screening	Limited use	Limited use	Widely used
Availability of material	Limited	Unlimited	Unlimited
Cell population	Heterogeneous	Highly heterogeneous	Homogeneous
Cell phenotype	Tissue selective	Cell selective	Not specific
Cell engineering	Difficult	Possible	Easy
Genetic stability	Diploid	Diploid	Aneuploid
Target expression	Native	Native	Artificial
Cell handling	Difficult	Difficult	Easy
Cell environment	Closer to *in vivo*	Closer to *in vivo*	Different from *in vivo*

and toxicology (ADMET) assays; and gene function and pathway analysis. The main advantages offered by primary cells are reported in Table 10.2.

10.4.1 Cell-Based Assays and Screening

Recent surveys indicate cell-based assays represent more than 50% of all screening campaigns performed every year in HTS laboratories for drug discovery purposes (Fox, 2006). Most of these assays are based on the use of transformed cell lines expressing a recombinant target, sometimes with a reporter gene to functionally detect the activation or repression of the target gene.

While the use of primary cells for developing cell-based assays is sometimes a matter of choice to provide a more native-like environment for the target of interest, in other cases the use of primary cells may be necessary as the only viable way to address completely the complexities of multiple subunit targets such as ion channels and multimeric G protein-coupled receptors (GPCRs). The recombinant expression of targets such as voltage-gated calcium channels is particularly challenging since multiple subunits must be co-expressed with the correct stoichiometry to fully resemble the native situation. Despite reports of the successful use of recombinant systems in screening for such difficult targets (Benjamin, 2006), the use of cells endogenously expressing these targets at the physiological level and stoichiometry is simpler and in most instances more reliable.

John McNeish (2007) described a primary screening campaign of one million compounds to search for modulators of the AMPA receptors (AMPARs) using mouse ES cells differentiated into neurons. These receptors are non-NMDA-type ionotropic transmembrane receptors for glutamate that mediate fast synaptic transmission in the central nervous system. AMPARs are composed of four types of subunits, designated GluR1 (*GRIA1*), GluR2 (*GRIA2*), GluR3 (*GRIA3*), and GluR4 (*GRIA4*), alternatively called GluRA–D, that combine to form tetramers. Most AMPARs are heterotetrameric, consisting of symmetric "dimers of dimers" of GluR2 and GluR1, GluR3, or GluR4 (Mayer, 2005; Greger, Ziff and Penn, 2007). The complexity of the stoichiometries of AMPARs makes them particularly challenging targets to study in recombinant systems since the subunit composition and expression levels may only partially reproduce the physiological situation. These differentiated cells demonstrated the same rank order pharmacology as primary rat neuronal cells, indicating their utility as valid and consistent cell models to measure the modulation of AMPAR functionality and leading to the successful identification of specific AMPAR modulators.

In the case of GPCRs, some targets or mechanisms studied in engineered recombinant cells may miss some naturally occurring protein partners. An example is the discovery of the role of receptor activity-modifying proteins (RAMPs) as accessory proteins. RAMPs are single transmembrane proteins that heterodimerize with a GPCR for the correct and appropriate localization and function of the calcitonin receptor (CT) and the calcitonin-like receptor (CLR) (Parameswaran, 2006). The discovery of RAMPs was facilitated by the efforts of several research groups to express CT and CLR in heterologous systems. This suggested the possible need of a cofactor for the correct expression of the receptor.

As a result, three different members of the RAMP family were identified as able to interact with these receptors. The diversity generated by interaction of two receptors with three accessory proteins or expressed alone, which respond to at least four endogenous ligands with differing affinities, depending on the exact receptor composition, gives rise to at least seven different receptor phenotypes that would not have been possible in a simple heterologous expression system unless all the necessary accessory proteins were also co-expressed at appropriate levels. Appropriately selected primary cells express most, if not all, of the cofactors and accessory proteins at the correct levels along with the receptor of interest.

Even if the concepts are good, the use of primary cells in screening laboratories is not yet widespread. Primary cells are mainly employed in secondary assays to validate effects of compounds identified in initial screens conducted with recombinant cells. Secondary assays using primary cells to validate positive hits found via primary HTS should aid in the selection of meaningful leads with enhanced probability of identifying successful new compounds (Gebrin-Cezar, 2007).

10.4.2   ABSORPTION, DISTRIBUTION, METABOLISM, EXCRETION, AND TOXICOLOGY (ADMET) ASSESSMENTS

The introduction and use of primary cells for ADMET assays may make a valuable contribution to the level and quality of information obtained from the tests. Absorption, distribution, metabolism, and excretion (ADME) encompass the disposition of a pharmaceutical compound within an organism. These four criteria influence the levels and kinetics of drug exposure to tissues and hence influence the performance and pharmacological activity of a compound as a drug.

Preclinical efficacy and toxicity testing of newly identified chemical entities have the main goals of validating the mechanism of action and predicting possible adverse effects on humans. These studies are mainly conducted in animal models such as rabbits and rats (Piersma, 2006) but despite the relative ease of performing such tests, the species differences of humans and animals and the differences in dose sensitivity and pharmacokinetic processing of compounds reduce the predictive power of this approach. At present these models are only 50% efficient in predicting the toxicity of a pharmaceutical to human heart, liver, and development processes (Greaves, 2004).

In vitro cell-based models have been developed to aid in the evaluation of ADMET properties of compounds to explore the influences of species differences. CaCo-2 cells, for example, constitute an immortalized line of heterogeneous human epithelial colorectal adenocarcinoma cells widely used to predict the absorption rates of candidate drug compounds across the intestinal epithelial cell barrier. Drug absorption rates are determined 21 days after CaCo-2 cell seeding to allow for monolayer formation and cell differentiation.

HepG2 cells are mainly used for hepatotoxicity studies and MDCK cells as blood–brain barrier models. CHO or HEK293 cells transfected with hERG are used as assays for monitoring cardiotoxicities of compounds. All these immortalized cell lines differ very significantly from their *in vivo* counterparts in the ability to accurately assess efficacy and toxicity (Mayne, 2006). Despite the availability of these models and their extensive use, one in three drugs fails in phase I trials and 10% fail in phase II trials due to pharmacokinetics, with an additional 25% failing for safety reasons (Schafer, 2008). Thus, a clear and significant need exists for developing cell-based assays and *in vitro* models that are more predictive of human physiology and clinical response. From this perspective, the use of primary cells and differentiated ES cells (e.g., hepatocytes, cardiomiocyte-like cells) is currently seen as the most promising technology expected to expand greatly in the next few years (McNeish, 2004; Davila, 2004; Doetschman, 1985).

Human hepatocytes, for example, are considered invaluable for testing hepatotoxicity and also the activities of drug transporters and the metabolism of xenobiotics by CYP450 enzymes. At present, ready-to-use frozen homogeneous preparations of human hepatocytes are commercially available for these purposes. Since primary human liver cells rapidly lose their functional properties when cultured *in vitro*, their usefulness relies on repeated sourcing, which raises the clear limitation of sample consistency due to donor–donor variability (Rodriguez-Antona, 2002).

Several available hepatic cell lines such as HepG2 cells can be propagated easily in culture but the levels of CYP450-metabolizing enzymes and other important proteins are substantially different from those of native hepatocytes (Wilkening, 2003).

Another possible and innovative approach to overcome the problem is the use of human or mouse stem cells differentiated into hepatocyte-like cells (Hai, 2007; Soderdhal, 2007; Soto-Gutierrez, 2007; Hamazaki, 2001). These cells display appropriate morphologies and express some of the typical hepatocyte markers such as albumin, α-1 antitrypsin, cytokeratins 8 and 18, and typical hepatic transcription factors such as HNF-1 and FoxA2. Functional analysis of the cells demonstrated glycogen accumulation, inducible cytochrome P450 activity, and production of urea and albumin. These models still need further evaluation of critical functions such as metabolic competence, biotransformation capacity, and transportation of exogenous substances, but preliminary results suggest that stem cell-derived hepatocyte-like cells are promising candidates for better prediction of hepatotoxicity and drug metabolism.

While primary human hepatocytes are commercially available, this is not the case for primary cardiomyocytes, mainly due to the lack of donor material and the problems associated with cell isolation procedures. The QT interval is the portion of an electrocardiogram representing the time from the beginning of ventricular depolarization to the end of ventricular repolarization. Prolongation of the QT interval is associated with the rare but potentially life-threatening type of ventricular arrhythmia known as torsade de points (Morganroth, 1993; Cubeddu, 2003). Blocking the cardiac hERG channel, which may lead to the prolongation of the QT interval, is the most common liability of small molecules. Therefore the early assessment of cardiotoxicity for any new chemical entity is of crucial importance. HEK293 or CHO cells recombinantly expressing hERG can monitor this danger adequately but other genes are also involved in prolongation of the QT interval.

The immortalized myocyte HL-1 cell line from a mouse atrial lineage (Claycomb, 1998; White, 2004) is one of the few cell lines available that continuously divides, spontaneously contracts, and maintains a differentiated adult cardiac phenotype through an indefinite number of passages in culture. In other respects, however, these lines do not resemble all the native and physiological characteristic of cardiac cells.

The most common method for obtaining cardiomyocyte-like cells from ES cultures is the induction of cell differentiation through embryoid body formation (Kehat, 2001). The technique is well established and reliable, if laborious. Murine ES cells and P19 embryonic carcinoma cells (Anisimov, 2002) also provide useful models for studying cardiomyocyte development and differentiation, but cardiogenesis in mice and humans clearly exhibits substantial differences. One major advantage of cardiomyocytes derived from stem cells is that they can be maintained in culture for extended periods without losing their spontaneous contractile capacity, allowing the use of the same cell preparation for several different analyses. Progress has been made in deriving functional cardiomyocytes from human ES cells for *in vitro* applications for drug development (Goh, 2005), which should allow the development of reliable *in vitro* models for human cardiac safety tests.

10.4.3 GENE FUNCTION AND PATHWAY ANALYSIS

Another field of application for primary cells is their use in target identification and validation studies. To fully understand the importance of a gene as a potential target for a drug discovery campaign, it is crucial to use the appropriate cell type or, when available, a clinically relevant, well defined tissue sample. For reasons of convenience, most pathway analysis utilizes transformed cell lines.

The combination of primary cells and modern gene knockdown technologies such as RNA interference (RNAi) provides effective tools to rapidly characterize the functions of genes *in vitro* and *in vivo* (Van Es, 2005; Kourtidis, 2007; Kurreck, 2009). RNA interference is a conserved biological process evolved to specifically and efficiently silence genes. The transduction of RNAi into appropriate cellular models allows rapid measurement of the phenotype observed upon abrogation of target expression. As previously noted, this can allow confirmation of the role of the suspected target in the interaction with an identified substance. The power of this new technology is fully realized when the biological material used for silencing the activity of a desired gene is relevant; the closer the cellular model resembles the disease situation, the better the target profile will be (Colombo, 2008; Van Es, 2005). Intuitively, primary cells would represent the ideal material, but executing experiments with primary cells can be difficult since they are hard to maintain in culture and transfection efficiency is often very low. A great deal of effort has been made to develop more efficient methods for delivering RNAi into cells. Progress has been made (Hartmann, 2007; Stewart, 2003) but delivery remains one of the most challenging aspects of the use of this technology (Kurreck, 2009).

10.5 CHALLENGES TO USE OF PRIMARY CELLS

While use of primary cells in the drug discovery process has clear advantages, their use presents challenges that have slowed their broader implementation in assays of other types. Typically, special requirements of assays that cannot be adequately addressed by transformed cells have served as the driving force behind the use of primary cells in assays. Among other considerations, in most instances primary cells are more expensive to acquire and use than comparable cell lines.

10.5.1 Availability of Relevant Primary Cell Types

The first challenge to the use of primary cells is identifying a primary cell type that is more relevant to the disease or condition under study than existing cell lines. This requires a thorough understanding of the disease and the tissues involved. Not all diseases can be adequately modeled at the cellular level and for those that can be primary cells are not always available in sufficient quantities.

The availability of sufficient quantities of relevant primary cells is a major limitation, especially if a large screening campaign is planned. The limited availability of primary cells often restricts their use to secondary testing of compounds or the hit validation phase in which smaller sets of compounds require fewer cells for testing. While large quantities of primary blood cells are readily available, this can be an insurmountable obstacle for other types of cells. At the 2007 Society for Biomolecular Sciences (SBS) meeting, Steve Rees from GlaxoSmithKline reported a successful effort to carry out a primary screen using peripheral blood mononuclear cells (PBMCs) isolated from a very large number of donors. The challenge then becomes ensuring the batch-to-batch reproducibility of the preparations, which requires careful quality control and decreases the throughput relative to immortalized cell lines.

Other primary cells, like cardiomyocytes or chondrocytes, can rapidly undergo de-differentiation to fibroblast-like cells when plated on a two-dimensional (flat) surface. In this case, three-dimensional matrices can help maintain the original features. BioLevitator by Hamilton Robotics, 3D Insert™ by 3dbiotek, and 3D cell culture by Invitrogen are only some of the products available to facilitate the correct growth and morphology of particular primary cell types.

High-throughput automated imaging systems, also sometimes referred to as high content screening (HCS) systems, enable the study of specific cell types within a mixed population of cells (Giuliano, 1997). Some primary cell types are dependent on interactions with other cell types, making their cultivation in isolation impossible, at least at the present time. Even when primary cells are not dependent on interactions with other cell types, primary cell preparations are almost never 100% pure, which could potentially lead to problems in assay quality and reproducibility. Systems such as Opera™ by Perkin Elmer, INCell™ by GE Healthcare, and ImageXpress by Molecular Devices enable the selective analysis of single cell types among a mixed cell population.

10.5.2 SIGNAL STRENGTH AND RELIABILITY

The existing technologies for cell-based assay generation and signal detection were mainly developed for use with transformed or recombinant cell lines. Some of these technologies must be adapted or modified for use with primary cells, considering that the protein of interest is often expressed at a lower level compared to equivalent recombinant systems. As a result, the available technologies often require the use of a large numbers of primary cells and even then sensitivity can still be a problem. New technologies able to generate and/or detect signals from fewer cells or higher sensitivity assay formats will offer advantages and increase the use of primary cells. For example, primary cells with limited survival time in culture, such as freshly isolated adipocytes, require an assay format that can be executed within hours. Simply optimizing the assay using a fluorogenic substrate instead of a chromogenic one will speed the process as a result of a greater sensitivity in detecting a fluorescent signal.

10.5.3 CELL ISOLATION AND QUALITY CONTROL

The methods and technology used for quality control and the quantity of parameters checked can vary, depending on the cell type preparation and the final use of the cells. Cell viability, morphology, and proliferative capacity are common, easily tested parameters. Immunofluorescence studies with panels of antibodies raised against specific markers of the isolated cell type are commonly employed to provide proof of the identity and health of the cells. Additional tests, for example, specific enzymatic activity or expression levels of particular genes, can be also performed as routine quality control parameters when they are crucial to the subsequent use of the cells.

10.5.4 NEGATIVE CONTROLS FOR PRIMARY CELLS

A screening campaign should always include at least a secondary screen of hit compounds against a negative control. This presents a challenge for primary cells since they lack a ready negative control, that is, a cell identical in every way except for the expression of the receptor of interest. For cell lines, the parental (untransfected or mock-transfected) cell line can be used to ensure that the measured response relates to the interaction of screening substance and the target under investigation. Without this control, it is difficult to be certain that a hit identified in a screen using primary cells is due to the receptor under investigation or another receptor also expressed by the primary cells. The use of well characterized antagonists can allow the confirmation of interactions for targets for which they are available but even then modulators acting through other binding sites may be incorrectly characterized as false positives.

As a result, a cell line expressing the target of interest is often required to confirm the compound–target interaction via secondary screening. Since one of the advantages of using primary cells is the time saved due to the endogenous expression of the target—avoiding the time-consuming cloning of a cell line—the requirement of a cell line for secondary screening neutralizes the advantage. RNAi technology, however, can be employed to selectively silence the expression of one or more components of the receptor of interest to prove that an observed interaction is dependent on the expression of the receptor of interest (Kurreck, 2009).

10.5.5 DONOR VARIABILITY

A further obstacle to more widespread use of primary cells is donor-to-donor variability, particularly for large compound collections. Differences can arise from variability in the isolation and preparation of the cells or from intrinsic differences among individual donors (e.g., age, gender, genetic background, and disease state). Standardization and automation of as many passages as

TABLE 10.3
Primary Cell Providers

Institution	Website
Lonza Group Ltd.	www.lonza.com
Innoprot	www.innoprot.com
Zen-Bio, Inc.	www.zen-bio.com
ScienCell Research Laboratories	www.sciencellonline.com
3H Biomedical AB	www.3hbiomedical.com
Millipore Corp.	www.millipore.com
Invitrogen Corp.	www.invitrogen.com
American Type Culture Collection	www.atcc.org
AllCells LLC	www.allcells.com
StemCell Technologies	www.stemcell.com
PromoCell GmbH	www.promocell.com
Sem Cells Sciences	www.stemcellsciences.com
Epithelix Sàrl	www.epithelix.com
European Collection of Cell Cultures (ECACC)	www.hpacultures.org.uk/collections/ecacc.jsp

possible during cell batch preparation is a crucial step for obtaining a homogeneous cell population of sufficient size. Rigorous standard operating procedures must be established before the start of any isolation; every possible operator-induced variability must be minimized. At the end of cell preparation, quality must be assessed to compare lot-to-lot variability.

The batch-to-batch variability introduced by biological and/or genetic differences among donors is difficult to assess and control. Whenever possible, preparations from single donors are used in parallel to assess whether possible differences in tested compound activity can be ascribed to a difference in the primary biological samples used for testing. While this clearly limits the number of compounds that can be investigated and limits throughput, this procedure is highly informative (Modriansky, 2000) and worth employing when subtle differences in primary cell performance can affect the results of the compound tested (Olsavsky, 2007).

Another possibility to minimize differences among donors is to pool cells from different donors to obtain a mixed but homogeneous population of cells. This approach provides more material to work with but must be carefully evaluated before use. Certain types of immune cells from different donors, for example, react with each other and cannot be pooled.

Many commercial suppliers (Table 10.3) have established protocols and procedures that address many of these issues. Primary human and animal cells of a variety of types are often available, sometimes as ready-to-use frozen stock aliquots or already added to 96-or 384-well plates. Specific culturing medium and detailed protocols for the correct maintenance of the cell line in culture are usually provided as well. This should facilitate the use of primary cells for primary HTS and compound validation.

10.6 FUTURE TRENDS

At present, the use of primary cells in the pharmaceutical industry is not as common as the use of recombinant cell-based assays. A rough estimate shows about 20% of all initial screening is carried out using primary cells and the rate is about the same for secondary screening. The use of primary cells is expected to increase in the next few years as emerging technologies start appearing on the market.

Improvements in the availability of growth factors and special media, isolation and culture methods, and signal generation and measurement expand the range of primary cells available for use in assays. Much work remains to be done in terms of new materials for supporting three-dimensional cultures and application protocols for a wider range of primary cells. High-throughput automated imaging systems will continue to improve to make the cultivation and analysis of mixed cultures simpler but more work on application protocols for all the available primary cell types is still needed.

As already described, substantial improvements have been made in transfection methods for primary cells, but in some cases it remains difficult to transfect them with reporter genes or RNAi without substantially altering their characteristics in undesirable ways. Label-free technologies preclude the need for cellular labeling or overexpression of reporter proteins, utilizing the inherent morphological and adhesive characteristics of cells as physiologically relevant and quantitative readouts for various assays (Xi, 2008; Atienza, 2006). These technologies utilize non-invasive measurements, allowing for time resolution and kinetics in the assay. The technology is very new and still needs to be validated on more cellular systems. In addition, the high costs of the instruments and consumables have slowed adoption. Those costs will presumably decline as the technology becomes more widespread.

The immense research effort focused on more functional culture and differentiation media has greatly facilitated the use of primary cells. Most of this research is now devoted to the formulation and validation of new media and protocols for the differentiation of stem cells into specific cell lineages. This relatively new field still needs support in terms of strongly validated and reproducible protocols for differentiation and related media formulation that will promote differentiation.

10.7 CONCLUSIONS

Assay design is inevitably a matter of finding the right balance of various, often interrelated parameters. Frequently an improvement of one parameter negatively impacts others. Until fairly recently, the necessary technology and techniques were not available to allow the use of primary or primary-like cells beyond small scale assays. New tools allow the use of primary cells in screening campaigns, secondary screening, ADMET assays, and pathway analysis.

Ideally any assay should model the biological system it represents as closely as possible and in this respect primary cells present a clear advantage. Transformed cell lines have, however, remained the work horses for most cell-based assay systems due to a number of limitations of primary cells. The first relates to the availability of suitable primary cells for the condition or disease state under study. Some primary cells, such as blood cells, are readily available in large numbers and made initial screening campaigns feasible, although slower and at higher cost than cell lines. In other cases, recent progress in the development of growth factors and media has allowed the cultivation of more demanding primary cells. Nevertheless a great deal of work must be done to expand the availability and improve the convenience of primary cells before they are likely to begin displacing recombinant cell lines from their places in screening laboratories.

Techniques to overcome the lack of a negative (parental) control cell have become available through the use of RNAi, at least in cases where RNAi enters or can be expressed in the primary cells. High-throughput automated imaging systems have helped lessen the need to fully purify cells by allowing analysis at the single cell level.

ADMET applications place particular demands on *in vitro* assays in that the objective is usually to detect biological processes that can influence the bioavailability or safety of a compound. These processes are often complex and involve more than a single interaction between a known target and a compound. Instead, the idea is to simulate a complex system *in vitro* to the degree possible. The lower throughput that is often a consequence of using primary cells is not a major disadvantage at this stage due to the relatively small number of compounds typically under investigation this late in the drug discovery process. As a result, the early use of primary cells for ADMET purposes was with primary hepatocytes and continues to drive developments in the field.

Gene and pathway analysis is another application of primary cells in which fidelity of the test system is crucial for reliable results. The use of RNAi technology in combination with primary cells has already simplified the process of *in vitro* analysis of loss of function phenotypes, although further progress is required to expand the range of primary cells that may be successfully treated with RNAi.

Improvements in assay technology will also facilitate the use of primary cells in assays. The limited availability of primary cells and the low signal strengths observed with normal expression (not overexpression) present significant challenges that may be overcome with more sensitive detection technologies. Developments in tissue culture have made possible the culturing of some primary cells dependent on three-dimensional matrixes, with more to come as the technology is refined and matures. Finally continuing work with growth factors and media will expand the range of cell types subject to culture *in vitro*, allowing primary cells for an ever-broader range of diseases and conditions to be used for assay development.

REFERENCES

Anisimov SV. 2002. SAGE identification of differentiation responsive genes in P19 embryonic cells induced to form cardiomyocytes *in vitro. Mech. Dev.* 117, 27–74.

Atienza JM. 2006. Dynamic and label-free cell-based assays using the real time cell electronic sensing system. *Assay Drug Dev. Technol.* 5, 597–607.

Benjamin ER. 2006. Pharmacological characterization of recombinant N-type calcium channel (Cav2.2) mediated calcium mobilization using FLIPR. *Biochem. Pharmacol.* 72, 770–782.

Carbone E. 1990. Ca currents in human neuroblastoma IMR32 cells: kinetics, permeability and pharmacology. *Pflugers Arch.* 416, 170–179.

Claycomb WC. 1998. HL-1 cells: a cardiac muscle cell line that contracts and retains phenotypic characteristics of the adult cardiomyocyte. *Proc. Natl. Acad. Sci. USA* 95. 2979–2984.

Colombo R. 2008. Target validation to biomarker development: focus on RNA interference. *Mol. Diagn. Ther.* 12, 63–70.

Cubeddu LX. 2003. QT prolongation and fatal arrhythmias: a review of clinical applications and effects of drugs. *Am. J. Ther.* 10. 452–457.

Davila JC. 2004. Use and application of stem cells in toxicology. *Toxicol. Sci.* 79, 214–223.

Doetschman TC. 1985. *In vitro* development of blastocyst-derived embryonic stem cell lines: formation of visceral yolk, blood islands, and myocardium. *J. Embryol. Exp. Morphol.* 87, 27–45.

Fico A et al. 2008. High-throughput screening compatible single-step protocol to differentiate embryonic stem cells in neurons. *Stem Cells Dev.* 17, 573–584.

Fox S. 2004. High-throughput screening: searching for higher productivity. *J. Biomol. Screen.* 9, 354–358.

Fox S. 2006. High-throughput screening: update on practices and success. *J. Biomol. Screen.* 11, 864–869.

Gebrin-Cezar G. 2007. Can human embryonic stem cells contribute to the discovery of safer and more effective drugs? *Curr. Opin. Chem. Biol.* 11, 405–409.

Giuliano KA. 1997. High-content screening: a new approach to easing key bottlenecks in the drug discovery process. *J. Biomol. Screen.* 9, 273–285.

Goh G. 2005. Molecular and phenotypic analysis of human embryonic stem cell-derived cardiomyocytes: opportunities and challenges for clinical translation. *Thromb. Haemost.* 94, 728–737.

Gonzalez JE. 1999. Cell-based assays and instrumentation for screening ion channel targets. *Drug Disc. Today* 4, 431–439.

Greaves P. 2004. First dose of potential new medicines to humans: how animals help. *Nat. Rev. Drug Discov.* 3, 226–236.

Greger IH, Ziff EB, and Penn AC. 2007. Molecular determinants of AMPA receptor subunit assembly. *Trends Neurosci.* 30. 407–416.

Hai DC. 2007. Direct differentiation of human embryonic stem cells to hepatocyte-like cells exhibiting functional activities. *Cloning Stem Cells* 9, 51–62.

Hamazaki T. 2001. Hepatic maturation in differentiating embryonic stem cells *in vitro. FEBS Lett.* 497, 15–19.

Herzenberg LA. 2002. The history and future of the fluorescence activated cell sorter and flow cytometry: a view from Stanford. *Clin. Chem.* 48, 1819–1827.

Kamihira M. 2007. Development of separation technique for stem cells. *Adv. Biochem. Eng. Biotechnol.* 106, 173–193.

Kehat I. 2001. Human embryonic stem cells can differentiate into myocytes with structural and functional properties of cardiomyocytes. *J. Clin. Invest.* 108, 407–414.

Kola I and Landis J. 2004. Can the pharmaceutical industry reduce attrition rates? *Nat. Rev. Drug Discov.* 3, 711–715.

Kourtidis A. 2007. RNAi applications in target validation. *Ernst Schering Res. Fdn. Workshop* 61, 1–21.

Kurreck J. 2009. RNA interference: from basic research to therapeutic applications. *Angew. Int. Ed.* Engl 48, 1378–1398.

Macarron R. 2006. Critical reviews of the role of HTS in drug discovery. *Drug Discov. Today.* 11, 277–279.

Mayer ML. 2005. Glutamate receptor ion channels. *Curr. Opin. Neurobiol.* 15, 282–288.

Mayne JT. 2006. Informed toxicity assessment in drug discovery: systems-based toxicology. *Curr. Opin. Drug Discov. Dev.* 9, 75–83.

McNeish J. 2004. Embryonic stem cells in drug discovery. *Nat. Rev. Drug Discov.* 3, 70–80.

McNeish J. 2007. Stem cells as screening tool in drug discovery. *Curr. Opin. Pharmacol.* 7, 515–520.

Modriansky M. 2000. Human hepatocyte model for toxicological studies: functional and biochemical characterization. *Gen. Physiol. Biophys.* 19, 223–235.

Morganroth J. 1993. Relations of QTc prolongation on the electrocardiogram to torsade de pointes: definitions and mechanisms. *Am. J. Cardiol.* 72, 10B–13B.

Müller-Hartmann H et al. 2007. High-throughput transfection and engineering of primary cells and cultured cell lines—an invaluable tool for research as well as drug development. *Expert Opin. Drug Discov.* 2, 1453–1465.

Naot D. 2005. Lactoferrin: A novel bone growth factor. *Clin. Med. Res.* 3, 93–101.

Olsavsky KM. 2007. Gene expression profiling and differentiation assessment in primary human hepatocyte cultures, established hepatoma cell lines, and human liver tissues. *Toxicol. Appl. Pharmacol.* 222, 42–56; Epub Apr. 21 2007.

Parameswaran N. 2006. RAMPs: the past, present and future. *Trends Biochem. Sci.* 31, 631–638.

Perriere N et al. 2007. A functional *in vitro* model of rat blood-brain barrier for molecular analysis of efflux transporters. *Brain Res.* 1150, 1–13.

Piersma AH. 2006. Alternative methods for developmental toxicity testing. *Basic Clin. Pharmacol. Toxicol.* 98, 427–431.

Rodriguez-Antona, C. 2002. Cytocrome P450 expression in human hepatocytes and hepatoma cell lines: molecular mechanisms that determine lower expression in cultured cells. *Xenobiotica* 32, 505–520.

Schafer S. 2008. Failure is an option: learning from unsuccessful proof-of-concept trials. *Drug Disc. Today* 13, 913–916.

Schuster D. 2006.Why drugs fail: a study on side effects in new chemical entities. *Curr. Pharm. Des.* 11. 3545–3559.

Soderdhal T. 2007. Glutathione transferases in hepatocyte-like cells derived from human embryonic stem cells. *Toxicol. in Vitro* 21, 929–937.

Soto-Gutierrez A. 2007. Differentiation of mouse embryonic stem cells to hepatocyte-like cells by co-culture with human liver nonparenchymal cell lines. *Nat. Protoc.* 2, 347–356.

Stewart SA. 2003. Lentivirus-delivered stable gene silencing by RNAi in primary cells. *RNA* 9, 493–501.

Van Es HH. 2005. Biology calls the targets: combining RNAi and disease biology *Drug Discov. Today* 10, 1385–1391.

White SM. 2004. Cardiac physiology at the cellular level: use of cultured HL-1 cardiomyocytes for studies of cardiac muscle cell structure and function. *Am. J. Physiol. Heart Circ. Physiol.* 286, H823–H829.

Wilkening S. 2003. Comparison of primary human hepatocytes and hepatoma cell lines HepG2 with regard to their biotransfomation properties. *Drug Metabol. Dispos.* 31, 1035–1042.

Xi B. 2008. The application of cell-based label free technology in drug discovery. *Biotechnol. J.* 3, 484–495.

Zhang JH, Chung TDY, and Oldenburg KR. 1999. A simple statistical parameter for use in evaluation and validation of high-throughput screening assays. *J. Biomol. Screen.* 4, 67–73.

Zhu ZR. 2007. Use of cryopreserved transiently transfected cells in high-throughput pregnane X receptor transactivation assay. *J. Biomol. Screen.* 12, 248–254.

11 Screening Automation

Wei Zheng and Catherine Z. Chen

CONTENTS

11.1 INTRODUCTION

Since its inception in the 1980s, high-throughput screening (HTS) has become a standard aspect of the drug discovery process in pharmaceutical and biotechnology companies and has also recently been explored in academia (Tolliday et al., 2006). Current drug discovery relies on screening of chemical libraries against various molecular targets to find lead compounds that serve as starting points for the drug development process. Technological development in HTS has been closely linked with advances in screen automation in the past two decades to increase throughput and reduce screen costs.

Recent advances in combinatorial chemical synthesis, improved isolation of natural compounds, and increased commercial availability of compound collections have expanded most compound libraries from hundreds of thousands to over one million. Concurrently, complete genome sequencing has provided novel targets for potential molecular intervention. The expansion of compound libraries and molecular target space brings a need to minimize cost and time spent per screen and maximize efficiency and reproducibility (Landro et al., 2000; Houston et al., 2008). Traditional test tube-based research strategies would simply be too costly, take too long, and introduce too many human errors to effectively screen current molecular libraries against the vast numbers of novel targets. These problems can be solved by assay miniaturization to conserve reagents and automation to increase speed and reproducibility (Rutherford and Stinger, 2001). The use of microplates allows multiple reagent additions and incubations to be carried out in parallel. Traditional biochemical and pharmacological drug discovery methods required 1-mL reaction volumes in individual test tubes—an approach that limited assay capacity to 20 to 50 compounds per week for a typical laboratory. The 96-well plate format was implemented in the mid-1980s at Pfizer for natural product screening (Pereira and Williams, 2008).

The use of microplates provided several immediate advantages including (1) simultaneous access to 96 samples, (2) compatibility with multichannel pipettes, and (3) reduced incubation space. Thus, screening capacity increased to 10,000 compounds per week by the early 1990s. The capacities of HTS have evolved from 96-well plates in the mid-1980s to 384- and 1536-well plates that are widely used today; ultra-HTS (uHTS) 3456-well plate formats are now available (Gwynne and Heebner, 2003, Mayr and Fuerst, 2008). Miniaturization requires increasing reliance on instrumentation. Major advances such as liquid handlers, signal detectors, and integrated robotic platforms have all been enabling factors in the miniaturization trend.

11.2 COMPONENTS AND INSTRUMENTS FOR SCREEN AUTOMATION

Several vendors manufacture liquid dispensers, plate storage and incubator devices, and plate detection instruments that are the three major components used in compound screens. Selection of equipment is usually based on several factors including need, previous experience, recommendations, cost, available funding, and quality of post-sale service provided by the instrument vendor. Many instruments from different manufacturers such as the photomultiplier tube (PMT)-based plate readers exhibit similar function and performance. Therefore, reliability, cost, and service quality become the dominating factors for purchase decisions.

11.2.1 LIQUID DISPENSERS

11.2.1.1 Types

Two types of liquid dispensers classified by their dispensing mechanisms are commonly used for compound screens. One is the tip-based type based on positive displacement. The other is nontouch-based and utilizes other dispensing mechanisms as shown in Table 11.1 and Figure 11.1.

Two groups of tip-based dispensers are available. One has a fixed tip usually made from metal materials coated with nonstick chemicals. The main advantages are (1) low operating cost since the

TABLE 11.1
Mechanisms and Classifications of Liquid Dispensers

Type of Dispensing	Dispensing Mechanism	Dispensing Range	Examples	For Details, See Websites:
Tip based:				
Fixed tip	Positive displacement	0.5 μL and up	Janus and Evo	www.perkinelmer.com; www.tecan.com
Changeable tip	Positive displacement	0.5 μL and up	Cybi-well	www.cybio-ag.com
Pin tool	Direct transfer	10 nL and up	Pin tool station	www.vp-scientific.com
Nontouch:				
Peristaltic	Mechanical force	2 μL and up	Multidrop	www.thermo.com
Solenoid valve	Air and valve	0.2 μL and up	FRD BioRAPTR	www.beckman.com
Acoustic	Acoustic beam	2.5 nL and up	Echo and ATS-100	www.labcyte.com; www.edcbiosystems.com
Piezoelectric	Piezoelectric sensor	1 nL and up	PicoRAPTR	www.beckman.com

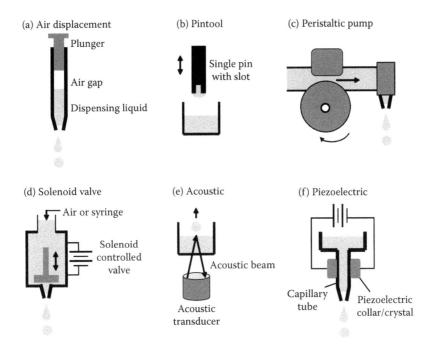

FIGURE 11.1 Liquid dispenser technologies. (a) Positive displacement devices utilize pistons to displace air in a column, forcing liquid into or from tip. (b) In pin tool dispensing, liquid is transferred via surface tension. Factors affecting transfer volume include pin diameter, pin shape, depth to which pin is submerged into liquid, surface tension, and speed of pin movement. (c) Peristaltic pumps use mechanical forces to eject liquid from a flexible tube. (d) In solenoid valve systems, positive pressure is placed on dispensing liquid via air or syringe. A valve moved by a solenoid controls the amount and timing of dispensing. (e) In acoustic dispensing, focused acoustic beams travel from transducer to sample. The energy of the beam is transferred to a liquid droplet that is ejected. (f) In piezoelectric systems, expansion of the piezoelectric collar induced by transient voltage creates a compression wave that propagates longitudinally in a capillary tube to force liquid from the tip.

user has no need to buy disposable tips and (2) flexibility to reach any well position in a plate. The disadvantages include low throughput due to the time required for tip wash, potential carryover of compounds, and lack of compatibility with 1536-well plate formats.

Another group of tip-based dispensers uses disposable tips made usually of plastic. This type operates faster (e.g., simultaneously transferring in 96-, 384- or 1536-well format), eliminates compound carryover, and is compatible with 1536-well plates. A disadvantage is the high consumable cost because the 384- and 1536-well tips are usually expensive. The authors have had good experience with the Cybi-well 98 and 384/1536-well dispenser (Cybio, Inc.) for 8 years and found that it is a very reliable instrument.

Four groups of non-touch liquid dispensers are available for compound screens (Taylor et al., 2002). The Multidrop is a simple instrument with relatively low cost. It is mainly used for dispensing liquid reagents and cell suspensions. The dispensing cassette is resistant to many solvents and usually lasts for several hundreds of plates before requiring replacement. The speed of dispensing is the fastest among all dispensers listed and the device requires less than 30 s per plate for dispensing in any plate format. However, only column or whole plate dispensing is possible because the Multidrop lacks the flexibility to allow users to define specific rows for dispensing. Another downside is that the machine reveals significant variability when dispensing less than 2 μL volume.

The solenoid valve-based dispenser is flexible for any well position in a plate and dispensing usually takes less than 1 min for a full plate with small dispensing volumes (Niles and Coassin, 2005). It is typically used for 1536- or 3456-well plates and for lower volume dispensing between 0.2 and 0.5 µL per well. The instrument costs more than the Multidrop and frequent clogging of dispensing valves may become an operational issue. Thus, the solenoid valve type is a good instrument for assay development and assay optimization, when the greatest flexibility is desired, but it is not good for high-throughput situations.

Acoustic-based dispensing has the advantage of accurately transferring small amounts of liquid (nanoliter volumes) from a compound source plate to an assay plate (Wong and Diamond, 2009). It is mainly used for compound dispensing between 2.5 and 50 nL per well. The disadvantages include the relatively slow dispensing speed, special requirements for source plate materials, and high instrument cost.

The piezoelectric-based dispensers such as the PicoRAPTR achieve the smallest dispensing volumes among all the liquid types and are capable of dispensing 1 nL and more. Their disadvantages include low throughput due to extensive tip wash requirements between dispenses and high instrument cost.

11.2.1.2 Applications in Screen Assays

Liquid dispensers are used in compound screens for dispensing reagents including cell suspensions and other reagent solutions. Due to the different needs of compound handling and cell and reagent dispensing, different instruments are recommended for each step.

11.2.1.2.1 *Compound Handling*

For preparing compound libraries, instruments with changeable tips are commonly used for dissolving powder samples to avoid compound carryover contamination. Instruments with fixed tips and those with changeable plastic tips can be used for compound dilution in combination with tip wash steps. Since dimethylsulfoxide (DMSO) is a standard solvent for dissolving compounds during compound library preparation, the tips are usually washed in DMSO solution and DMSO in combinations with dH$_2$O, ethanol, and methanol to eliminate residual compounds and dry the tips. Performing tip washes in a sonicating bath in combination with blotting on filter paper can further improve the effectiveness of tip refreshments. Compound libraries are usually prepared and stored in DMSO solution in millimolar concentrations (2 to 10 mM). Dispensers with changeable tips are usually used to transfer compounds from mother/master plates to daughter/source plates to avoid potential crossover contamination.

For transferring compound from a source plate to an assay plate that is used repeatedly, dispensers with fixed and changeable tips, pin tools, and acoustic dispensing can be chosen based on the screen throughput needs and availability of dispensers (Cleveland and Koutz, 2005). The pin tool is highly recommended due to its relative high-throughput (usually 1 min per plate at any plate density including tip wash time) and low consumable cost as long as compound carryover can be effectively managed by adequate washes between dispensings. The wet dispensing method should be applied with a pin tool or tip-based dispenser that transfers DMSO solution from a source plate to an assay plate whose bottom is covered with assay solutions. Only acoustic or piezoelectric dispensing can be used for dry dispensing in which the compound solutions are dropped directly onto an empty assay plate.

11.2.1.2.2 *Reagent and Cell Dispensing*

Nontouch dispensers are the common choices for dispensing cell suspension and reagent solutions into plates. Cells can be dispensed quickly by various instruments. We highly recommend the Multidrop (96- and 384-well) and Multidrop-Combi (1536-well) because of their high dispensing speeds and low costs. Issues such as tip clogging and dispensing cassette failure should be carefully monitored when the Multidrop instruments are used. Other reagent solutions can be dispensed by nontouch dispensers including solenoid valve-based instruments and Multidrops. For cleaning and sterilizing of dispensers before and after experiments, we recommend the use of 70% ethanol or 10% bleach solution followed by rinses in dH$_2$O.

11.2.2 Compound Plate Storage and Assay Plate Incubators

Compounds for primary screens and confirmation tests are usually dissolved in DMSO and stored in 96-, 384-, and 1536-well plates. The plates should be kept in –70°C freezers for long term storage and can last up to 10 years. The compound plates, covered with air-sealing lids (Kalypsys, Inc.) or heat-sealer (Velocity-11), are kept in a plate storage unit at room temperature in a robotic screen system ready for use. The compound plate storage units are usually made by robotic integration companies or can be converted from plate incubators. Usually, the compounds in DMSO solution can be stored at room temperature for 3 to 6 months without significantly affecting the quality of the compounds. Frequent freeze–thaw cycles and storage at 4°C to –40°C should be avoided for compounds in DMSO solution as these conditions accelerate compound decomposition. Compounds in aqueous solutions should be used when freshly diluted and should not be stored under any conditions because many compounds will decompose.

A manual or robotic assay plate incubator is an important part of a compound screen system. The incubators in robotic systems usually consist of multiple shelves that can be rotated to accommodate frequent access by the robotic arm. For cell-based assays, a CO$_2$-injected incubator should control temperature between 25°C and 50°C. In addition, since many cell-based assays require long incubations (1 day or more), the incubator should have a humidity control that can achieve 98% humidity at a given temperature. The humidified environment is needed to prevent liquid evaporation from assay plates during prolonged incubation, especially in a miniaturized format (1536-well plate). The edge effect—uneven evaporation of liquid with more evaporation occurring at the edges of a plate—is usually observed when plates are kept in under-humidified incubators or incubated for more than 3 days. We found that the incubation time of cell plates in a well humidified incubator can be extended up to 3 days without significant edge effects.

11.2.3 Plate Readers

11.2.3.1 Typical Plate Readers

After assay plates are incubated with compounds and detection reagents, they are transferred to a plate reader for the measurement of results. The selection of detection mode in a plate reader is determined by the nature of the screen assay. Currently, over 95% of screening assays use the macro well-based detection systems that measure the signal from the entire well in an assay plate. Fluorescence intensity, fluorescence polarization, time-resolved fluorescence, absorbance, and luminance are common screen assay detection methods. The multimodality plate reader is a common choice for a screening laboratory that requires frequent changes of detection modes. Based on the signal detection device, plate readers can also be grouped into two types: PMT-based and charge-coupled device (CCD)-based (Figure 11.2). The PMT-based plate readers have relatively low cost and provide higher detection sensitivity. The CCD-based devices are considered "high end" instruments that allow fast detection speed and lower well-to-well variability because they record an image for an entire plate. This instrument is a good choice for miniaturized (1536-well) assays.

Based on optical devices used for selecting the spectra of exciting and emitting lights, the PMT-based plate readers fall into two categories; filter-based and monochromator-based (Henshall and Held, 2008). They differ mainly in their methods of specifying excitation and mission wavelengths. Filter-based detectors utilize filters to select light of a specific wavelength band. Monochromators use diffraction gratings that reflect white light. Filter-based detection is more common than monochromator-based detection because of relatively lower instrument cost, better light transmission efficiency, higher sensitivity, and faster ability to switch between two detection wavelengths.

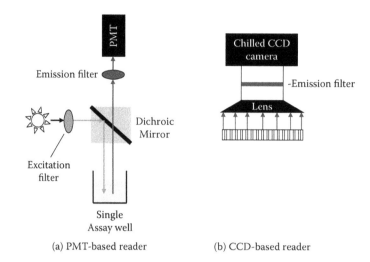

(a) PMT-based reader   (b) CCD-based reader

FIGURE 11.2 Photomultiplier tube (PMT) and charge-coupled device (CCD) imaging systems. (a) In PMT-based readers, the excitation beam travels through an excitation filter and is reflected to the sample well via a dichroic mirror. The emission beam travels through the dichroic mirror and emission filter, and is then read by the detector. This sequence repeats for each sample well. (b) For CCD-based readers, a camera records an image of an entire plate.

Monochromator-based detection, however, can perform spectral scanning that is useful for the assay development and for spectral determinations of fluorophores and compounds. Monochromator types eliminate the need to buy different filter sets for each wavelength. Our laboratory uses several filter-based plate readers and one monochromatic type to accommodate the needs of assay development and compound screening (Table 11.2).

11.2.3.2 Other Plate Readers

In addition to the above plate readers, several other types are used to perform specific tasks in screening laboratories. Fluorescence kinetic plate readers can measure an assay plate quickly and

TABLE 11.2
Common Plate Detectors

Type of Reader	Detector	FI	FP	TRF	Lum	Abs	Alpha Screen	For Details, See Websites:
Monochromator-based:								
Safire	PMT	+	+	+	+	+		www.tecan.com
Synergy	PMT	+	+	+	+	+		www.biotek.com
SpectraMax M5	PMT	+	+	+	+	+		www.moleculardevices
Filter-based:								
Envision	PMT	+	+	+	+	+	+	www.perkinelmer.com
POLARstar	PMT	+	+	+	+	+	+	www.bmglabtech.com
ViewLux	CCD	+	+	+	+	+		www.perkinelmer.com

FI = fluorescence intensity; FP = fluorescent polarization; TRF = time-resolved fluorescence; Lum = luminance; Abs = absorbance.

kinetically (usually at 1 Hz). They are used to measure intracellular calcium release in G protein-coupled receptors (GPCRs) and ion fluxing for ion channels. Both the FDSS-7000 (Hamamatsu) and FLIPR Tetra (Molecular Devices/MDL) can be used for kinetic measurements in 96-, 384-, and 1536-well plates.

Imaging plate readers are used for high content screening (HCS) (Carpenter, 2007). These instruments are usually equipped with fluorescence microscopes with automated plate-moving stages. The details of a field or several fields in a well of an assay plate can be imaged and recorded with high resolution for analysis of subcellular structures. However, the quantitative data analysis of the cell phenotypes and the storage and retrieval of results are still huge challenges for HCS because one compound library screen can easily produce one to several terabytes of data. Current HCS instruments include the Opera (EvoTech/Perkin Elmer), ImageXpress Micro/Ultra (Molecular Devices/MDL), Pathway Bioimager (BD Biosciences), and ArrayScan (Cellomics/ThermoScientific).

11.3 COMPOUND SCREEN PLATFORMS: WORKSTATIONS AND FULLY AUTOMATED ROBOTIC SYSTEMS

Based on the numbers of daily compound screens, a screening laboratory may choose a workstation or fully automated screen system (Figure 11.3) (Hamilton, 2002). A workstation-based screening platform can be set up quickly and requires less capital funding. The screening throughput with a workstation platform is relatively small—usually 20 to 100 plates per day in 8 hr. A fully automated robotic screen system requires 6 to 12 months for implementation and a minimum investment of

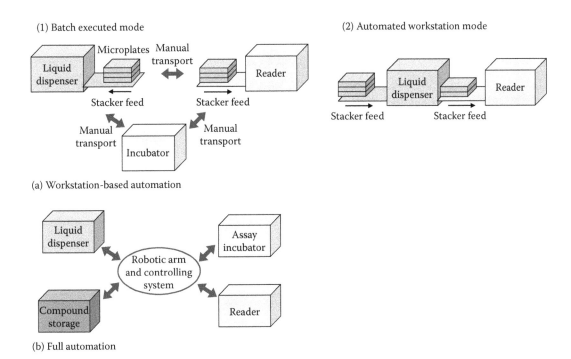

FIGURE 11.3 Levels of screen automation. (a) Workstation-based automation can be categorized as (1) batch automation in which plate stackers feed assay plates into each device and plate stacks are transported between devices manually and (2) automated workstations that include liquid dispensers and readers and can transfer plates between the two devices automatically. (b) In a fully automated system, transport of plates between devices is carried out by robotic arms and scheduling software.

millions of dollars. Fully automated systems operate continuously (24 hr per day, 7 days per week) and achieve throughputs of 100 to 600 plates in 24 hr.

11.3.1 WORKSTATION-BASED SCREEN PLATFORMS

Workstation-based screening is easy to set up and flexible in terms of operation (Figure 11.3A) (Menke, 2002). Plate stackers are commonly used in this type of operation and can typically handle 20 to 50 plates per batch semi-automatically. Frequent manual operation is usually needed.

11.3.1.1 Batch Executed Semi-Automated Screen Format

Liquid dispensers and plate readers equipped with plate stackers are needed for this type of screening operation. Usually, 20 to 50 plates are executed sequentially as a batch for a screening assay. For example, in a fluorogenic enzyme assay, the Multidrop-Combi takes empty assay plates from a plate stacker, dispenses 2 μL enzyme solution to each of 1536 wells and places the plates in another plate stacker. The operator manually transfers assay plates filled with 2 μL per well of enzyme solution from the stacker in the Multidrop-Combi and compound plates to a pin tool station with stackers (Kalypsys, Inc.) that then adds 20 nL per well of compound solution to the assay plates. The operator then generates a tracer file in Excel format using a barcode reader that pairs a compound plate with an assay plate.

Subsequently, a 1 μL per well substrate solution is dispensed to the assay plates in another Multidrop-Combi equipped with plate stackers. After 30-min incubation at room temperature, the operator manually moves the assay plates to a plate detector with stackers and executes the measurement of assay plates in a specific detection mode (fluorescence intensity in this case). The data file from the plate reader and the tracer file will be copied to a computer for data analysis after the experiment. This screening platform is very useful for laboratories that screen small compound collections, assay validation with the LOPAC collection, and follow-up screens for hit confirmation and lead optimization.

11.3.1.2 Automated Workstation-Based Screen Format

Several automated work stations including the Biomak FX (Beckman), Janus (Perkin Elmer), BioCel (Velocity-11), and Freedom EVO (Tecan) are available for this type of screening operation. The automated workstation consists of one or two liquid dispensers, plate layout stations, plate stackers, and plate readers. The process can be automated using the software at the workstation. It can execute small numbers of compound screens with 20 to 100 plates in each batch. It can also be used for the determination of EC_{50} and IC_{50} measurements during hit follow-up stage and lead optimization. This platform is also commonly used for compound plate preparation and other analytical screening experiments.

11.3.2 FULLY AUTOMATED SCREENS IN A ROBOTIC SYSTEM

A fully automated robotic system is composed of liquid dispensers, plate storage, incubators, plate detection readers, and other accessories including de-lidders, plate centrifuges, and plate sealing and piercing devices integrated in a platform with one or more robotic arms and controlled by customized computer software (Figure 11.3b) (Michael et al., 2008). All robotic systems are customized to fit the special needs of a laboratory and are usually capable of continuous operation with high-throughput.

Special equipment can also be integrated to meet the new challenges of screening operations. Several systems are available for consideration including those of Kalypsys (www.kalypsys.com), RTS (www.rts-group.com), HRE (www.highresbio.com), and CRS/Thermo (www.thermo.com). The advantages of an automated system include high screening throughput, even time scheduling between plates, elimination of most human errors, and continuous (24/7) screening. These systems usually require special operators who have the engineering capability to operate and maintain them.

The operator's attention is still necessary during continuous screening modes, and he or she can be alerted via email or telephone of problems arising during off hours. Cooperation between assay development scientists and robotic engineers is required for reagent preparation and trouble shooting. Some of the operational issues of fully automated robotic systems include:

1. Minimize dead volume. Approximately 20% to 30% more reagent is needed because of larger dead volumes in automated reagent dispensing instruments. This should be considered before implementation of HTS. To reduce the reagent dead volume, the reagent needed for overnight runs should be carefully calculated and freshly prepared reagents should be added to the small amount remaining the following morning to "top off" reagent levels if possible. One common mistake is underestimating the volumes of reagents needed for overnight robotic runs. Insufficient reagents waste assay plates and robotic time.

2. Determine reagent stability before robotic screening. All reagents should be stable for at least 8 to 12 hr. Certain reagents must be kept at 4°C or shielded from light during screening.

3. High quality assays are keys to successful lead identification. The assays should yield greater than threefold signal-to-basal ratio (above twofold may be acceptable for ratiometric assays such as fluorescence polarization, TR-FRET, and kinetic assays) and Z' factors greater than 0.5.

4. Simplifying assay steps. A screening assay should be simplified; ideally it should require only three to five reagent and/or compound addition steps. Cell wash and medium aspiration steps should be avoided, especially in the 1536-well format because they increase well-to-well variability. The plate incubation time is usually 10 min to 3 hr and can lengthen to 1 to 3 days for cell-based assays. Higher humidity (98% or more) should be maintained in plate incubators to avoid uneven evaporation of liquid in the assay plates for prolonged incubation (up to 3 days).

5. Cell dispensing. Online dispensing of cells to assay plates represents a challenge because most cell types survive only 2 to 4 hr in suspension. Continual cell suspension replacement at 2- to 4-hour intervals becomes a tremendous burden for a 24/7 continuous robotic screen operation. Thus, offline batch dispensing using a Multidrop or Multidrop-Combi with stackers is a more practical alternative to prepare cells for automated screens. The cell plates from the stackers can be loaded to the robot after the cells are dispensed to batches of 200 to 400 plates at a time. Two to three batches of cell plates per day can be prepared to fill the needs of a 24-hr robotic screening system. We found that in most cell-based screens this batch method worked very well. Readout as a percentage increase or decrease of assay signal resulting from compound treatment is usually fairly stable with this method.

6. Automated online determination and reporting of screen quality including signal-to-basal ratio and control compound activity should be applied. The robotic system operator should be notified immediately via automated phone call or email if significant changes of parameters and/or robotic malfunctions occur. The problems should be resolved immediately to avoid screen failure that may lead to expensive waste of reagents. Always remember that fully automated screening does not ensure trouble-free operation. Frequent checking of screen results from detectors and continuous system monitoring are critical measures for the success of robotic screens.

In summary, we reviewed all the components and instrument choices for screen automation. To set up a new screen lab, one should decide whether to implement semi-automated or fully automated systems based on throughput needs and available capital funding. Since both instrumentation and screen assay technologies advance quickly, continual testing and evaluation of new technologies are necessary to stay abreast of new developments in the field and continue to improve throughput and quality, and also reduce screening costs.

REFERENCES

Carpenter, A.E. 2007. Image-based chemical screening. *Nat. Chem. Biol.* 3, 461–465.

Cleveland, P.H. and P.J. Koutz. 2005. Nanoliter dispensing for uHTS using pin tools. *Assay Drug Dev. Technol.* 3, 213–225.

Gwynne, P. and G. Heebner. 2003. Laboratory automation: bursting through the bottlenecks. *Science* 299 (5606), special advertising section.

Hamilton, S. 2002. Introduction to screening automation. In *High-Throughput Screening: Methods and Protocols*, 190, Janzen, W.P., Ed. Totowa, NJ: Humana Press, 169–189.

Henshall, M. and P. Held. 2008. Assay: hybrid bridges microplate reader gap: Synergy 4 combines features of filter- and monochromator-based systems. *Genet. Eng. Biotechnol. News* 28 (2). http//genengnews.com/chtitem.aspx?tid=2332&chid=1.

Houston, J.G. et al. 2008. Case study: impact of technology investment on lead discovery at Bristol-Myers Squibb, 1998–2006. *Drug Disc. Today* 13, 44–51.

Landro, J.A. et al. 2000. HTS in the new millennium: role of pharmacology and flexibility. *J. Pharmacol. Toxicol. Meth.* 44, 273–289.

Mayr, L.M. and P. Fuerst. 2008. The future of high-throughput screening. *J. Biomol. Screen.* 13, 443–448.

Menke, K.C. 2002. Unit automation in high-throughput screening. In *High-Throughput Screening: Methods and Protocols*, 190, Janzen, W.P., Ed. Totowa, NJ: Humana Press, 195–212.

Michael, S. et al. 2008. A robotic platform for quantitative high-throughput screening. *Assay Drug Dev. Technol.* 6, 637–657.

Niles, W.D. and P.J. Coassin. 2005. Piezo- and solenoid valve-based liquid dispensing for miniaturized assays. *Assay Drug Dev. Technol.* 3, 189–202.

Pereira, D.A. and J.A. Williams. 2008. Origin and evolution of high-throughput screening. *Br. J. Pharmacol.* 152, 53–61.

Rutherford, M.L. and T. Stinger. 2001. Recent trends in laboratory automation in the pharmaceutical industry. *Curr. Opin. Drug Disc. Dev.* 4, 343–346.

Taylor, P.B. et al. 2002. A standard operating procedure for assessing liquid handler performance in high-throughput screening. *J. Biomol. Screen.* 7, 554–569.

Tolliday, N. et al. 2006. Small molecules, big players: the National Cancer Institute's Initiative for Chemical Genetics. *Cancer Res.* 66, 8935–8942.

Wong, E.Y. and S.L. Diamond. 2009. Advancing microarray assembly with acoustic dispensing technology. *Anal. Chem.* 81, 509–514.

12 Compound Library Management

Dalin Nie

CONTENTS

12.1 INTRODUCTION

For the past 15 to 20 years, a new multidiscipline function has emerged in the pharmaceutical arena: compound management. Thanks to the advancement of high-throughput screening, compound management has evolved as a critical component of drug discovery (Gosnell et al., 1997; Chan and Hueso-Rodriguez, 2002). Today, compound management is cited as a central and enabling function in the discovery process, from early phases involving high-throughput screening (HTS) to late-stage lead optimization screening cascades (Archer, 2004; Nie and Hartman, 2005).

What is compound management? Before HTS appeared, a chemist synthesized some chemical samples and handed them to the biologist in the next laboratory for testing; compound management did not exist. When the number of samples to be tested increased, so did the need for a simple technique to collect samples from makers and distribute them to analysts in their original forms. Drug discovery organizations set up pharmacy-type operations with shelves full of bottles to house the samples and balances for weighing and dispensing and considered those steps constituted compound management. Many people still hold that view when they think of compound management.

The introduction and growth of high-throughput screening (HTS) late in the 20th century revolutionized compound management. HTS, which screens literally tens of thousands to hundreds of thousands of samples daily, set completely different requirements for compound management. First, the total number of samples in inventories increased by several orders of magnitude from thousands to millions (Webb, 2004). A compound collection ("the crown jewels") became a primary asset of a drug discovery organization (Yates, 2003) and demanded totally new storage and retrieval mechanisms with high capacity and the capability to quickly access samples. Second, library samples in liquid form had to be delivered quickly to analytical laboratories for screening. This required all samples in a library to be solubilized and stored in solution and solid forms. Third, while the number of samples needed daily reached the magnitude of hundreds of thousands, the volume required for each sample had to be as low as a few nanoliters. Also delivery formats varied from one HTS campaign to another: one customer may have required delivery of 2 µL in a 384-well format and another may have requested 50 nL in a 1536-well format. The compound management task now requires maintenance of millions of chemical samples in both solid and liquid inventories, and more importantly, delivering up to hundreds of thousands of compounds upon request in a large variety of different formats, volumes, and concentrations daily.

Clearly, the traditional operation simply could not meet these types of demands and the drug discovery industry had to devise and invest in totally new compound management operations to cope with the dramatic changes in demand due to HTS. The field of compound management was recognized in organizational structures (Ray, 2001; Holden, 2003). Large-scale storage and retrieval robotic systems have been developed and implemented to expand storage capacity and expedite access (Spencer, 2004; Wood and Keighley, 2004; Schopfer et al., 2005; Fayez, Morin, and Paslay, 2008). A wide variety of liquid handling technologies have been adopted to enable low volume transfer. New logistics and workflows have been defined and refined to handle the demands (Sofia, Stevenson, and Huston, 2005; Diratsaoglu, 2008). Today, compound management has developed into a process-oriented, technologically intensive, multidiscipline function that integrates multiple fields, including chemistry, biology, process engineering, robotics automation, informatics, product quality assurance and quality control, and customer service.

12.2   COMPOUND MANAGEMENT PROCESS

Compound management encompasses acquiring, archiving, storing, and maintaining a library of chemical samples, along with preparing and processing samples upon request from scientists who screen and test the compounds. A modern compound management function must provide compounds for all drug discovery activities, from HTS to lead optimization, in forms most desirable for the testing, from solid or neat samples to solutions dissolved in solvents such as dimethylsulfoxide (DMSO). While different compound management operations may have different customer bases and their activities can vary from one group to another, support to HTS customers involves almost the full spectrum of compound management services. The main focus of this chapter is compound management for HTS.

Since HTS is the driving force for modern compound management, the two functions are often well integrated. Most samples in a large compound library are maintained for purposes of HTS and the success of HTS depends greatly on the timely supply of quality compounds. While exact processes of compound management for HTS can vary among organizations, HTS generally needs

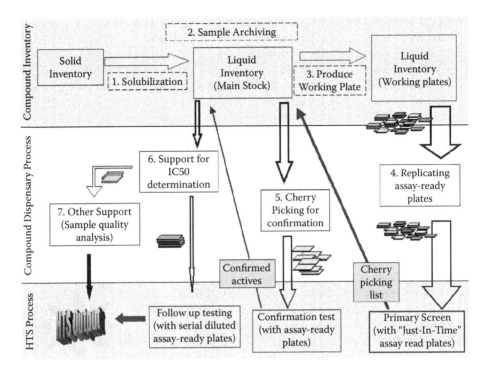

FIGURE 12.1 Typical compound management processes supporting HTS. Top (gray area): inventory management. Middle: compound dispensing and distribution. Bottom (gray area): HTS activities.

support from compound management for primary screening, confirmation retesting, and follow-up testing with concentration response curves. Figure 12.1 illustrates typical compound management processes that fully support HTS.

12.2.1 Compound Solubilization

A typical HTS campaign may require hundreds of thousands of samples delivered daily. Thus it is necessary to have all neat (solid) samples in liquid form in order to prepare and deliver them quickly. The first process in compound management to support HTS is dissolution of all solid samples into solutions—a process known as solubilization that makes samples readily available for biological testing. Another advantage to centralized compound solubilization, even for post-HTS activities, is to preserve the samples and reduce waste of the "crown jewels." Compound solubilization usually requires (1) selection of solvent, (2) determining concentration and volume, (3) weighing of solids, and (4) solubilization.

DMSO is the most common solvent in compound solubilization because of its physicochemical properties, high solvent power, low chemical reactivity, and low toxicity (Balakin, 2003). However, DMSO presents some challenges. First, neat DMSO has a freezing point of 18°C (Catalan, Diaz, and Garcia-Blanco, 2001). This must be taken into account in sample storage and processing environments because frozen liquid does not mix well with liquid handlers. Second, DMSO is hygroscopic and aggressively absorbs moisture from the air. Exposure of neat DMSO in a 30% relative humidity environment for 30 min can increase volume 5% or more (Nie, Hilton, and Gosnell, 1998). This can lead to quality concerns related to changes in concentration, solubility, and long-term stability (Di and Kerns, 2006). In addition, cell-based and other types of assays may exhibit limited tolerance to DMSO. Nevertheless, most HTS libraries are solubilized and stored in neat DMSO or in mixtures of DMSO and water, although other solvents may be used.

Determining target concentration is important because it must work for most processes after the sample is solubilized to a certain concentration. After solubilization, a lower concentration solution may be achieved by dilution, but it is no longer possible to fill a request for a higher concentration. Therefore, a compound management laboratory must ensure the concentration is high enough to cover the needs of its customers. Further, more solubility issues may arise as the concentration increases.

Excessively high concentrations require extra dilution processes. Concentrations from 10 to 20 mM are common in the drug development industry. The proper volume to solubilize should be determined based on the projected amount needed during the life expectancy of the library for the operations, considering storage conditions and testing processes. This is important because the volume must be sufficient not to require unnecessary resolubilization that may be labor-intensive and time-consuming. Conversely, the volume should not be so large as to result in significant unused (and wasted) amounts at the end of a compound's life expectancy. It is also important to consider that dead volumes in containers may be significant as well.

The next step is to retrieve solid samples (usually stored in barcoded glass vials) from storage and weigh the desired amount. The calculation of desired amount is based on targeted concentration, individual sample molecular weight, and target volume. After weighing, the stock vials are returned to storage.

Sample retrieval and re-storing may be automated as many HTS supporting organizations do, but weighing is a labor-intensive and time-consuming manual process. Although some automated weighing instruments are commercially available, most weighing processes are still manual operations—a scientist removes the desired amount of solid sample from stock vials or tubes and weighs it into a solubilization vial using an analytical balance. This seemingly simple task requires people who can concentrate on the task and pay great attention to detail. An error at this point may be difficult to find and will void every process downstream for the compound. The chance of error may be greatly reduced if the weighing step can interface in real time with the inventory database, the weighing balance, and a barcode tracking and printing system.

After the solid compounds are weighed into the vials, the vials can be moved to the solubilization station. An automated liquid handler can add the correct amount of solvent to the vials to achieve the desired concentration. Although the amount of solids needed is calculated to yield standard amounts for all compounds, some fine-tuning of the amount of solvent during solubilization is essential to achieve accurate concentration and allow for the weighing step to be slightly less precise in achieving the desired amount. It is very desirable to verify the barcode to ensure the right amount of solvent is added to the right vial, and more importantly, to ensure the correct identity of the liquid sample if it is to be transferred to another container. To assist solubilization of the sample, some mixing, shaking, or sonication is usually applied after the addition of solvent. The liquid stock is then transferred to long-term storage containers.

12.2.2 SAMPLE ARCHIVE

While sample archiving generally refers to the process of storing samples (solids and solutions) in inventory, we will focus on the system and process of storing solubilized samples. How a compound is stored is critical to the success of compound management. The key elements of sample archiving are accuracy, accessibility, and quality preservation. Accuracy involves the record of a sample in storage that must match exactly the physical properties of the sample, such as location, availability, etc. Accessibility allows a sample to be retrieved quickly for dispensing when requested. Quality preservation means the sample will maintain its integrity during storage.

Depending on the size of the library and the needs for testing, storage systems for solubilized samples can vary from a simple refrigerator, to a walk-in freezer, to a large-scale, environmentally controlled, fully automated storage and retrieval system. In the past decade, many fully automated systems have been customized to fit the compound management processes required by various organizations. Because these systems often require investments of millions of dollars of capital, modular systems that allow scalability according to the needs of a particular organization have been developed.

FIGURE 12.2   (See color insert following page 114.) Example of large-scale automated storage and retrieval system. Left: environmentally controlled store with tube-filled trays inserted in slots of shelves. A transporting robot pulls a tray and sends it to a picking robot (upper right) where tubes are picked and placed in a rack with SBS standard plate footprint. The rack with tubes is sent for processing via a conveyance (lower right).

The storage container is another key component of a sample archive. Early liquid libraries stored samples in microtiter plates in 96- or 384-well formats that matched the format of the biological screening process. This works well for primary screens when all samples in the plates are needed. The issues with storing samples in plates arise during follow-up testing when only small portions of samples are needed. First, the speed of "cherry picking" the needed samples from large sets of plates becomes a bottleneck. It can take weeks to pick the 1% of active compounds from a deck size of millions. Second, sample quality is compromised because most of a library is subjected to unnecessary exposure to dispensing conditions. In recent years, individually barcoded and sealed storage vessels and/or tubes and fast cherry-picking robots have been introduced. Systems based on these new technologies can select tens of thousand of individual tubes in a matter of hours. Figure 12.2 shows a large-scale storing and retrieval system.

If plates are chosen as storage containers, it is necessary to isolate each sample from cross contamination and interaction with ambient air. This can be done by attaching lids or sealing the plates. A lid can be as simple as a plastic cover or a proprietary device that allows automated placement and removal. Sealing techniques range from cap mats, to adhesive seals, to heat seals. Traditionally, a sealed plate requires a manual step for seal removal. Recently a new technology provides automated seal removal.

Two types of strategies are involved in storing main liquid stocks. One is to set up a master storage for multiple uses; the other is to have multiple copies made and have each copy serve a one-time use. Both strategies have pros and cons. With the multiple copy–single use strategy, each copy is stored presumably under optimum conditions until retrieval for screening. This system presents little concern for sample integrity but requires far more storage capacity and more initial processing time. Imagine the effort required to make 30 copies of a library containing millions of compounds and provide the space needed to store them. This approach is also less flexible for adopting changes. For example, 2-μL copies made 2 years ago may no longer be suitable for present requirements

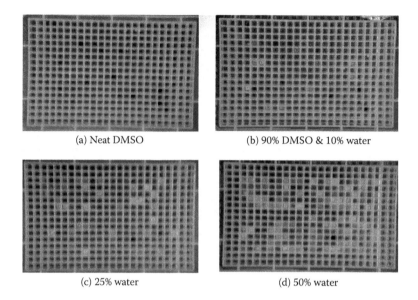

(a) Neat DMSO   (b) 90% DMSO & 10% water

(c) 25% water   (d) 50% water

FIGURE 12.3 (See color insert following page 114.) Effect of water on compound precipitation. A set of selected samples that previously showed signs of precipitation was placed in a 384-well plate. Four identical copies were made and the copies were diluted to sample volume; each had a different DMSO-to-water ratio. After 1 week in an ambient environment, the number of wells showing precipitates increased with the increase of water.

for 50 nL. By contrast, the master storage approach requires smaller storage capacity and is easier to adjust for changes. It requires no "mad rush" to create large numbers of copies and appropriate amounts can be taken as needed. However, it may cause quality concerns every time a container holding total stock must be retrieved for processing.

Storage conditions represent another challenging aspect of compound management. No single consensus dictates the best way to store and process samples. However, the following major factors are commonly considered when determining storage conditions.

Humidity — General agreement surrounds the effects of humidity on sample quality. Water in a sample is believed to be detrimental to quality in a number of ways. First, it affects the concentration of the solution. Second, it reduces the solubility of the sample. More compounds tend to precipitate and fall out of solution as water content increases, as demonstrated in Figure 12.3. The hydrophilic nature of DMSO makes controlling humidity an important consideration. Humidity is generally the greatest concern among all the factors that influence sample integrity.

Temperature — Temperatures for storing liquid samples can vary from ambient to –80°C in the compound management community. Some believe that samples should be stored at –20°C or lower to reduce chemical activities and slow sample degradation (Cheng, et al., 2003). Others demonstrate that storing samples at higher temperatures is adequate (Bowes et al., 2006). The argument is that drug-like compounds should be stable at ambient temperature and storing at sub-freezing temperatures should not be necessary. Keep in mind that the purpose of HTS is not to find a drug; rather it is to identify compounds with interesting properties that may serve as starting points for developing a drug.

Freeze-and-thaw cycle — Evidence indicates that sample decomposition is accelerated at the solid–liquid interface (Lipinski, 2004). Many believe that a freeze-and-thaw cycle negatively affects compound quality, especially when the thawing temperature is significantly higher than the ambient. Kozikowski et al. (2003) found ~1.7% compound loss for every freeze-and-thaw cycle in their study of over 25 cycles. Some companies set a limit on the number of cycles a compound can undergo before it is discarded.

Oxygen — Lower oxygen levels are believed to reduce chemical activity, and therefore the desire is to store and process samples in an oxygen-free environment. Some organizations displace the ambient air in a storage container with inert gas before a sample is added to the container. Others have tried to conduct sample processing operations in inert enclosures, but no studies document the effectiveness of such practices.

Light — It is generally believed that light causes even stable samples to degrade; therefore most storage should be in a dark environment. The processing area, particularly for automated operations, should also be dark.

While it may seem simple to consider these factors individually, the effects of a combination of some or all may be more difficult to assess. This may be one reason some study results are contradictory. For example, which procedure is better for sample quality: storing samples at 18°C to avoid freezing and thawing or storing them at −80°C and then subjecting them to 30°C or higher temperatures (Lund, 2004) for an hour every time they are processed? In some cases, samples are requested several times a day and study results support both processes. In reality, the industry employs a wide variety of storage conditions.

12.2.3 PRODUCTION OF WORKING PLATES

Working plates are compound plates created from master liquid stock to supply needed samples for multiple HTS screens. Working plates are needed for two main reasons. First, although the samples are in liquid form, it is very difficult to supply from the main stock on demand. For example, the fastest tube-based master inventory system reported to date can process no more than 50,000 samples a day, far short of demands for HTS campaigns. Furthermore, multiple screens may be run simultaneously. Creating a set of working plates with the same format as HTS enables fast replication of assay-ready plates. The second reason is that the screening process may require full integration of the compound supply because the use of pre-made assay-ready compound plates may be impossible or undesirable for the assay. In this case, the working plates are integrated into the screening process (Yasgar et al., 2008) and the desired amounts of compounds added to the assay plates "on the fly" from working plates in the screening process. The working plates can then be re-used for the next screen.

Working plates are created in a reformatting system (or workstation) with a traditional syringe type liquid handler; volumes range from a few to tens of microliters. Figure 12.4 shows an example of a fully automated reformatting system. Basically the desired volume is taken from each container storing the main stock (whether in high-volume plate or tube format) and placed into the working plates.

Working plates today generally contain 384 or 1536 wells that match the formats of HTS screens. Working plates can contain an entire library or only a subset of a library. They can be at a single concentration or multiple concentrations for each compound. The life expectancy is usually a few months to about a year—significantly shorter than the main stock. Since the working plates will last for months, they generally must be sealed or lidded and stored. The sealing techniques and storage conditions described previously for the main stock apply here. Because of the shorter life expectancies of working plates, some organizations relax their storage requirements slightly for working plates.

12.2.4 REPLICATE ASSAY PLATES FOR PRIMARY SCREENING

When screeners request compounds on a per-screen basis, it is necessary for compound management to replicate assay plates from working plates. Depending on the number of plates needed, replication can be performed by workstations requiring human intervention and/or fully automated systems that run unattended, like the one illustrated in Figure 12.5. Because assay plates are normally copies of the original working plates from which they are made, but in lower volumes, the working-plates and assay plates are sometimes called mother and daughter plates, respectively. Replicated assay plates are generally used for a single screen.

FIGURE 12.4 In-house developed, fully automated reformatting system. Compounds stocked in individually barcoded and sealed tubes are transported (A). A robot (B) delivers the tubes in a 96-tube rack to a liquid handler (C), where samples can be added to 96-, 384-, or 1536-well plates.

Assay plates are traditionally replicated in advance. Compound management anticipates the coming screens and produces assay plates in advance so that the compounds are already in plates for screening. To increase efficiency and reduce the costs, multiple copies of assay plates are often replicated at one time. This ensures the timely delivery of compounds to the screening area, but presents a few challenges. First, the requirements for compounds with respect to volume, concentration, and

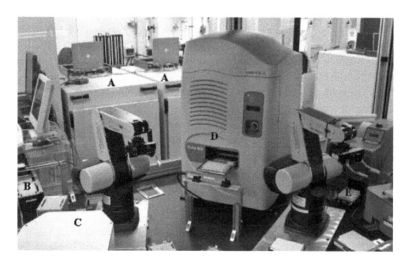

FIGURE 12.5 In-house developed, fully automated replicating system. Working plates are automatically retrieved from module working plate stores (A) and presented to the system individually by a handoff arm (B). A robot carries the working plate through a centrifuge (C) and brings it to a liquid handler that utilizes acoustic technology (D) to transfer compounds in low nanoliter volumes. The assay-ready plate is then sealed in a heat sealer (E) after it is replicated.

types of plates may change from one screen to another; pre-generated assay plates may not exactly fit the screen requirements. The scientists must modify the assay or perform extra steps before the plates may be used in the screen. Second, more storage capacity is required. Third, the volume of samples in the assay plates is much smaller and their life expectancy is shorter. A delay in screening may mean having to produce plates again and money and effort are wasted when the unused plates are trashed.

Some organizations adopt a just-in-time assay-ready plate (ARP) or assay-ready plates on-demand type of operation (Nie, 2007). The screener requests the compounds exactly as they are needed and specifies the type of plate, volume and concentration in each well, and number of plates needed for each day of screening. The compounds in the assay plates will be generated right before they are needed ("just in time") for the screen. This technique presents a number of advantages. The assay plates are customized for each screen and fit perfectly. Since plates are produced on-demand, storage is unnecessary, plates are not wasted, and less concern surrounds sample quality. However, the requirements concerning the capacity and reliability of replicating plates are rigorous.

Miniaturized assay formats constantly introduce challenges to assay-ready-plate production. In the days of the 96-well format, traditional syringe-type liquid handlers, even those using disposable tips, could deliver compounds with adequate results. The move to 384 and 1536 wells or more required the delivery of nanoliter volumes of compounds with high accuracy and precision to the right well positions. New types of instruments also replicate assay plates. The most notable ones are pin tools and capillary tube-based instruments. Pin tools utilize a set of pins that may vary in diameter and shape to deliver a targeted volume. Instead of aspirating and dispensing the liquid, the pin is dipped into the source liquid and a small portion remains in the pin when it leaves the source. The pin is then put in contact with the destination well and leaves a small amount of liquid there. The amount of liquid it carries depends on factors such as the diameter and shape of the pin.

The capillary tube-based system employs a set of tiny tubes of fixed length. When a tube is dipped into a source liquid, the tube is filled with liquid via capillary action, then compressed air blows the liquid in the tube into the destination well. Both types of tools can deliver nanoliter volumes with reasonable accuracy and precision. The limitation is that they can only deliver a fixed volume without changes of pin or tube settings. New syringe-type liquid handlers capable of delivering 100 nL or less are now on the market.

The introduction of acoustic technology into the liquid transfer arena revolutionized compound transfer, especially for replicating assay-ready plates. As illustrated in Figure 12.6, this technology utilizes a transducer placed beneath a source (working) plate (Heron, Ellson, and Olechno, 2006). The transducer emits a defined amount of acoustic energy focused on the liquid surface. This propels a small droplet to jump from the liquid surface. The droplet is then captured in the destination (assay-ready) plate.

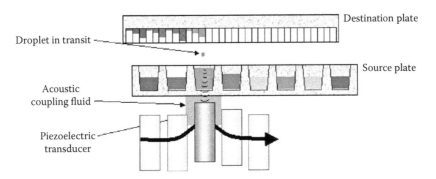

FIGURE 12.6 Liquid transfer by acoustic droplet ejection. A transducer travels beneath a source plate and emits a pulse of acoustic energy focused on the liquid surface, causing a droplet of liquid to jump upward. The droplet is captured by an inverted destination plate. (Photo courtesy of Labcyte.)

The droplet can be as small as 2.5 nL and exhibit amazingly good accuracy and precision, allowing delivery of compounds in assay-ready condition in any format available. Since nothing is in contact with the source compound, cross contamination is not a concern. The non-contact nature of the instrument also saves time and money—no need for tip washing or buying disposable tips. An added benefit is the fairly accurate measure of water content in the well (Ellson et al., 2005).

12.2.5 CHERRY PICKING FOR CONFIRMATION

Unless replicates of each sample are tested in the primary screen, screeners will want to run a confirmation test for compounds showing certain activities in a primary screen. The number of samples to be included in a confirmation test can range from a few thousand to a small percentage of total samples screened in the primary step and scattered in the inventory. The task is to produce assay plates containing only the compounds shown as active in the identical format as the primary screen. This normally involves picking the active samples from the library, reformatting into the screening format with intermediate plates, and replicating assay-ready plates.

Although the number of samples is smaller than a few percent, the time required to prepare cherry picking plates for confirmation may be longer than the time needed to replicate a full copy of the library if samples are stored in plates. Imagine that you must check every plate in the working set. Instead of making a complete copy in one transfer, the liquid handler must move to the individual wells of interest—a process that can take weeks or even months, depending on library size. Unlike a primary screen for which compounds can be prepared in advance, cherry picking cannot start until the primary screen reveals actives and constitutes a bottleneck by holding up compound supply during screening. Thanks to faster technology, cherry picking now can be completed within a day or two when performed by a tube-based system.

12.2.6 SUPPORT FOR FOLLOW-UP TESTING

Follow-up testing performed after confirmation is usually to determine IC_{50} or EC_{50} values. These tests require preparation of each compound in a series of different concentrations according to twofold diluting schemas (the concentration of each consequent point is half the concentration of the previous point in the series; with the half log, every other point would differ in a magnitude of concentration). The number of concentration points in a series can range from 2 to 12 (Quintero et al., 2007; Turner and Charlton, 2005).

If the concentration span is small, the sample may be prepared by direct transfer of different amounts of the source into destination wells to achieve the different concentrations. However, most traditional liquid handlers can only cover a narrow concentration range so the direct transfer is impractical. The most common way to prepare these samples is to add samples at the highest concentration in the first column (or row) as a starting point. The next point would be made by taking an aliquot of the solution in the first point and adding the proper amount of solvent to achieve the desired concentration. The third point is made by taking an aliquot of the solution in the second point and adding the proper amount of solvent and continuing until the last point is made. In other words, every point is diluted down from the previous point; this is often called serial dilution and may be performed by column or by row within the same plates. Serial dilution can also be performed across plates. The serial diluted plates can be replicated to make assay-ready plates at lower volumes such as nanoliters.

Liquid handlers generally face uncertainty when performing transfers. The transferred amount can be a few percent off from the targeted amount, depending on the type of instrument. When serial dilution is used, instrument uncertainty is an accumulating effect. Since the source of the compound in the second point is the first point, the uncertainty of the second point constitutes the uncertainty of the instrument in performing the transfer plus the uncertainty of whatever is already

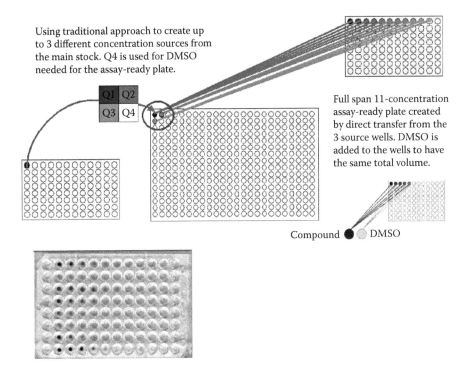

Using traditional approach to create up to 3 different concentration sources from the main stock. Q4 is used for DMSO needed for the assay-ready plate.

Full span 11-concentration assay-ready plate created by direct transfer from the 3 source wells. DMSO is added to the wells to have the same total volume.

Compound ● ○ DMSO

FIGURE 12.7 (See color insert following page 114.) New methods of creating IC$_{50}$ assay-ready plate via direct transfer. Top: achieving 11 point, half log serial dilution by first creating a three-concentration source plate with a traditional approach and obtaining a full span assay-ready plate. Bottom: visual inspection of the plate.

in the first point. For example, for a liquid handler with an accuracy of 5% (transfer error is within ±5% of the target volume, which is widely accepted in the industry), the uncertainty may be as high as 19% by the fourth point.

An instrument utilizing acoustic technology makes direct transfer more practical. Since the instrument can deliver low nanoliter volumes, it is possible to cover a concentration span of two magnitudes. However, most IC$_{50}$ tests require a concentration span of four to five magnitudes. To address this, Cesarek and Nie (2005) proposed a new combined method that can cover a whole range of concentrations and greatly reduce the uncertainties of traditional methods. The new method utilizes the traditional way to create two or three source concentrations based on the required concentration span, then utilizes the acoustic liquid handler to create assay-ready plates by directly transferring different amounts from the two or three sources. This method allowed the authors to cover a concentration range from 0.0001 to 10 mM as shown in Figure 12.7, while reducing the uncertainty by 5% to 30%, compared to the traditional method.

12.2.7 ADDITIONAL SUPPORT

Depending on the operation, HTS users can request additional support such as analysis of sample quality and request for solids.

Analysis of sample quality — The customer usually wants to know physical qualities such as purity and concentration when determining the IC$_{50}$ of a compound. This request is normally made when IC$_{50}$ compounds are requested. A separate set of plates should be made for the analytical testing discussed later in this chapter.

Request for solids — With improved storage conditions, better processing practices, and delivery of correct sample quality data (purity and concentration), requests for solids from HTS investigators have decreased to almost none. Nevertheless, solids or freshly solubilized solutions are requested occasionally for analyzing concentration response curves, for example, to determine the borderline purity of an active sample. If a solid is requested, the process is simple and similar to preparing a solid for solubilization. The sample is retrieved from the solid inventory and the requested amount is weighed from stock.

12.3 CHALLENGES OF COMPOUND MANAGEMENT

While compound management has overcome many obstacles and resolved a great number of issues during its evolution, new challenges arise steadily because of larger corporate collections, greater complexity in lead generation processes, and industry pressures to optimize time and cost efficiency. Assay technologies continue to evolve rapidly and demand wider varieties of plastic ware, plate formats, and dispense volumes. To succeed, compound management must address issues related to sample integrity, delivery accuracy and efficiency, storage capacity, process capability, and flexibility.

12.3.1 QUALITY OF SAMPLES

Sample quality is critical to the success of screening and drug discovery; it serves as the basis of all testing. It may be argued that the results of a screen can only be as good as the quality of the samples. This highlights the importance of attention to sample quality on the part of compound management personnel because the first question when screening results are unsatisfactory relates to compound delivery. Three aspects of compound quality are important for HTS.

12.3.1.1 Chemical Quality

Chemical quality of a sample is determined by structure characteristics. Some compounds never reach market because of their structures. Certain structural characteristics are more likely to enable a compound to "move up the pipeline" and are known as drug-like or lead-like properties. Lipinski et al. (1997) proposed a classic criterion to determine whether a compound is drug-like and it is known as the rule of five: molecular weight not exceeding 500; ClogP value below 5; no more than 5 hydrogen bond donors; and no more than 10 hydrogen bond acceptors. Many drug discovery organizations have adopted this rule or a modified version to select compounds for their libraries. Another factor to consider is the diversity of the samples in the library to enable HTS staff to cover as much chemical ground as possible in discovering leads.

12.3.1.2 Physical Quality

Physical quality relates to physical condition of a sample. The main aspects of physical quality are the presence of a compound and the purity of the sample. If the sample contains substances other than the desired compound, they can cause the compound to fail to produce the activity it would produce if pure. Conversely, the inclusion of other substances may enable a compound to show activity and lead to questions about whether the activity arose from the compound or one or more of the other substances present. Another aspect of physical quality is the concentration of the compound in solution form. An incorrect sample concentration yields misleading results and may lead to incorrect decisions. The physical quality of a sample may change over time; thus it is important for compound management to assess physical quality of a sample during its life cycle. This is achieved by three types of assessments:

Initial assessment — The physical quality of a sample should be checked when it is first submitted for inventory. This ensures that only high quality samples enter the library.

Periodic assessment — The intent is to determine how the physical quality of the entire library changes over time. The results can be used to guide storage and processing practices and also determine the life expectancy of the library. Generally, this assessment involves the analysis of a small set of samples that represent the library rather than analyzing every sample.

Project assessment — This analysis aims to examine compounds of interest to specific projects to aid managers in making informed decisions about compounds. Many organizations assess active compounds resulting from HTS; others perform analysis when a project focuses on a particular compound.

The physical quality of samples is assessed via analytical methods. Liquid chromatography with mass spectrometry (LC-MS) is the most common instrumentation used to determine sample presence and purity (Kerns et al., 2005). Chemiluminescent nitrogen detection (CLND), charged aerosol detection (CAD) and other techniques have also been used to determine concentration (Popa-Burke et al., 2004; Gamache et al., 2003).

12.3.1.3   Process Quality

In the process of preparing compounds for screeners, uncertainty can be unintentionally introduced in several areas. For example, a wrong compound may be delivered; an empty container may be delivered if a delivery tip is clogged. Some measures of process quality are:

- Samples are stored where they should be stored.
- Amounts of samples in inventory are the amounts required.
- Samples delivered to customers are correct.
- Amounts delivered are exactly as requested.

Compound management must take all measures to ensure high process quality by eliminating human error and minimizing instrument uncertainties.

Barcode — This procedure is the most effective way to avoid loss of sample identity. It should be mandatory to barcode all containers that can be moved independently and the barcode should be checked every time a container is in a situation where its identity can be lost or misunderstood. Furthermore, sample and container identifiers should be implemented in real time as soon as a sample is placed in a container. Each barcode should be unique and should include a label readable by humans. Alphanumeric barcodes are preferred because pure numerical codes tend to lose their leading zeros in some situations. Although a barcode is intended to provide a unique identifier to a container, it may be of great value in improving process quality if systematically and intelligently generated. For example, a barcode may identify a container and also carry certain information about its purpose; the first character of a barcode may identify whether the content of the container represents main stock or is for intermediate use such as preparation of working plates or delivery to a customer as an aid in system quality checking. A user can also identify an error when a container is in the wrong place at the wrong time.

Quality assurance and quality control — Quality assurance (QA) activities ensure that processes will perform their tasks as designed and produce high quality products. Since compound management is a process-oriented discipline, establishing robust processes and pursuing rigorous quality assurance are crucial steps; every process should be validated when first established, periodically thereafter, and whenever a change is made. Other examples of procedures include periodic checks of instrument performance and checking the effectiveness of washing processes to ensure they eliminate carryover.

Quality control — Also known as QC, these activities verify product quality before delivery to customers. Compound management for HTS, in a sense, manufactures sample plates for screening and it is necessary to establish rigorous QC to ensure the quality of deliverables before they are delivered. A few examples would be checking for missing wells in assay plates (as shown in Figure 12.8), inspection of plate orientations, and examinations of barcode clarity among other reviews.

Liquid handler performance — While this check may be considered an aspect of process QA, liquid handler QC warrants a separate discussion (Hentz, 2008). It is critical in compound management to perform regular checks of instruments based on vendors' recommendations to ensure they meet specifications because deliveries from compound management to HTS customers rely heavily on the performances of liquid handlers.

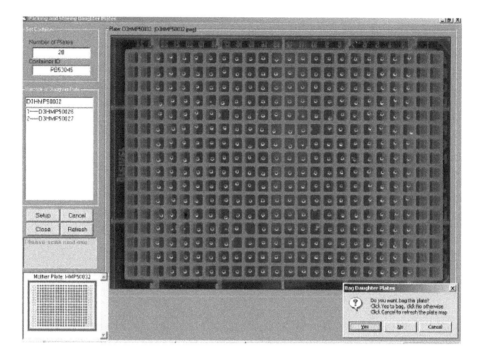

FIGURE 12.8 Quality control of assay-ready plate production: checking empty wells. An image of the assay-ready plate is captured during replication. The plate map is brought up for comparison to the actual image to identify missing wells.

Instrument accuracy is a measure of an instrument delivery. For a liquid handler, accuracy measures how close the instrument-delivered volume is to the targeted amount. It requires a standard for comparison, for example, a gravimetric approach to compare the weight of a liquid delivered by the instrument to the expected weight.

Instrument precision is a measure of reproducibility, i.e., how consistent an instrument is in delivering volume. This is generally measured as a coefficient of variance (CV). Reproducibility for multichannel (96- or 384-tip) pipetting instruments has two aspects. Within-plate variability measures the consistency from one channel to another in the same transfer; cross-plate variation measures the consistency of the same channel through multiple transfers. A good liquid handler should be consistent in both functions.

12.3.2 SPEED OF DELIVERY

Although the capability, capacity, and speed of processing customer compound requests have increased greatly over time, meeting delivery timelines remains a challenge. First, customers are more demanding as their capability and capacity increase, and this increases their expectations for compound management. For example, the cherry picking of 5000 compounds in a few weeks may have been considered very fast 5 years ago; current expectation is to deliver 5000 compounds the next morning. Second, compound management builds its capability based on average demands from customers, but peak demands can reach three or four times the average.

For example, if your HTS customers plan 20 to 25 screens annually, you may build the capacity to supply assay-ready plates on demand for two or three screens per month. What may happen is a slow period for a few months followed by demands for products for six to eight screens at one time. Finally order processing for HTS depends on automated systems that may break down at times of heaviest demand. To meet this challenge, it is essential to engage customers in developing clear and realistic expectations for deliverables, applying customer insight for short and long term needs, establishing mechanisms for timely communications at all levels, and developing contingency plans for unexpected scenarios.

12.3.3 CHANGES IN REQUIREMENTS

Drug discovery groups continually search for new ways to find potential drugs faster. HTS, while new, continues to change and evolve constantly. HTS is now far more productive, cost effective, and efficient due to the years of process improvement, new technologies, and miniaturization. However, advances introduce significant challenges for compound management because the process requires good aim at a moving target. Since compound management for HTS is capital intensive, it takes time to implement new functionalities. A new system may become obsolete soon after the site acceptance test; that has occurred in recent decades.

12.3.4 AUTOMATION

The world of automation is amazing when a system works as it should. However, maintaining such systems is challenging when the robots do not cooperate. Since modern compound management depends on automated systems, it becomes very difficult to meet customer timelines if a system is down and a vendor does not repair the system in a timely manner. Some steps can mitigate this negative impact (Nie, 2008). Backup processes are essential, at least for processing urgent requests. If possible, including individuals with automation skills within the group is the best option.

12.4 INFORMATION SYSTEMS FOR COMPOUND MANAGEMENT

A well designed, robust, and reliable information system is crucial for the success of a compound management group. Some companies consider e-commerce systems suitable for compound management because the e-commerce model is very similar to some aspects of compound management. A customer places an order; compound management receives the order, fills it, and delivers it. However, in e-commerce, material ordered by a customer is already in the inventory so no real intelligence is needed to fill an order. However, in most compound management situations, material in inventory represents only a raw ingredient of an ordered product and it is necessary to deal with production of different products from raw materials when filling orders.

This adds a new layer of complexity to a compound management system. Such systems must be integrated or interfaced with many other systems beyond the compound management area. Full descriptions of the design and architecture of such systems are beyond the scope of this chapter, but it is useful to review the major functionalities of information systems required for effective compound management operations.

12.4.1 SAMPLE REGISTRATION

Compounds arrive at a compound management facility from a number of sources: in-house synthesis by internal chemists, acquisition from external collaborators, and purchases from commercial vendors. Regardless of the source, all compounds must be registered with information properly describing the compound, such as its chemical structure and formula. Information on compound registrations must be recorded and kept in a database that is accessible company-wide. A unique identifier should be assigned to each sample during registration so it can be distinguished from others. Compound registration may or may not be the responsibility of the compound management team.

12.4.2 SUBMISSION

When a compound arrives at the compound management operation, the initial information must be verified and recorded in an inventory database. First, the sample must be associated with its container. This means a unique container identifier such as a barcode must be assigned to the sample. The container must carry the identifier throughout its lifecycle. Then initial data about the sample must be recorded. This includes the initial amount (weight for solid samples and volume for liquid solutions), concentration if in solution, container data such as barcode, type, and tare weight (if weighing will be part of the dispensing process). Submission creates or initiates the record of the container in the inventory.

12.4.3 INVENTORY MANAGEMENT

The purpose of inventory management is to maintain the integrity of data to ensure that it accurately reflects the physical inventory (location of compound, amount available, and condition). Inventory management should be associated with all transactions such as archiving, order fulfillment, and sample removal so that the inventory always reflects samples in real time. Inventory management should also allow system auditing and error correction.

12.4.4 ORDER PLACEMENT

Order placement enables the customers to place orders for what they need. A customer should be able to place an order anywhere in an organization by specifying the desired final product without worrying about where the "raw material" is or how compound management will deliver the final product needed. All necessary information should be recorded and sent to compound management. It is helpful for compound management to indicate the availability of samples in inventory and update order status to allow customers to track their orders.

12.4.5 ORDER ADMINISTRATION

After a customer submits an order, it should follow certain steps (order administration) before it can be fulfilled. Order administration can combine human intervention with automatic administration by an information system. An order must be checked against business rules first. The next step is to determine the workflow needed to take an order from starting inventory stock to final product by selecting systems and processes required to achieve the final outcome specified by the request. The workflow may involve any combination of automated systems, stand-alone workstations, and manual operations.

For a simple example, suppose a customer from HTS orders 3000 samples of 40 nL of 5 mM solution each in 384-well assay-ready plates for confirmation after a primary screen. The workflow to fill this order can involve retrieval from main stock of 10 mM in microtubes from the storage system, creation of intermediate plates by diluting the samples to 5 mM with 4 μL stock and 4 μL DMSO, then producing final assay-ready plates by using a replication system capable of making 40-nL transfers. Up to three systems may be involved, depending on the set-up, so more than one workflow may be required to fill some orders.

It is necessary to determine at the outset how much "inventory" is needed to fill the order by temporarily reserving (making unavailable) the required amount in inventory. Note that the amount ordered by a customer does not necessarily equal the amount to be withdrawn from stock. In the above example, the order amount was 40 nL of 5 mM concentration for each compound; that required 4 μL of 10 mM stock. The next step is to create jobs for the system(s) in the selected workflow to fill the request and send data to the system(s) to execute the order.

Project chemists often place dispensing restrictions on samples in inventory and must be consulted for approval before dispensing these samples. When an order for restricted samples is received, compound management is responsible for seeking and obtaining approval in the order administration step before dispensing samples.

12.4.6 ORDER EXECUTION

Order execution consists of steps to process an order to achieve the final outcome. As noted earlier, this may involve a combination of an automated system, workstations, or manual operations. Management should include information services (IS) applications for all automated systems and workstations. Recording activities is necessary even with manual operations; thus an IS application provides great benefit. An execution application should interface with other IS systems in compound management to handle job information and update progress in real time via a defined interfacing mechanism. However, the execution application should accurately perform the tasks independently, not as part of an inventory or order administration system to minimize the impacts of any execution system changes on the business process and the IS systems.

12.4.7 ORDER FINALIZATION

After a final product is made and an order fulfilled, some further steps must be performed to finalize the order before delivery to the customer. First, QC should be performed to check the product for defects. Then it is necessary to make sure all systems are properly updated and that all data associated with the product (e.g., plate maps) are ready. It may also be necessary to inform the customer that an order is completed and ready for delivery or pickup or an order may need to be packed and shipped.

12.5 COMPOUND MANAGEMENT TEAM

Effective compound management operations can exert significant positive impacts on drug discovery. The key to the success of an operation is the compound management team. A good team must be very customer focused, willing and capable to deliver whatever customers need. Moreover, they must fully understand drug discovery processes, know where they can make an impact, and bring new ideas to enhance their relations with customers. They should have technical proficiency along with all necessary skills in chemistry, process engineering, QA and QC, customer service, automation solution development and equipment services, data management, and information system development and support.

It is interesting to observe where different companies place the compound management function within their organizations. Some companies include the function in chemistry groups since its task is to manage chemical samples; other companies place it in the biology and screening areas because the function provides samples for screening. Still others consider compound management part of the IS/IT area because of the emphasis on automation and information systems. Compound management requires a multidisciplinary team that fits well with other parts of an organization. Regardless of organizational placement, the importance and value of compound management to a drug discovery organization are unquestionable.

12.6 SUMMARY

Compound management has developed as a discipline that represents a core component of the drug discovery process. It will continue to play a critical role in the race to discover promising leads for potentially marketable drugs. In the future, compound management will face challenges in supplying higher quality, faster delivery, more cost-effective operations, and the ability to adapt quickly to changes in customer requirements. The success of these groups depends on a clear understanding of the drug discovery process and a strong customer focus with adaptable approaches to meet project needs.

ACKNOWLEDGMENTS

The author wishes to thank Manori Turmel and Irene Pappas of Compound Management Automation at AstraZeneca, Wilmington, Delaware, for their comments and suggestions.

REFERENCES

Archer, J.R. 2004. History, evolution and trends in compound management for high-throughput screening. *Assay Drug Dev. Technol.* 2, 675–681.

Balakin, K.V. 2003. DMSO solubility and bioscreening. *Curr. Drug Disc.* 8, 27–30.

Bowes, S. et al. 2006. Quality assessment and analysis of Biogen Idec compound library. *J. Biomol. Screen.* 11, 828–835; Epub Sept. 6, 2006.

Catalan, J., Diaz, C., and Garcia-Blanco, F. 2001. Characterization of binary solvent mixtures of DMSO with water and other cosolvents. J. Org. Chem. 66, 5846–5852.

Cesarek, J. and Nie, D. 2005. New tool for automating serial dilutions for activity confirmation experiments. Poster presented at Laboratory Automation Exhibition and Conference, San Jose, January 2005.

Chan, J.A. and Hueso-Rodriguez, J.A. 2002. Compound library management. *Meth. Mol. Biol.* 190, 117–127.

Cheng X. et al. 2003. Studies of repository compound stability in DMSO under various conditions. *J. Biomol. Screen.* 8, 292–304.

Di, L. and Kerns, E.H. 2006. Biological assay challenges from compound solubility: strategies for bioassay optimization. *Drug Disc. Today* 11, 446–451.

Diratsaoglu, J. 2007. Compound management workflow and best practices in vendor selection and collaboration. *Am. Drug Disc.* 3, 18–19.

Ellson, R. et al. 2005. In situ DMSO hydration measurements of HTS compound libraries. *Comb. Chem. High-throughput Screen.* 8, 489–498.

Fayez, H., Morin, J., and Paslay, J. 2008. Compound storage evolution: unlocking the right combination. *Drug Disc. World* 9, 62–68.

Gamache, P.H. et al. 2003. HPLC analysis of nonvolatile analytes using charged aerosol detection. *LC/GC N. Am.* 23, 150–155.

Gosnell, P.A. et al. 1997. Compound library management in high-throughput screening. *J. Biomol. Screen.* 2, 99–102.

Hentz, N.G. 2008. The importance of liquid handling quality assurance through the drug discovery process. *Drug Disc. World,* Spring 2008, 27–31.

Heron, E., Ellson, R., and Olechno, J. 2006. Acoustic droplet ejection in drug discovery. *Drug Plus Int.* 5, 22–25.

Holden, K. 2003. Significance of effective compound management. *Curr. Drug Disc.* 9, 9–10.

Kerns, H.E. et al. 2005. Integrity profiling of high-throughput screening hits using LC-MS and related technology. *Comb. Chem. High-Throughput Screen.* 8, 459–466.

Kozikowski, B.A. et al. 2003. The effect of freeze/thaw cycles on the stability of compounds in DMSO. *J. Biomol. Screen.* 8, 210–215.

Lipinski, C.A. 2004. Solubility in water and DMSO: issues and potential solutions. In Borchardt, RT. et al., Eds. *Pharmaceutical Profiling in Drug Discovery for Lead Selection: Biotechnology.* Arlington: AAPS Press, pp. 93–125.

Lipinski, C.A. et al. 1997. Experimental and computational approaches to estimate solubility and permeability in drug discovery and development settings. *Adv. Drug Del. Rev.* 23, 3–25.

Lund, K.O. 2004. Solving the HTS compound thawing bottleneck. *Am. Biotechnol. Lab.* April 2004, 10–12.

Nie, D. 2007. Provide assay-ready plates on demand at AstraZeneca: practice, benefits and challenges. In *IQPC Compound Management and Integrity.* London.

Nie, D. 2008. Toys or tools? Automation in the drug discovery process. *Am. Drug Disc.* 3, 32–36.

Nie, D. and Hartman, D.S. 2005. The AstraZeneca Automated Compound Management Facility (ACMF) in Wilmington. *Eur. Pharm. Rev.* 10: 49–53.

Nie, D., Hilton, A., and Gosnell, P. 1998. DMSO in HTS: the effect of water absorption and evaporation. Paper presented at 4th annual conference of Society for Biomolecular Screening, Baltimore, September 1998.

Popa-Burke, I.G. et al. 2004. Streamlined system for purifying and quantifying a diverse library of compounds and the effect of compound concentration measurements on the accurate interpretation of biological assay results. *Anal. Chem.* 76, 7278–7287.

Quintero, C. et al. 2007. Quality control procedures for dose–response curve generation using nanoliter dispense technologies. *J. Biomol. Screen.* 12, 891–899.

Ray, B.J. 2001. Value your compound management team. *Drug Disc. Today* 6, 563.

Schopfer, U. et al. 2005. The Novartis Compound Archive: from concept to reality. *Comb. Chem. High-Throughput Screen.* 8, 513–519.

Sofia, M.J., Stevenson, J.M., and Huston, J. 2005. Compound management: integrated chemistry, biology and technology in the modern drug discovery environment. *Pharm. Disc.* 5, 22–31.

Spencer, P. 2004. Challenges of managing a compound collection. *Eur. Pharm. Rev.* 9, 51–57.

Turner, R.J. and Charlton, S.J. 2005. Assessing the minimum number of data points required for accurate IC_{50} determination. Assay Drug Dev. Technol. 3, 525–531.

Webb, T.R. 2004. Current directions in the evolution of compound libraries. *Curr. Drug Disc.* 4, 28–30.

Wood, T. and Keighley, W. 2004. Automated sample supply for high-throughput screening. *Eur. Pharm. Rev.* 9, 68–73.

Yasgar, A.L. et al. 2008. Compound management for quantitative high-throughput screening. *JALA* 13, 79–89.

Yates, I. 2003. Compound management comes of age. *Drug Disc. World*, 2, 35–42.

13 Unique Discovery Aspects of Utilizing Botanical Sources

Susan P. Manly, Troy Smillie, John P. Hester,
Ikhlas Khan, and Louis Coudurier

CONTENTS

13.1 INTRODUCTION

Based on a long tradition of use in humans, botanicals present many unique advantages as sources of natural products with pharmaceutical influences, especially in terms of opportunities for the development of diverse botanical products. This chapter outlines their use in screening programs, including distinctive informatics needs and certain paths and modes of development inherent to botanicals versus other sources of natural product leads.

The uses of botanically derived product continue to change and grow from traditional medicinal uses to nutraceutical functions such as dietary supplements, functional foods, food additives, sports and energy foods and drinks, and drugs (small molecules and complex mixtures). The perspectives and goals of practitioners of botanical research are very diverse. Many research groups seeking to identify particular activities in plant extracts have adopted technologies developed by pharmaceutical companies for empirical discovery, high-throughput screening (HTS) and data management that underpins HTS.

Natural products continue to be hailed as the unassailable best source for diversity and novelty (Harvey, 2000; Rouhi, 2003; McChesney, Venkataraman, and Henri, 2007). Although 50% of all drugs approved from 1981 to 2008 were of natural product origin (Figure 13.1; Newman and Cragg, 2007), during the same period, pharmaceutical companies largely divested themselves from natural products to focus on small molecule synthetics as resources for their discovery programs (McChesney, Venkataraman, and Henri, 2007; Rouhi, 2003).

The origins and processes of screening samples have always followed trends. Synthetic libraries (commercial and in-house proprietary) constituted one trend. Others include diversity synthetic libraries, peptide libraries (often viral peptide expression libraries), combinatorial synthetic libraries, fragment libraries, natural product source libraries and others. Natural products from various sources such as soil microorganisms and marine plants have gone in and out of fashion but were always recognized as more expensive and labor intensive than synthetic and molecular biological entities. Plants are distinct from all other natural product sources in that they have a very long history of use in humans and continue to distinguish themselves as sources of medicinal care for about 80% of the world's population (Shanmugam, Manikandam, and Rajendran, 2009; http://www.siu.edu/~ebl/leaflets/juand9.htm).

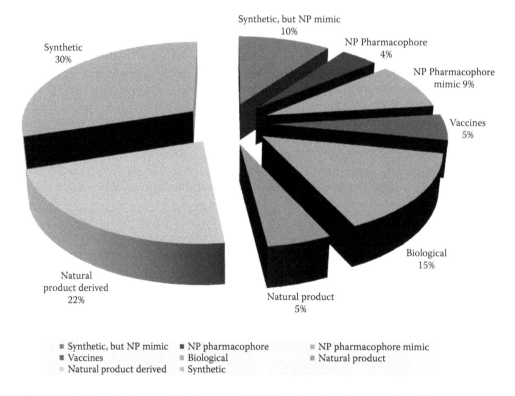

FIGURE 13.1   (See color insert following page 114.) Pie chart showing genesis of all drugs approved (*n* = 1272) from January 1, 1981 through October 12, 2008. (*Source:* Adapted from Newman, D.J. and Cragg, G.M. 2007. *J. Nat. Prod.* 70, 461–477.)

Plants are not different from other natural product samples in that they too tend to interfere with various screening formats in nonspecific ways as "nuisance" compounds displaying unwanted color, inherent fluorescence, "promiscuous" or aggregate behaviors, detergent-like activities, or toxicity (Feng et al., 2005; Appleton, Buss, and Butler, 2007). Biochemical assays (cell-free defined systems) are notoriously sensitive to such interference by natural product extracts. Cell-based reporter assays and cell-based so-called phenotypic screens always require parental cell controls to determine extract toxicity. However, plants contain their own sets of components that are problematic to screening assays and compound identification.

How traditional screening programs dealt with these challenges is discussed further in Section 13.4. The use of natural products in traditional pharmaceutical company screening programs was considered too expensive and time consuming. Through the expanding use of laboratory automation, chemistry became the rate-limiting factor in drug discovery and since the mid 1990s, most natural product programs have been eliminated (Rouhi, 2003). Academic institutions established natural product screening programs in response to the much touted "chemical genetics" approaches embraced by various research universities and by the Roadmap Initiative of the National Institutes of Health (NIH; http://nihroadmap.nih.gov/overview.asp; Kugawa, Watanabe, and Tamanoi, 2007).

More recently, a plethora of small companies and biotechnology ventures were established to handle "biologics" as well as natural products, many of them with proprietary screening platforms designed to make natural product screening more straightforward and quicker. Some developed proprietary collections of particular natural product sources for extract production such as organisms isolated from extreme environments, leaf litter, caves, unique water environments, and endophytes from marine and terrestrial plants.

A partial list of companies that sell natural products can be found at http://www.rechemical.com/natural-compounds. Several such companies have entered agreements with the same large pharmaceutical companies that were so eager to abandon in-house natural product groups. Such deals usually involve the supply of natural product samples or the provision of certain screens and targets as identified by the large companies against their natural products libraries to find new and novel natural product leads (Bouley, 2007; Rouhi, 2003).

Other companies developed proprietary screening platforms that facilitate the identification of natural product leads. For example, WellGen's transcriptional analysis technology helps identify natural product extracts that are active in eliciting certain healthy changes in cell expression aimed at eliminating diabetes and obesity. BioPlanta and AnalytiCon Discovery combine plant tissue culture, bioengineering, and high-throughput structure elucidation technology to generate large libraries of novel natural products. With Cetek's fluid-based screening technology, the separation of each component of an extract is separated from the others and tested individually in the assay of choice. The goal of these companies is to identify and develop pharmaceuticals, functional foods, medicinal foods, and food additives (WellGen: http://www.wellgen.com/products/php; Cetek: http://www.cetek.com/drug_disc_plat.html; AnalytiCon: www.ac-discovery.com; BioPlanta: www.bioplanta-leipzig.de).

Academic institutions are assembling both synthetic libraries and natural product repositories. The National Cancer Institute (NCI) operates a well-received resource providing natural product extracts, mostly from marine sources (http://www.dtp.nci.nih.gov/branches/npb/repository.html).

This chapter will outline the use of botanical samples in traditional screening environments and the unique opportunities for development of botanical products.

13.1 PLANT ACQUISITION AND STORAGE

13.1.1 PLANT SOURCING

The National Center for Natural Products Research (NCNPR) has contracts in place with various outside collectors and botanical collaborators. Samples are classified by authority, genus, and species by botanists in the field. These collections are gathered via the best botanical practices, taking the necessary precautions needed to minimize impact and to ensure continued integrity and health of the sampled populations. Samples from countries outside the United States are collected with explicit agreements from countries of origin in full compliance with all international biodiversity agreements for material transfer. Care has been taken to work with the collectors to ensure diversity of samples. Criteria such as phytochemical activity, availability, and rarity are used to assess collection strategies for samples.

In keeping with good practices, each species collected includes the collection of a "voucher" sample along with material to be processed for activity testing. This sample consists of representative plants or plant parts including stems, leaves, flowers and/or fruits where possible. The voucher samples are pressed and dried using standard equipment and each sample is given a unique identifying number by the collector. The collection number protocol is the standard method used for uniquely identifying plant collections. The voucher samples are maintained at botanical garden herbaria or at University of Mississippi's Pullen Herbarium.

Information about each collection site is recorded as well. Latitude, longitude, and altitude are determined by a global positioning system (GPS) unit. Habitat descriptions follow a general format including land use characteristics (open grassland, disturbed or undisturbed forest, wetland, alpine meadow, etc.); soil composition, color, moisture; sun exposure (full sun, part sun, part shade, full shade); associated plant species; and other notes including vegetation zone, slope, etc. Plant descriptions are prepared for each species collected and note plant size, flower presence, colors, unique aromas, etc. All this information is entered into the NCNPR database.

13.2.2 PROCESSING PLANT SAMPLES

The collected plants intended as sources of phytochemicals to be screened are received frozen and freeze dried in a VirTis 24Dx48 general purpose freeze dryer. The purpose of freeze drying is to remove moisture content without affecting physical and chemical characteristics. The lyophilization of natural products requires special care to ensure that the integrity of the plant material is not harmed. To complete this task effectively, frozen samples are placed in a temperature-controlled chamber where they are maintained in their frozen state. The chamber is subjected to a vacuum (~100 millitorr range) and a condensing unit maintains a temperature of −40°C. Over a week, the temperature is slowly increased from subzero to room temperature. The key in this process is to increase the temperature slowly enough to not melt the plants. NCNPR's freeze drier has a programmable interface allowing the protocol to run automatically and ensure consistent treatment of samples. In one week, it can completely process about 30 to 45 samples.

Only live, clean plants free of insects, disease, molds, bryophytes, etc. are included. The collection of sample parts depends on plant habit, characteristics, and abundance. Woody species may be divided into several samples of leaves, twigs, bark, wood, roots, and fruits. Herbaceous plants, when abundant, are divided into aerial and below-ground parts; otherwise they are collected as whole-plant samples. Samples are weighed fresh, before freezing.

The dried samples are ground in a Retch SM100 grinder with a 1-mm mesh. The grinder is cleaned thoroughly between each processed sample to eliminate crossover. Grinding time is usually 10 min followed by a 20-min cleaning. Each plant sample is ground, weighed, packaged, and labeled prior to initiation of the next sample to ensure the integrity of the samples and their taxonomic data. Each sample is packaged in a 950-mL opaque amber HDPE wide-mouth jar and labeled with collection information (collector, collection number, plant part, hazard warnings, plant species, and date processed). The jarred samples are barcoded and stored in a well-ventilated, temperature- and humidity-controlled environment.

The ground plant samples are weighed (typically 7 to 20 g, depending on plant part) and extracted in a semi-automated process utilizing a Dionex ASE300 accelerated solvent extractor system. The samples are placed in stainless steel cells and sealed. Approximately 33 to 66 mL of 95% EtOH are added and the cells are placed under 1500 PSI, preheated for 5 min to 40°C and held at this temperature for 10 min. The sample extract is flushed through the cell by an equal volume of 95% EtOH and this process is repeated twice. The extracts are combined and concentrated down to the remaining organic material. A portion of the residue is resuspended in DMSO and transferred to microtiter ("daughter") plates for distribution to biologists for various screening assays. Deep well "parent" or "mother" plates are stored to replenish screening plates as needed. All plates are stored at −80°C.

It is possible to use multiple solvents to generate samples with varying solubility characteristics from the same ground plant material [sequential extraction of a single ground plant sample with solvents of increasing polarity: (1) hexane, (2) EtOAc, (3) EtOAc/MeOH, (4) MeOH/H_2O]. This sequential solvent use is a crude form of pre-fractionation of extract samples. More sophisticated protocols for fractionating extracts are often proprietary (Appleton, Buss and Butler, 2007; natural product company list; personal communications: Kip Guy, St. Jude; Ikhlas Khan, NCNPR).

Generating partially purified fractions prior to screening has several clear advantages including more accurate assessment of the potencies and selectivities of active compounds from biological results and better comparisons of activities of controls and synthetic compounds early in the screening process. Although the generation of purified fractions, including those of inactive samples, seems time consuming in its early phases, it provides a clear productivity and quality boost that encompasses all stages of screening and lead optimization. This is especially true in industrial settings where the time saving allows natural product samples to become competitive against synthetic samples in the deck.

13.3 USE OF BOTANICALS AS DISCOVERY SCREENING SOURCES

13.3.1 HIGH-THROUGHPUT SCREENS: VALIDATION FOR NATURAL PRODUCT SAMPLES

As important as validation studies for synthetic samples are in running a successful discovery program, they are even more important for running natural product samples and especially plant extracts, which are notoriously difficult to run in biochemical or enzyme-based screens. Tannins in particular are responsible for nonspecifically inhibiting activities in such screens. Other nuisance compounds include chlorophyll, melanin, lipids, and waxes.

Traditional screening programs had to be adapted to respond to issues arising from the interference of plant extracts in bioassays and new assay formats often had to be developed. Whole crude extract hits from screens were routinely subjected to de-replication studies to identify known nuisances and uninteresting compounds. De-replication schemes were developed to quickly identify extracts that appeared active in screens due to spurious interfering compounds (Van Middlesworth and Cannell, 2008; Lang et al., 2008).

It was important to eliminate these extracts from consideration so that they would not join the queue of extracts slated for isolation studies, effectively diluting the efforts and resources of isolation chemists. De-replication is de-convoluting a complex mixture of a crude extract in a timely fashion to eliminate nuisance compounds; many of these protocols are proprietary (Carter, 1998). This usually depends on separation steps, such as high performance liquid chromatography (HPLC) of the crude extract and compound identification by ultraviolet (UV) or mass spectrometry, coupled with UV or light scattering detectors (Carter, 1998). Fortunately, several commercial databases allow searches based on UV/visible spectra and molecular weight to identify natural products. Historically, crude extracts were utilized and provided many new drugs; however, today's drug discovery environment demands shorter timelines. Identification of chemotypes and liabilities much earlier in the developmental process is fueling trends such as pre-fractionating samples and developing quicker chromatography methods (Manly, Lowe and Padmanabha, et al. 2002).

Inherent fluorescence in natural product samples rules out the use of frank fluorescence as an assay readout. In the early 1990s, the development of homogeneous time-resolved fluorescence (HTRF) allowed the use of fluorescence with such samples (Kolb, Yamanaka, and Manly, 1996; Kolb and Manly, 1997; Bazin, Trinquet, and Mathis, 2002; Mathis, 1999). This technique

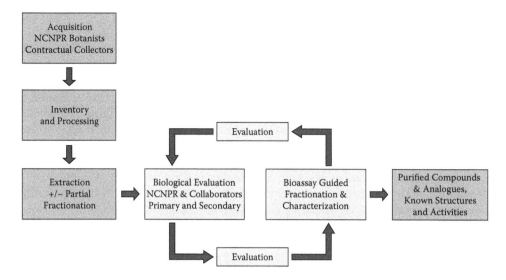

FIGURE 13.2 Flowchart of work for plant to extract through discovery studies to pure compounds with known structure and biological activities.

incorporates a time delay in reading of the signal capitalizing on the short fluorescent lifetimes of extract components and capturing only the long fluorescent lifetime of the lanthanide-labeled target complex. These assays are homogeneous, robust and enable widespread adaptation of fluorescence to analyze natural product samples.

13.3.2 VALIDATION AND PLANT EXTRACT USE

Figure 13.2 is a flowchart for a traditional screening platform utilizing botanical extracts. Previous literature outlined the validation protocols for utilizing both plant and microbiological extracts in screening assays (Manly, Lowe and Padmanabha, 2002; Appleton, Buss, and Butler, 2007). However, screen validation for plant samples is usually more difficult to achieve. Sensitive assays may require removal of tannins from crude extracts by de-fatting with a nonpolar solvent such as hexane or petroleum ether or by polyamide chromatography (Wall et al., 1996; Phillipson, Zhu, and Cai, 1998). The hexane layer (fats, waxes, chlorophylls) is discarded (Silva, Lee, and Kinghorn, 1998). Hexane extraction can be accomplished directly in daughter extract plates by adding hexane, then mixing and freezing the plates. The hexane can be removed by gently inverting frozen plates.

Another useful practice for HTS of plant extracts is to empirically determine the concentration of extract that works best for a particular screen. This is accomplished by running 3000 or 4000 crude extracts at several concentrations through the screening assay to establish a suitable concentration for the screening campaign. The dilution series data are treated to generate a normal curve by establishing the range of activity in the screen, i.e., the x axis is assay signal and the y axis represents test sample activity value relative to control. The concentration of samples yielding the tightest normal distribution will be the concentration of choice. Running the screen under these conditions ensures confidence that the outlying points are genuine actives and that the interfering components are neutralized by dilution. Curves that are shallow and wide indicate nonspecific effects of the extracts and the extracts should be further purified.

13.4 BOTANICAL DIFFERENCES

Nearly as diverse as phytochemistry are the perspectives and goals of today's botanical researchers. Whether the goal is nutraceutical, herbal compound, supplement, or drug product, many of these practitioners adapted the high-throughput technology developed by pharmaceutical companies to maximize the likelihood of identifying particular and novel activities. This section describes the evolution of and need for database management strategies for high-throughput endeavors and the best methods of incorporating natural product samples into screening campaigns. These database management approaches developed as drug discovery strategies are now used for discovering and developing botanical products that are not drugs per se.

13.4.1 PLANT EXTRACT TO PURE COMPOUNDS: ACTIVITIES, TRACKING, AND CHARACTERIZATION

All natural product samples have more pronounced tracking and purification requirements than those of synthetic and defined sources such as small molecules. This puts additional pressure on the information management capabilities traditionally maintained in the pharmaceutical industry [chemistry registration system, laboratory information management system [LIMS], and basic sample tracking]. Plant genealogy, geographic references (location, season, environmental details), ambient parameters (soil pH, salinity, marine or terrestrial habitat, etc.), plant parts, extraction methods (nature of extraction, solvent, etc.), and subsequent treatment and fractionation studies must be carefully recorded for later cross referencing to determine biological activity and for confirmation studies.

The need to maintain this data induces complex multidimensional information systems that few firms will even attempt to tackle. Clearly, this complex multidimensionality of source data exerts a profound effect on viewing and interpreting biological results correctly and efficiently. With natural products, it becomes very obvious that the traditional structure–activity relationship (SAR) rendition of results, so important to biochemists, becomes pointless due to the lack of a structure and the addition of multidimensional relationships between and within studied samples.

Furthermore, natural product research organizations are likely to include synthetic chemistry functions as part of their portfolio. Ultimately, the idea is to integrate the natural product aspects of research and its small molecule chemistry counterpart on the basis that, to some degree, synthesized molecules are derived from natural substances.

Traditionally, laboratory information systems followed two directions: screen-centric and chemistry-centric systems. Screen-centric activities focus on protocols, tests, and biological results. Chemistry-centric laboratories deal primarily with compound registration and structure manipulation. The two systems only meet when an SAR table is finally generated as a report that biochemists can study. Other fields such as gene therapy, assay development, target identification, and others usually have their own satellite computer systems, more or less integrated with the rest of the research organization, and are often treated as local rather than global research systems. The same principle applies to natural product research whenever an information system is developed or acquired to support it. Natural product research simply adds another silo of information with questionable integration into other research informatics. The basic problem is that the need for across-screen comparisons holds regardless of the sample types screened; the nature of natural products makes such cross comparisons rather difficult.

From a purely biological perspective, data analysis of small molecules and natural product screening differ only slightly. Certain screen formats such as enzyme-based screens are vulnerable to interference by extracts and can exhibit high false positive rates. The ability to "flag" troublesome extracts is vital to allow the identification of genuine active samples. Conversely, the identification of false negatives is afflicted by the very complex natures of natural products due to their inherent properties and extraction methods.

13.4.2 INFORMATICS AND DATA MANAGEMENT

"Amateurs talk about tactics, but professionals study logistics," as noted by General Robert H. Barrow, Commandant of the U.S. Marine Corps, in 1980—logistics is precisely the informatics layer that can allow both the integration of chemistry and natural products and their distribution throughout the entire drug discovery engine. Today, most organizations possess some kind of LIMS that stores all their biological results. Most research centers also have some degree of chemo-informatics system in which chemical structures are stored—a simple file system or a far more sophisticated chemical structures database cartridge technology. What is rarely recorded and/or rarely usable is what happens between the registration of a chemical structure and the recording of its final biological results. In other words, the entire logistical chain that crosses the discovery landscape is typically lacking. This problem is exacerbated by natural products since the tracing of their genealogy in combination with biological activity is paramount.

13.4.3 EVOLVING NATURE OF DRUG DISCOVERY INFORMATICS

The current lack of integration of the specialty areas within discovery is not the result of chance alone and can be traced directly to the nature of drug discovery and its evolution over time. In the early 1990s, automation and databases had not yet been implemented and biologists were usually inputting their results in Lotus 123 or later in Excel on a Windows 3.1-equipped personal computer. At the same time, the pharmaceutical world faced an ever-decreasing pipeline of new drugs and future predictions were dire. The industry had exhausted the small molecule bonanza that started in the early 1970s. Since most pharmaceutical companies had their roots in hard core chemistry, a chemistry-oriented solution was seen as a "life saver" to rescue diminishing drug research productivity. If we could no longer efficiently hit a biological target with a handful of painfully crafted molecules, maybe a very large number of industrially produced chemicals would work. "Carpet bombing" a target with millions of molecules was therefore perceived as having great potential in solving the diminishing pipeline problem.

This led to the birth in the mid 1990s of combinatorial chemistry in tandem with HTS (Pisano, 2006) and the combination was expected to produce an industrialized version of drug discovery based on a simple hits filter engine that could process hundreds of thousands of molecules against multiple targets in record time. This should have enhanced the chances to uncover hits, and therefore resupply the drying drug pipelines (see Figure 13.3). It is important to understand that the level of informatics and automation in discovery laboratories was not very advanced then and certainly could not support the huge quantities of molecules and results that HTS and combinatorial chemistry were expected to generate.

HTS became the de facto driver of automation in drug discovery. This, in turn, set into motion what would soon become the fundamental structure of informatics in drug discovery. The simple filter design, essentially a binary cluster of industrial biology and chemistry, was almost exactly reproduced from an informatics perspective (see Figure 13.4). Time was of the essence, resources were strained, and key technologies were in their infancy. These factors led to the extremely fast construction of core databases and automation engines. It was the time of rapid application development (RAD) and the only focus was making the industrial wheels of combinatorial chemistry and HTS turn and turn again. At that time, natural products and small nonindustrial laboratories were addressed as side activities and their integration was not a priority. Ultimately, drug discovery informatics delivered screen-centric solutions, and between 1994 and 1998, total HTS LIMS systems were implemented. Data flowed back and forth from chemistry to biology within the so-called new leads generation engine.

A side effect of this informatics focus on a screen-centric industrialized view of research further entrenched drug discovery, and most of life science for that matter, as a patchwork of islands (or silos) of information. This condition was enhanced by the high degree of specialization that underlines each drug research activity. As a result, the greatest challenge today is to allow researchers to build bridges between these islands. To date, information technology (IT) solutions usually remain adequate in solving problems for one department or activity, but fail to address overall research objectives, therefore further enhancing the silo effect.

Another effect of the industrialization of drug research is the difficulty of displaying and comprehending the huge quantities of results generated (Figure 13.5). This led drug discovery operations

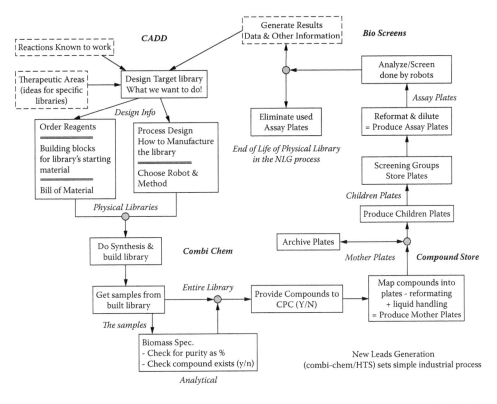

FIGURE 13.3   New Leads Generation (combinatorial chemistry and HTS) system sets simple industrial process.

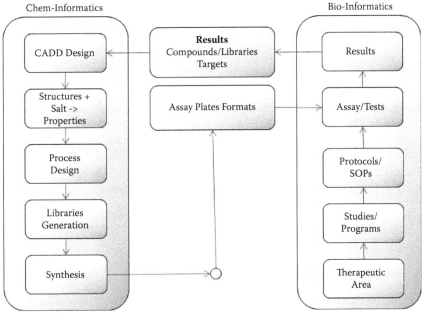

FIGURE 13.4   New Leads Generation (combinatorial chemistry and HTS) system drives bipolar information engine.

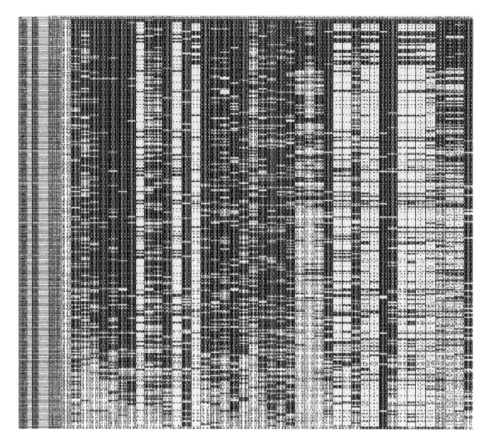

FIGURE 13.5   (See color insert following page 114.) Partial view of screen in New Leads Generation (combinatorial chemistry and HTS) generating a huge quantity of hard-to-analyze results.

to exhibit short term data myopia—past information is quickly forgotten (or untraceable) and the focus is on the latest results.

This created the problem in today's drug discovery informatics landscape. Drug research is anamorphic in nature. The inclusion of different perspectives from various research vantage points provides the broad characterization of new molecules or natural products. Again, the silo nature of discovery and its informatics components make it very difficult to build cross referencing tools or bridges. Natural products and their constant need for cross referencing and genealogy constructs are particularly afflicted by the silo dynamic. Many plants may nonspecifically interfere with assays and will be exposed by data comparison and cross referencing activities. This is especially true for biochemical or enzyme-based screens.

13.4.4   LOGISTICAL SOLUTION TO LOGISTICAL PROBLEM

General (and later President) Dwight D. Eisenhower said, "You will not find it difficult to prove that battles, campaigns, and even wars have been won or lost primarily because of logistics." Linking chemicals, plants, or other natural products to biological results to allow cross referencing and data integration occurs in the real physical world. Substances originated from a chemistry department or natural product center do not find their ways to fluorescent scanners or scintillation readers alone.

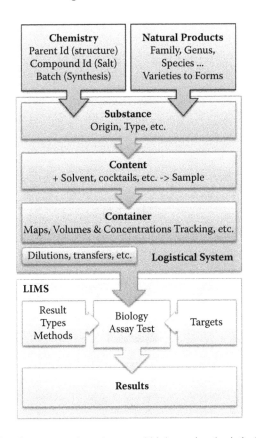

FIGURE 13.6 Merging chemistry, natural products, and biology via a logistical system.

In fact, a plethora of activities takes place between the recording of chemical structures and natural products and the results generated by biologists. This set of activities, in fact, constitutes the core of discovery logistics. Therefore, a very good starting point for integrating natural products, small molecules, and biology is designing a logistical chain that connects all the drug discovery laboratories together (Figure 13.6).

A logistical system in discovery can be described as a set of relationships of substances, contents, containers, carriers, storage units, and locations (Figure 13.7). In fact, maintaining strong logistics within discovery allows the tracing of every bit of information generated. Therefore, logistics is truly at the core of all data reduction, summary, tracking, and analysis activities within research laboratories. However, building or integrating a logistical system is a demanding, resource-consuming, and painstakingly difficult task. Meanwhile, the existing chemistry registration systems and biological LIMS continue to be used and raise the level of difficulty in retroactively inserting a logistical system into the current informatics scene. The growth of natural products research exacerbated the complexity of the task by introducing an additional layer of specific taxonomic information.

13.4.5 KEY DESIGN CHANGES FOSTERED BY NATURAL PRODUCTS

When dealing with natural products, the starting point is core definition; thus, the first information to be gathered relates to taxonomy. Taxonomic information is not proprietary and must adhere to officially sanctioned data sets. Existing species information resources such as the National Center for Biotechnology

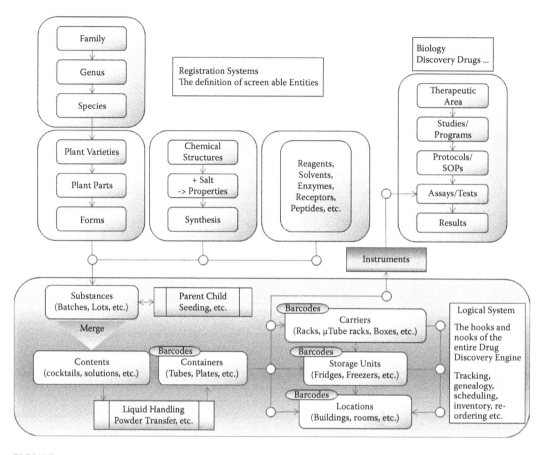

FIGURE 13.7 Logistics serves as the rails of the discovery research engine.

Information (NCBI) taxonomy databases help in that regard. However, if chemical structures are very straightforward and do not change over time, thus providing a very stable registration system, taxonomic information is different. Taxonomy relies on systematics—the study of the diversity of living organisms. Systematics is broken down into classification and nomenclature fields. Classification is the process of defining systematic groups (taxa); nomenclature is a system of allocating names to taxa.

The problem arises because the same name can be used for different organisms or different names can be applied to a single organism. This situation is the consequence of the misapplication of names generally caused by ignorance of previous studies. Only one name can be applied, and the various names created over time are recorded as synonyms. This, in turn, leads to a complex history of name changes (Sutherland et al., 1999). These name changes create a data repository's worst nightmare. Complete tracking of all objects and events in a research environment relies on proper naming conventions, and if the names have the potential to change, the entire logistical system of a research organization may be jeopardized. The drug discovery industry never had to face this situation when working with small molecules and their reliable structures, but systematics is a challenging issue when dealing with natural products.

Systematics produces database design ramifications. In designing databases, the use of surrogate keys within a third normal form is no longer optional but must be diligently observed (Figure 13.8). Another issue arising from natural product research is the need to track the genealogy of each specimen in association with its extracted samples. The genealogies of natural

Table Name	Owner
BIO_SAMPLE	dbo

Field Name	Type
SAMPLE_KEY	bigint
ENVIRONMENT_KEY	bigint
TAXONOMY_KEY	bigint
SAMPLE_ID	char
SAMPLE BATCH	char

Simple update statement on the identifier field or other data entry fields should not affect the Entity Relationships.

The use of Surrogate Keys enhances scalability & Sustainability.

From Original Sample Records

SAMPLE_KEY	ENVIRONMENT_KEY	TAXONOMY_KEY	SAMPLE_ID	SAMPLE LOT
1	1	1	XYZ012	78FR56G43
2	1	2	XYZ034	7Y912GT3
3	1	3	XYZ055	56HJ1
4	1	3	XYZ055	81DR45L91
5	1	4	XYZ134	T67E21O87

All Primary Keys & Foreign Keys Relationships are **Preserved**

To New Modified Sample Records

SAMPLE_KEY	ENVIRONMENT_KEY	TAXONOMY_KEY	SAMPLE_ID	SAMPLE LOT
1	1	1	A00001	A00001–001
2	1	2	A00002	A00002–001
3	1	3	A00003	A00003–001
4	1	3	A00003	A00003–002
5	1	4	A00004	A00004–001

RESULT_KEY	SAMPLE_KEY	RESULT_TYPE	MY_RESULT
67564	4	P-INH	45.67

FIGURE 13.8 The use of surrogate keys in designing databases.

products extend beyond simple parent–child relationships and involve the management of items such as seeds and seed collections. Of course, mapping the logistics of natural products from an informatics perspective can yield very complex designs. Nonetheless, the integration and management of natural products and small molecules require the implementation of a logistical system for any laboratory.

13.5 BOTANICAL PRODUCTS IN THE UNITED STATES

13.5.1 DSHEA: Framework for Regulating Dietary Supplements and Definitions of Foods, Dietary Supplements, and Drug Products

The United States Congress passed the Dietary Supplement Health and Education Act (DSHEA) in 1994. It describes a dietary supplement as a product (other than tobacco) that (1) is intended to supplement the diet, (2) contains one or more dietary ingredients (including vitamins, minerals, herbs, other botanicals, amino acids, and other substances) or their constituents, (3) is intended to be taken orally as a pill, capsule, tablet, or liquid, and (4) is labeled as a dietary supplement on the front panel. This means that dietary supplements are regulated as foods within the meaning of the act and such products do not require pre-market notification or registration except for new dietary ingredients (NDIs).

In terms of botanical products, where does a dietary supplement fit into the spectrum of use in the United States? See Figure 13.9. The same botanical parent can yield derivative products that can be marketed and therefore regulated in many categories (Figure 13.10). In the United States, botanicals can be regulated as (1) foods: conventional foods, functional foods, spices, dietary supplements; (2) drugs:

Dietary Supplement

Food   Drug

Where do you want to go with your product?

FIGURE 13.9   Spectrum of botanical products.

over-the counter (OTC) and prescription; (3) biologicals such as vaccines used to ameliorate allergies; (4) cosmetics and shampoos; and (5) devices such as dental alginates, poultices, and adhesives.

The way botanicals and their derivative products are marketed depends on both their intended uses and their safety categorization. Figure 13.11 illustrates the intended use of a substance. Its intended use directly influences the route of administration and level of regulation. Unless a substance is "generally regarded as safe" (GRAS) by qualified experts (and, often, by history of human use), any substance added to food must attain pre-market approval of the U.S. Food and Drug Administration (FDA; http://www.cfsan.fda.gov/~dms/opa-noti.html). Pre-market approval is attained when a company issues a GRAS notice with which the FDA may or may not concur. After a

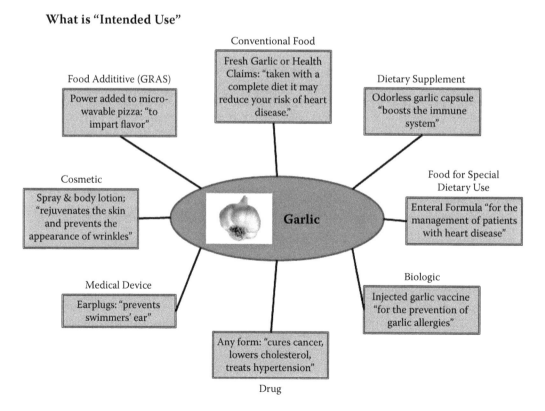

What is "Intended Use"

Conventional Food
Fresh Garlic or Health Claims: "taken with a complete diet it may reduce your risk of heart disease."

Food Addititive (GRAS)
Power added to micro-wavable pizza: "to impart flavor"

Dietary Supplement
Odorless garlic capsule "boosts the immune system"

Cosmetic
Spray & body lotion; "rejuvenates the skin and prevents the appearance of wrinkles"

Garlic

Food for Special Dietary Use
Enteral Formula "for the management of patients with heart disease"

Medical Device
Earplugs: "prevents swimmers' ear"

Biologic
Injected garlic vaccine "for the prevention of garlic allergies"

Any form: "cures cancer, lowers cholesterol, treats hypertension"

Drug

FIGURE 13.10   Marketing a botanical product can span the available spectrum of product types and depends on the intended use. (From Hoffman, F.A. and Garvey, T.N: *Textbook of Legal Medicine*, 5th ed. 2001.)

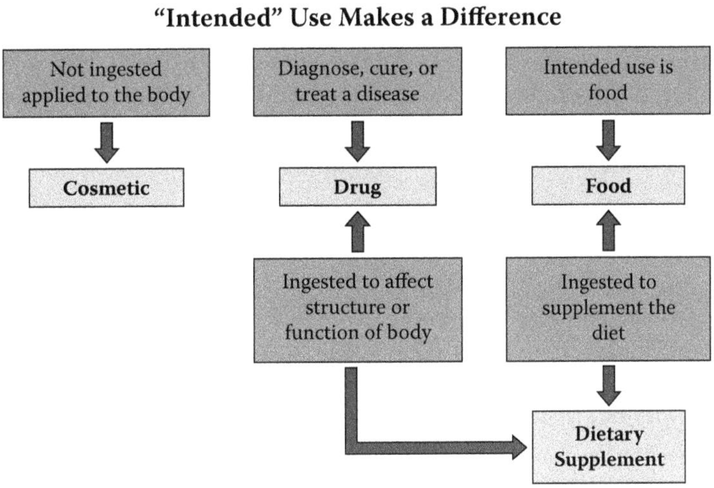

"Intended" Use Makes a Difference

FIGURE 13.11 In terms of approval, marketing, and regulation of a botanical product, the details of application, ingestion and expected benefits dictate possible niches of product development.

substance is GRAS, it may be used as a component in dietary supplements or as a food additive. A proposal to use a substance or an extract as a drug will result in a health benefit versus risk evaluation by the FDA. The burden of proof as to the safety of a new drug is on the submitter since a cautionary principle is de facto applied here.

Are expectations of quality and level of standardization different for marketing a botanical product for various categories of uses? Drugs, dietary supplements, and food products are required to meet Good Manufacturing Practices (GMPs), but dietary supplements and food products are required only to be relatively free of contaminants and adulterants. Drugs must also demonstrate consistency, potency, and purity.

While drug manufacturers can claim that their products will diagnose, cure, mitigate, treat, or prevent a disease, dietary supplement manufacturers cannot legally make such claims. What claims are they permitted to make? A dietary supplement, food product, or food additive may make one of three types of claims, (1) a health claim: "diets high in calcium may reduce the risk of osteoporosis," (2) a nutrient content claim: "a good source of…," or (3) a structure–function claim: "calcium builds strong bones" and "antioxidants maintain cell integrity" (http:cfsan.fda.gov/~dms/hclaims.html).

Three types of health claims relate to dietary supplements: (1) health claims authorized by the Nutrition Labeling and Education Act of 1990 (NLEA) stating that a dietary supplement must meet a "significant scientific agreement standard," (b) health claims based on an authoritative statement issued by a scientific body of the U.S. government or the National Academy of Sciences as provided by the FDA Modernization Act of 1997 (FDSMA), or (3) qualified health claims that contain qualifying language to reflect level of scientific support and are not misleading to consumers. FDA guidance on these claim definitions and their use can be reviewed at http://www.cfsan.fda.gov/~dms/hclaims.html.

The FDA requires the manufacturer of a botanical product designed for human use to submit a new dietary ingredient (NDI) notification for any new ingredient that was not marketed in the United States before October 15, 1994 (DSHEA). Notifications should be submitted to the FDA 75 days prior to the marketing date and require information establishing a reasonable expectation of safety for products containing the NDI. The FDA does not approve or disapprove an NDI; it may post objections. Examples of substances disallowed for use by the FDA include androstenedione, Ephedra, steroidal precursor substances, and aristolochic acid.

13.5.2 BOTANICAL DRUG PRODUCTS

The FDA published its *Guidelines for Botanical Drug Products* on June 9, 2006 (www.fda.gov/cder/guidance). Botanical drugs can be developed in the United States through two mechanisms: (1) a new drug application (NDA) submitted to the FDA to cover an OTC drug or a prescription drug, or (2) a monograph for an OTC product. The following list summarizes the guidelines:

- Identification of active constituents is not essential.
- Purification is not required.
- Chemistry, manufacturing and control (CMC) extend to raw materials.
- Nonclinical evaluations may be reduced.
- Same level of clinical efficacy and safety requirements for standard drugs apply.

Generally, the FDA utilizes historical safety information to expedite early stage testing and evaluation of botanical products.

What are some considerations influencing a botanical investigational new drug (IND) application? A botanical IND application should be filed if the proposed product contains crude plant extracts, partially purified plant extract fractions, or a combination of highly purified compounds from different plants. A botanical IND application should also be filed if a proposed product contains a single herb, multiple herbs, or a botanical alone or with additional active components such as vitamins, minerals, or animal parts.

A botanical review team (BRT) from the Center for Drug Evaluation and Research (CDER) will provide guidance to applicants throughout the pre- and post-application processes. The CDER website states, "The BRT participates in all phases of review, meetings and decision-making processes for all botanical pre-Investigational New Drug (IND) applications, INDs, and New Drug Applications (NDAs)" (http://www.fda.gov/Cder/). The review process is clearly detailed in the manual on the CDER website, http://www.fda.gov/cder/regulatory/default.htm#CDER%20Policies%20&%20Procedures.

If the plant(s) from which a proposed botanical product is derived have been in prior human use, the BRT review of the IND relies heavily on data from the previous experiences. BRT reviewers compare the doses and durations of the botanical product cited in the IND with previous human uses. They determine whether a product or trial appears reasonably safe and whether previous data reveal any side effects or potential safety issues. BRT reviewers expect IND applications to address the relationship between prior human use and the proposed indications.

The BRT will review clinical data in relation to the scope of the proposed clinical studies: (1) preliminary studies (Phases I and II; marketed product versus not-marketed products or products marketed with safety concerns), (2) expanded studies (Phase III), end-of-phase II meeting. The BRT will provide feedback and guidance at the pre-NDA meeting, then review the data in the NDA in relationship to safety, efficacy, quality, and therapeutic consistency.

Initial botanical IND submissions exhibit some common issues including (1) incomplete information about raw materials such as scientific name or botanical parts not specified and (2) safety information gaps regarding botanical raw materials and products such as not disclosing yields of extracts from raw materials not known or providing unreasonably high doses (in terms of weight of raw herb). Additionally, if the information concerning prior use of the material in man is vague or contradictory, the duration of clinical trials will be longer, as is the case with small molecules previously untested in humans.

Chen et al. (2008) reviewed the numbers and types of botanical IND applications received by the FDA from 1999 to 2007. A total of 225 botanical IND applications covering 12 therapeutic indications were received by the agency. The most common indication/target is oncology, followed by rheumatological analgesic and endocrine metabolic indications. Chen also noted that the proposed studies of 13% of the INDs submitted were placed on clinical hold due to unresolved safety concerns.

13.6 SCIENTIFIC ISSUES IMPACTING REGULATION OF BOTANICAL PRODUCTS

As the popularity of new botanical products increases, they are used in concurrence with modern and traditional medicine—a practice known as complementary and alternative medicine (CAM). A survey in the United States shows an increase of CAM use from 34% to 42% of the population in 1997 (Eisenberg et al., 1998). The same report estimated Americans spent some $5.1 billion on herbal medicines (in a $20 billion global market) and another $27 billion on alternative medical treatments (Eisenberg et al., 1998; Dev, 1997, 1999).

13.6.1 GOOD AGRICULTURAL PRACTICES AND GOOD MANUFACTURING PRACTICES: IDENTIFICATION, SUBSTITUTION, PURITY, AND QUALITY ISSUES

Khan (2006) wrote an excellent review citing multiple sources for the complexity of addressing quality for botanical products, including species differences, organ specificities, diurnal and seasonal variations, environment, field collection and cultivation methods, wild collection, contamination, substitution, adulteration, and processing and manufacturing practices (Reichling and Saller, 1998; Simon, 1999; McChesney, 1999; Flaster, 1999; Busse, 1999). The diverse development process undertaken by (usually) multiple parties reinforces the importance of oversight and compliance to ensure both the safety and the consistency of the content and efficacy of a final botanical product. The Good Manufacturing Practice (GMP) and Good Agricultural Practice (GAP) guidelines are outlined at various websites (http://www.ienica.net/policy/goodagpracherbs.htm; http://www. actahort.org/members/showpdf?booknrarnr=249_16; http://www.inaro.de/Deutsch/ROHSTOFF/ industrie/HEILPFL/GAPengl.htm; http://whqlibdoc.who.int/publications/2003/9241546271.pdf).

13.6.2 PRODUCT EFFICACY, ACTIVE COMPONENT, AND STANDARDIZATION ISSUES

It is often the case that a purported active ingredient for a botanical product is not fully characterized. This leads to the use of surrogate "marker" compounds as de facto identification criteria for these products. Unfortunately, this often leads to the selection of non-unique compounds that do not represent the true activity or potency of the traditional therapy. The commonality of some of these markers can lead to potential adulteration or "spiking" of the product.

13.6.3 NEW DIETARY INGREDIENT (NDI) SAFETY ISSUES

NDI status is granted by the BRT (Section 6.2, Botanical Drug Products) when the team does not contest the application submitted by the advocating commercial or private interest. The application must contain evidence ensuring the safety of the NDI. Typically two thirds of the applications are rejected due to unresolved safety and identity issues. The burden of proof of reasonable safety of the NDI falls directly on the applicant. Unfortunately, applicants typically provide only anecdotal safety information. Typically, it is based on folkloric or traditional uses and does not include appropriate or relevant clinical data.

Unfortunately, most applications do not include sufficient supporting information regarding the proper identification of the specified material, including Latin binomial nomenclature, plant parts utilized, processing specifications, extraction procedure (if applicable), or method of identification. The failure to provide this basic information is a common oversight on the part of applicants. The industry's counterpoint is that some of the required information is proprietary and that the FDA does not provide sufficient protection of NDI applications that would prevent a competitor from utilizing the NDI after the initial applicant invests all the work and expense.

13.6.4 INTERACTIONS OF DRUGS AND OTHER DIETARY INGREDIENTS

Marketed dietary supplements are not required to contain identified active components; therefore they often exert their pharmaceutical influence via unknown mechanisms and it is difficult to predict potential adverse events. After a supplement or dietary ingredient has been consumed for a long enough time by a large number of people, a volume of empirical evidence about possible interactions and side effects begins to build. This is why practitioners of herbal medicine, regulatory agencies, and informed lay people favor botanicals with long records of use in humans. If a botanical extract or herb has a long history of safe ingestion in humans, it has likely been consumed in many contexts.

Industry-sponsored and advocacy-related adverse event reporting (AER) agencies address safety-related events to the FDA for dietary supplements (www.SafetyCall.com). These groups submit AERs to the Food and Drug Administration's Center for Food Safety and Applied Nutrition's Adverse Event Reporting System (CAERS) at (http://www.fda.gov/opacom/backgrounders/problem.html).

13.7 SUMMARY

This chapter offered a review of some of the unique aspects of utilizing plants as sources for screening samples for drug discovery programs. Botanical products from such programs can be marketed as foods, dietary supplements, and drugs.

ACKNOWLEDGMENTS

The authors wish to acknowledge the critical review of the manuscript by Drs. Melissa Jacob and Larry Walker and would also like to acknowledge other members of the NCNPR natural products community including Steve Duke, Aruna Weerasooriya, Maria Bennett, Mary Heather Martin, Derek Oglesby, and Ed Lowe. This research is funded in part by the U.S. Department of Agriculture's Agricultural Research Service Specific Cooperative Agreement 58-6408-2-0009.

REFERENCES

Appleton, D.R., Buss, A.D., and Butler, M.S. 2007. A simple method for high-throughput extract prefractionation for biological screening. *Nat. Prod. Drug Disc.* 61, 327–331.

Bazin, H., Trinquet, E., and Mathis, G. 2002 Time resolved amplification of cryptate emission: A versatile technology to trace biomolecular interactions. *J. Biotechnol.* 82, 233–250.

Bouley, J. 2007 A natural fit. *Drug Disc. News*, Internet version.

Busse, W. 1999. The processing of botanicals. In Eskinazi, D.P. (Ed.). *Botanical Medicine: Efficacy, Quality Assurance and Regulation.* Larchmont, NY: Mary Ann Liebert, pp. 143–145.

Carter, G.T. 1998. LC/MS and MS/MS procedures to facilitate dereplication and structure determination of natural products. In *Natural Products Drug Discovery II: New Technologies to Increase Efficiency and Speed.* Sapienza, D.M. and Savage, L.M., Eds. Southborough, MA: IBC Communications, pp. 3–19.

Chen, S.T. et al. 2008. New therapies from old medicines. *Nat. Biotechnol.* 26, 1077–1083.

Dev, S. 1997. Ethnotherapeutics and modern drug development: the potential of Ayurveda. *Curr. Sci.* 73, 909–928.

Dev, S. 1999. Ancient–modern concordance in Ayurvedic plants: some examples. *Environ. Health Persp.* 107, 783–789.

Eisenberg, D.M. et al. 1998. Trends of alternative medicine use in the United States, 1990–1997: results of a follow-up national survey. *JAMA* 280, 1569–1575.

Feng, B.Y. et al. 2005. High-throughput assays for promiscuous inhibitors. *Nat. Chem. Biol.* 1, 146–148.

Flaster, T. 1999. Shipping, handling, receipt, and short-term storage of raw plant materials. In Eskinazi, D. et al., Eds. *Botanical Medicine: Efficacy, Quality Assurance and Regulation.* Larchmont, NY: Mary Ann Liebert, pp. 139–142.

Harvey, A. 2000. Strategies for discovering drugs from previously unexplored natural products. *Drug Disc. Today* 5, 294–300.

Hoffman, F.A. and Garvey, T., IV. *Textbook of Legal Medicine*, 5th ed. 2001.

Kahn, I.A. 2006. Issues related to botanicals. *Life Sci.* 78(18): 2033–2038.

Kolb, J.M. and Manly, S.P. 1997. Adaptation of time-resolved fluorescence to homogeneous screening formats. In Janzen, B., Ed. *High-Throughput Screening: The Discovery of Bioactive Substances*, New York: Marcel Dekker, pp. 377–388.

Kolb, J.M., Yamanaka, G.Y., and Manly, S.P. 1996. Use of novel homogeneous fluorescent technology in high-throughput screening. *J. Biomol. Screen.* 1, 203–210.

Kugawa, F., Watanabe, M., and Tamanoi, F. 2007. Chemical biology/chemical genetics/chemical genomics: importance of chemical library. *Chem-Bio Inform.* J. 7, 49–68.

Lang, G. et al. 2008. Evolving trends in the dereplication of natural product extracts: New methodology for rapid, small-scale investigation of natural product extracts. *J. Nat. Prod.* 71, 1595–1599.

Manly, S.P., Lowe, S., and Padmanabha, R. 2002. Natural products or not? How to screen for natural products in the emerging HTS paradigm. In Janzen, B., Ed. *High-Throughput Screening: Methods and Protocols.* Totowa, NJ: Humana Press, pp. 153–168.

Mathis, G. 1999. HTRF® technology. *J. Biomol. Screen.* 4, 309–313.

McChesney, J.D. 1999. Quality of botanical preparations: environmental issues and methodology for detecting environmental contaminants. In Eskinazi, D. et al., Eds. *Botanical Medicine: Efficacy, Quality Assurance and Regulation.* Larchmont, NY: Mary Ann Liebert, pp. 127–131.

McChesney, J.D., Venkataraman, S.K., and Henri, J.T. 2007. Plant natural products: back to the future or into extinction? *Phytochemistry* 68, 2015–2022.

Newman, D.J. and Cragg, G.M. 2007. Natural products as sources of new drugs over the last 25 years. *J. Nat. Prod.* 70, 461–477.

Padmanabha, R., Cook, L.S., and Manly, S.P. 1996. Use of equilibrium dialysis to estimate sizes of active materials in natural product extracts. *J. Biomol. Screen.* 13, 131–133.

Phillipson, J., Zhu, M., and Cai, Y. 1998. Biological testing of plant extracts: should polyphenols be removed? *Polyphen. Act.* 18, 22–25.

Pisano, G.P. 2006. Can science be a business? *Harvard Bus. Rev.* October, 114–125.

Reichling, J. and Saller, R. 1998. Quality control in the manufacturing of modern herbal remedies. *Q. Rev. Nat. Med.* Spring, 21–28.

Rouhi, A.M. 2003. Natural products redux *Chem. Eng. News* 81(41), 77–107.

Shanmugam, S., Manikandan, S., and Rajendran, K. 2009. Ethnomedicinal survey of medicinal plants used for treatment of diabetes and jaundice among villagers of Sivagangai District, Tamil Nadu. *Ethnobot. Leaf.* 13, 189–194.

Silva, G.L., Lee, I.S., and Kinghorn, A.D. 1998. Special problems with the extraction of plants. In *Natural Products Isolation*, Vol. 4, Methods in Biotechnology, Cannell, R.J.P., Ed. Totowa, NJ: Humana Press, pp. 343–363.

Simon, J.E. 1999. Domestication and production considerations in quality control of botanicals. In Eskinazi, D. et al., Eds. *Botanical Medicine: Efficacy, Quality Assurance and Regulation.* Larchmont, NY: Mary Ann Liebert, pp. 133–137.

Sutherland, I. et al. 1999. LITCHI: knowledge integrity testing for taxonomic databases. Eleventh International Scientific and Statistical Database Management Conference.

Van Middlesworth, F. and Cannell, R.J.P. 2008. Dereplication and partial identification of natural products. In *Natural Products Isolation*, Vol. 4, Methods in Biotechnology, Cannell, R.J.P., Ed. Totowa, NJ: Humana Press, pp. 279–327.

Wall, M.E. et al. 1996. Effects of tannins on screening of plant extracts for enzyme inhibitory activity and techniques for their removal. *Phytomedicine* 3, 281–285.

WEB SITE REFERENCES

Medicinal plant use: http://www.siu.edu/~ebl/leaflets/juand9.htm

NIH Roadmap Initiative: http://nihroadmap.nih.gov/overview.asp

Companies selling collections of natural product compounds: http://www.rechemical.com/natural-compounds

Companies with unique platforms for natural products:

WellGen: http://www.wellgen.com/products/php

Cetek: http://www.cetek.com/drug_disc_plat.html

AnalytiCon Discovery: www.ac-discovery.com & BioPlanta, www.bioplanta-leipzig.de

NCI Natural Products Repository: http://www.dtp.nci.nih.gov/branches/npb/repository.html

DSHEA: http://www.fda.gov/opacom/laws/dshea.html#sec3

GRAS notification: http://www.cfsan.fda.gov/~dms/opa-noti.html

Dietary supplement health claims guidance: http://www.cfsan.fda.gov/~dms/opa-noti.html

Botanical drug product guidance: www.fda.gov/cder/guidance

CDER: http://www.fda.gov/Cder/

To report safety-related adverse events for dietary supplements to FDA: www.SafetyCall.com

FDA's Center for Food Safety and Applied Nutrition's Adverse Event Reporting System (CAERS): http://www.fda.gov/opacom/backgrounders/problem.html

Good Manufacturing Practices and Good Agricultural Practices: http://www.ienica.net/policy/goodagprac-herbs.htm; http://www.actahort.org/members/showpdf?booknrarnr=249_16; http://www.inaro.de/Deutsch/ROHSTOFF/industrie/HEILPFL/GAPengl.htm

http://whqlibdoc.who.int/publications/2003/9241546271.pdf

14 Screening Informatics

Stephan C. Schürer and Nicholas F. Tsinoremas

CONTENTS

14.1 INTRODUCTION

High-throughput screening (HTS) and ultra-high-throughput screening (uHTS) have evolved into industrialized processes that are fundamental to identifying novel starting points for the development of drugs and chemical probes. New technologies in robotics, liquid handling, sensitive detectors, material science, software, and information technology have driven continuous miniaturization and improvements in efficiency, precision, and ultimately throughput. It is now possible to routinely screen several hundred thousand to a million samples.

Our increasing understanding of human biology and the natures of many diseases as a result of advances in human genetics, functional genomics, and molecular biology led to the discovery of many novel molecular targets and pathways for therapeutic intervention that in many cases are amenable to HTS or uHTS approaches (Hertzberg and Pope, 2000). Conversely, enormous growth in parallel synthesis methodologies for the production of chemical libraries and technologies for high-throughput purification and analytics have made available more than ten million small molecule compounds that are commercialized for biological screening.

The advances in parallel synthetic methodology and high-throughput screening capacity have reinforced themselves in an environment where the pharmaceutical industry is under pressure by declining numbers of approved drugs in combination with patent expirations and, of course, the quest to develop drugs in novel and challenging therapeutic areas and individualized medicines (Gribbon and Sewing, 2005; Posner, 2005). A recent case study of the impacts of high-throughput technologies at Bristol Meyers Squibb provides some insight (Houston et al., 2008); two other insightful articles review the evolution of HTS and high-throughput drug discovery at Pfizer (Gribbon and Sewing, 2005; Pereira and Williams, 2007). Bender et al. (2008) published an analysis of which HTS campaigns are more likely to generate lead optimization projects as a function of target classes, assay technologies, assay types, library formats, and other factors at Novartis.

The required capital and operational costs for an HTS facility including robotics, compound libraries, and informatics have largely limited access to HTS resources to the pharmaceutical and biotechnology industries until recently. The Molecular Libraries Initiative of the National Institutes of Health (NIH; http://mli.nih.gov/mli/) has made screening resources available to academic research (Austin et al., 2004; Lazo, 2006). The screening data endpoints generated through NIH funding are in the public domain and accessible via PubChem (http://pubchem.ncbi.nlm.nih.gov/; Baker, 2006)—a fast evolving and growing chemical genomics database closely integrated with various NCBI databases and informatics tools (http://www.ncbi.nlm.nih.gov/).

The goal of an HTS campaign is to identify tractable chemical starting points for drug discovery programs for the development of small molecule chemical probes. Figure 14.1 is a simplified illustration of the HTS-facilitated lead discovery process. It begins with screening a diverse or focused set of commercially available or proprietary compounds to identify screening hits that are then confirmed by replication or concentration response assays. Appropriate secondary assays are then used to eliminate undesired hits, for example, assay- or detection-specific artifacts, undesired mechanisms, lack of specificity, etc.

The design of a screening campaign includes the order in which confirmatory, secondary, and concentration response screening are carried out in conjunction with various intermediate data analysis and prioritization steps with the goal to maximize efficiency to identify suitable lead series of interest. While it is common practice to run the initial primary screen at a single concentration, a recent innovative method of quantitative high-throughput screening (qHTS; Inglese et al., 2006) in which each compound is screened at multiple concentrations was developed by NIH at the Chemical Genomics Center (NCGC). One advantage of this approach is that structure–activity series can be identified directly without the need for confirmatory and concentration response follow-ups. Based on identifying chemical series suitable for further optimization, Figure 14.1 also illustrates where in the process structure the clustering of hit compounds and structure–activity

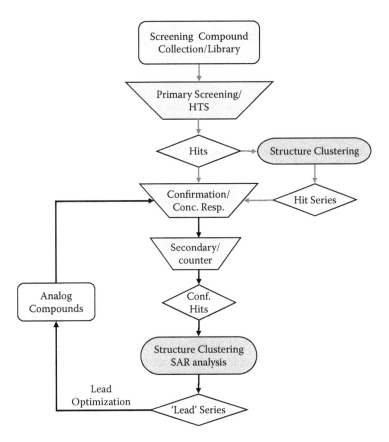

FIGURE 14.1 Lead discovery process facilitated by HTS.

relationship (SAR) analysis are carried out. For example, hit selection by a simple threshold-based activity cut-off is less efficient than considering chemical series in the primary data analysis.

14.2 SCREENING INFORMATICS OVERVIEW

Every step of the HTS-driven lead discovery process outlined in Figure 14.1 depends on informatics. The required HTS informatics tools and processes can be categorized as (1) operational/transactional informatics and (2) data analysis informatics classified as statistical data analysis and discovery cheminformatics. In practice these functional categories are tightly integrated. Operational components include all aspects of HTS data processing: (1) capturing the chemical structures of screening libraries, (2) inventorying and tracking associated samples and plates, (3) processing raw screening data and associating structures with calculated screening endpoints, and (4) software integration of these informatics components and data types with HTS instrumentation to optimize plate and sample handling, automatically monitor screening progress, and process reader output data.

Based on the quantity of data and the complexity of the process, an enterprise operational informatics system must be robust and scalable. It should also integrate with analytical instrumentation such as high performance liquid chromatography (HPLC) MS to verify compound identity. A recent review of HTS informatics provides a compact overview of the key informatics elements of HTS operations and data handling (Ling, 2008).

Statistical HTS data analysis and discovery cheminformatics are integral components of an HTS-driven lead discovery process. Various statistical and visualization techniques are routinely used to validate and assure data quality and screening efficiency in HTS campaigns, often without considering chemical structures. Cheminformatics methods including HTS data modeling, SAR analysis, and visualization of results are required to explore screening endpoints in the context of chemical structures in order to identify the best chemical series from the screening campaign and effectively follow-up on these series.

Transactional screening informatics, statistical data analysis, and cheminformatics as discussed in this chapter are important components of global discovery informatics—an increasingly important component of the drug (or chemical probe) discovery pipeline—with the goal of integrating and mining the heterogeneous data related to drug discovery in a distributed environment (Augen, 2002; Claus and Underwood, 2002).

This chapter discusses in more detail the operational informatics systems and their integration with compound management and HTS instrumentation. It then covers various aspects of screening data analysis such as statistical methods to define hits, quality considerations, reporting, visualization, and cheminformatics-driven analysis and mining of HTS data. We start from the assumption that the HTS assay has been optimized and miniaturized on an automation platform and that the assay has been validated to deliver robust and reproducible high quality results—the prerequisite of a successful HTS campaign.

From a screening informatics perspective, we should emphasize that established processes, procedures, and business rules are as important for a successful operation as the underlying informatics software, database, and hardware components. We assume the perspective of an end user of screening results with the goal of acquiring and using high quality data in the lead discovery process. We describe the object and data relationships in a screening informatics environment to allow a screener, informatics implementation manager, medicinal chemist, and scientific data analyst to conceptualize, design, or evaluate a screening informatics solution in the context of specialized individual and organizational requirements. A successful informatics environment must also maintain and organize data over time and across discovery projects while allowing easy data access to scientists of different disciplines.

Figure 14.2 illustrates the main transactional and data analysis components required for HTS. Prior to a screening campaign, compound libraries are entered into the operational informatics system, which includes compound registration and inventorying all samples and plates—critical tasks of compound management. Integration of the operational informatics system with the HTS laboratory involves synchronization of the sample and plate database with the physical inventory on the compound management instrumentation to enable real-time sample tracking and conversely real-time access of screening readouts by the operational assay data system to process and monitor screening results when generated.

Similarly the plate and sample logistics system can be integrated with analytical instrumentation to verify structures of screening hits or identified series (not shown). The data analysis environment includes tools for statistical HTS data analysis and visualization that should be integrated with the (operational) assay data system to facilitate interactive data processing and analysis and quality control. The data analysis environment also includes cheminformatics. The sample-based organization of data as generated in the screening operation is not suitable for the analysis of screening results by chemical structure or across different assays and endpoints. It is therefore most practical to reorganize and aggregate the screening endpoints into a compound-centric data warehouse. Results reorganized in such a format may be reported conveniently and queried by chemical structure or screening endpoints. Data analysis in the context of chemical structures (SAR analysis, HTS data modeling, visualization) can be applied to an individual screening campaign, but cheminformatics methods are particularly useful to analyze data across various campaigns and screening endpoints. To allow efficient data analysis, it is most practical to integrate the cheminformatics systems via the compound-centric assay data warehouse that holds the aggregated assay data endpoints.

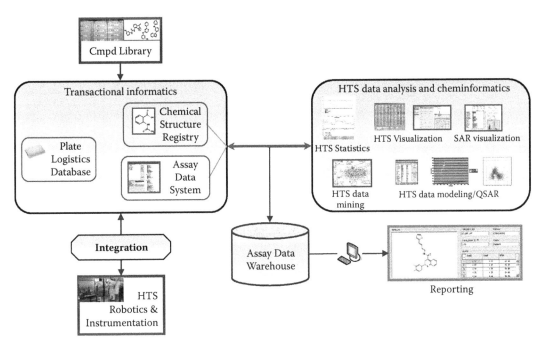

FIGURE 14.2   (See color insert following page 114.) Components of transactional and data analysis (statistical and cheminformatics) environment in the context of HTS campaign.

14.3   OPERATIONAL SCREENING INFORMATICS

14.3.1   COMPONENTS OF THE OPERATIONAL SYSTEM

Operational enterprise HTS informatics systems enable automated HTS operations by tracking objects, data, transactions, and relationships during a campaign and associate the chemical structures and screening data endpoints without error. We differentiate the operational screening informatics system from a laboratory information management system (LIMS) that addresses the complete laboratory process lifecycle to improve business performance. The informatics infrastructure described here is specialized for processing uHTS data on small molecules; it does not necessarily track pre-screening efforts of assay development, optimization, or miniaturization.

Key operational software components include a chemical compound registration system (chemical structure database), a plate and sample (container) tracking system (plate logistics), and a system to track and process all assay and associated screening data. All data are stored in a relational database that must be robust and scalable—Oracle is the industry standard. Although the compound registration, plate logistics, and assay data management systems perform distinct functions, they must be closely integrated. An enterprise operational informatics solution also includes software to integrate the plate and sample logistics database with HTS and analytical instrumentation. Operating procedures and business rules are also critical components of a functional informatics environment. Figure 14.3 illustrates the components, object relationships, and data flow of an operational screening informatics environment in the context of screening instrumentation and data warehousing.

New chemical libraries and individual compounds not already tracked in the system are entered into the chemical registration system for assignment of unique IDs—usually after pre-processing to assure all compound files conform to the business rules of the organization. The source plate and sample files are associated with the structure IDs assigned by the chemical registration system and loaded into a plate logistic system that tracks all physical transactions involving plates and samples.

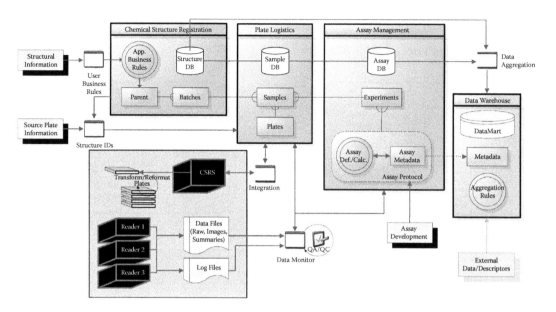

FIGURE 14.3 Operational informatics components, object relationships, and data flow.

All well samples remain associated with the structure IDs and the real molecular weight and other data (such as initial purity) of the original compounds.

The physical inventory of the plates on an HTS robotics system must be synchronized with the plate logistics system to ensure that all physical transactions are tracked and any transformation requests can be directed to the instrumentation via the plate logistics software. Any analytical instrumentation should also be integrated here, because analytical data relates to specific samples in specific plates. The HTS plate reader output files are loaded into the assay data system via software that integrates the HTS reader instrumentation with the operational assay data management system.

This system manages all data related to assays including description and metadata, variables, calculation instructions, and data objects required to process raw reader files into the screening endpoints for which the assay was designed. The assay data management system should be integrated with statistics and visualization software to monitor stability and data quality over a screening run and facilitate interactive post-processing of results by the screener. Screening endpoints are aggregated by chemical structure, assay metadata, and aggregation rules into a data warehouse where they can be readily queried and analyzed by the organization's user community. The warehouse and/or multiple data marts also allow integrating external data.

14.3.2 CHEMICAL COMPOUND REGISTRY AND BUSINESS RULES

14.3.2.1 Function of Compound Registration System

The chemical structure database and registration procedure play dual roles in an enterprise screening informatics environment. These components are essential to the transactional HTS informatics system and serve as the basis for cheminformatics data analysis. While capturing the chemical structures for association with the final screening endpoints, this system also captures and applies the organizational chemical business rules to validate and standardize chemical structures and canonicalize their representations. Structural representation is a critical prerequisite for any subsequent cheminformatics analysis.

The chemical business rules of a compound registration solution define the individual identities of chemical structure representations in order to meaningfully assign a unique corporate ID to each unique structure. After applying these rules, standardized parent structures are stored in a main structure table. The (structure) records from the original compound library data file entering the system are saved into a batch table on which each batch entry is associated with its unique parent structure. All data relating to the physical compound sample remain uniquely associated with this batch entry and thus are also associated with one unique parent structure.

14.3.2.2 Chemical Structure Representation and Standardization

Currently no universally accepted validation and standardization protocols exist for chemical structures and a number of factors should be considered before selecting or building a compound registry. Two-dimensional chemical structure representations are almost exclusively used to communicate and depict structures in a screening and medicinal chemistry environment. Chemical structure file formats differ conceptually by including two-dimensional (2D) coordinates such as v2000 or v3000 SDFile formats or storing structure connectivities without coordinates such as the SMILES format (Weininger, 1988). InChI (IUPAC International Chemical Identifier) was developed by the International Union of Pure and Applied Chemistry (IUPAC) and the National Institute for Standards and Technology (NIST) to identify chemical structures as unique strings of characters (2005; http://wwmm.ch.cam.ac.uk/inchifaq/). InChI is an open format and software to generate InChI codes is freely available (http://www.iupac.org/inchi/).

Representing chemical structures including coordinates allows capture of specific orientations or conformations. More compact connectivity-based formats rely on software to depict structures accurately and esthetically for viewing and reporting. While 2D depiction has dramatically improved, it can still be challenging to depict structures with complex ring systems or transition metal complexes with multiple components in a familiar and esthetically pleasing way for chemists. It is also important to remember that aromaticity in the SDFile format is encoded by bond definitions, but in SMILES by atom definitions. This may result in different outcomes of certain substructure queries, depending on the query language and canonicalization algorithm.

The heterogeneity and diversity with which chemical structures from various sources are represented usually call for a pre-processing step in which user business rules are applied and may include manual curation for exceptional cases before the structures are processed by the registration system (Figure 14.3). At this step it is also appropriate to perform some level of quality control if the structures can be compared to an external source. Quality control (QC) can include verifying chemical structure correspondences to external identifiers or simply comparing molecular weight correspondences to identify potential mistakes in data files before they are loaded into the operational system. For small numbers of identified mismatches, a manual review is often the most efficient. Thorough and even manual quality control of compound data files can be justified because it is only required when additional external compound libraries are added to the institutional database and can save more work later. Registration of individual compounds is usually performed interactively with implicit (expert) pre-processing and QC.

When developing and implementing an organization's chemical structure standardization processes, several important points must be considered beyond the common canonicalization rules such as fixing bond orders and mesomeric representations. The most important issues for typical small molecule screening compounds are representation of stereochemistry, mixtures of stereoisomers, geometric isomers of carbon–carbon and carbon–hetero double bonds, regio-isomers or isomers of other structures, tautomers, non-isomeric mixtures, salt forms or other addend structures (such as water), and ionization states. Another consideration is whether the original structural representations should be kept along with the final standardized structures. From operational, reporting, and data analysis perspectives, it is desirable to represent each compound in an institutional library as a one-component canonical (parent) structure. For compounds of multiple components such as salts or isomeric mixtures, the parent structure should ideally represent the biologically active component.

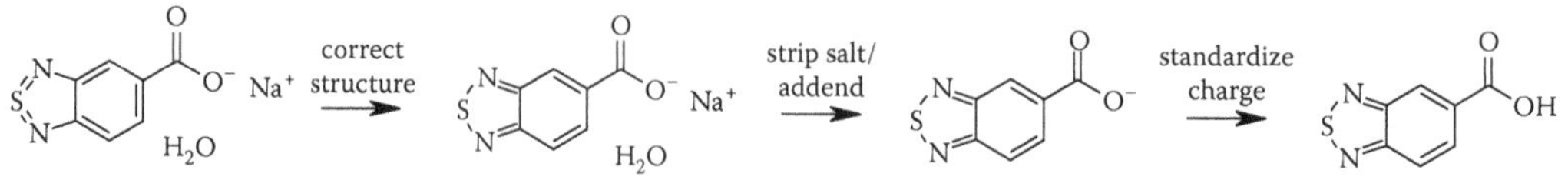

FIGURE 14.4 Example of simple sequence of structure standardization.

In some cases, single component representation is unsuitable. Figure 14.4 shows a simple sequence of structure standardization executed by a compound registration system including structure correction of the benzothiadiazole, salt/addend stripping and neutralization to depict the canonical parent structure representation. Any alternative salt form or structure representation of this compound will be standardized into the same parent structure and thus recognized as identical (at that level).

The databases of salt and other (usually small) addend structures are important components of standardization business rules. A starting set of most common salt forms and addend structures should be optimized in the context of the institutional library. It is not recommended to simply keep the largest (by molecular weight) fragment; this can be misleading, particularly for real mixtures. In some cases, the definition of a salt addend depends on the context of the registered structure. For example, in cisplatin (cis-diaminedichloridoplatinum(II) a platinum-based chemotherapy drug), the ammonia and chloride components are clearly parts of the structure and should not be stripped. In most organic compounds, however, components like ammonia and chloride serve only as counterions (salts) of a parent. If specific salt forms are considered critical for the identity of a compound (e.g., solubility or formulation is the subject of studies), the standardization rules may not remove these components. However, it is still important to standardize the ionization states of the structures and their stoichiometries in the representation of the standardized (parent) structure. The addend components and their exact stoichiometric data are always kept in the batch structure table of the database to compute the exact molecular weight and ratio of each component or regenerate a salt form of the parent if required.

Another important issue is tautomer standardization (Sayle, 1999; Szegezdi and Csizmadia, 2007). Figure 14.5 shows 11 possible tautomers without considering the geometric isomers of the two possible exocyclic double bonds. A tautomer canonicalization algorithm generates a unique structure representation (ideally a physiologically preferred low energy form) from any possible tautomeric starting form. It is likely in the illustrated case that the actual compound is a mixture of different forms. Any tautomer can potentially be responsible for biological activity and, depending on the equilibrium constants, the tautomeric ratios may shift under assay conditions and in the presence of a target. Regardless of the actual (unknown) biologically active tautomer, a canonical representation is important to ensure that all tautomeric (input) representations are recognized by the registry as the same (physical) compound (mixture).

The preferred representation of the Figure 14.5 structure may be as a mixture of the E and Z geometric isomers because they can interconvert via other tautomeric intermediates. Related to tautomer canonicalization is the ability to identify tautomers by exact or substructure querying of the

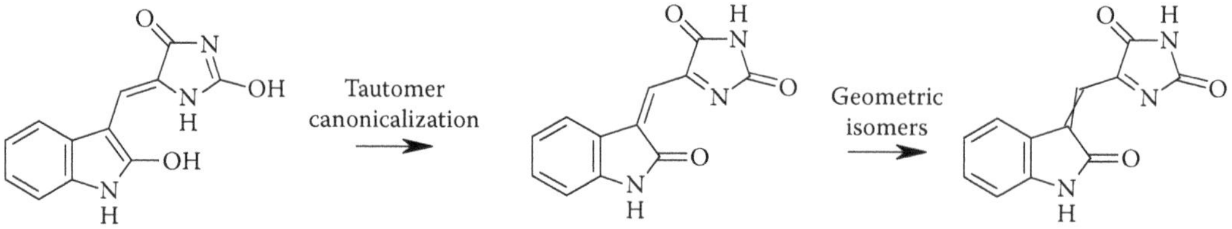

FIGURE 14. 5 Example of tautomer structure canonicalization. In reality tautomers can exist as mixtures.

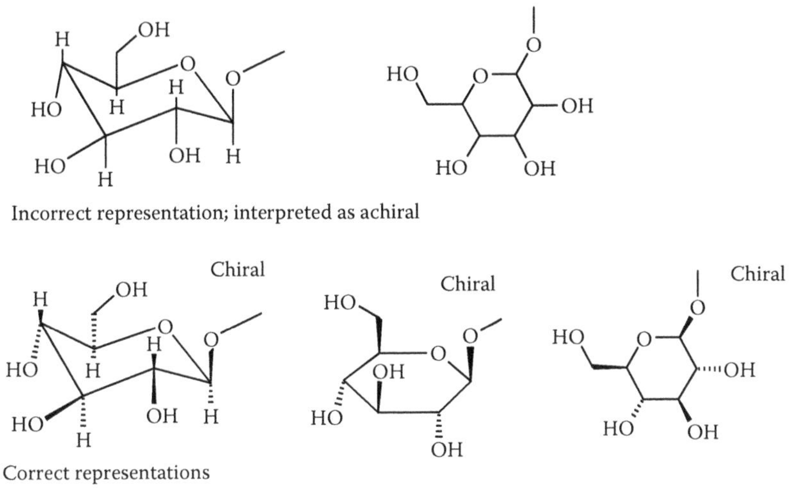

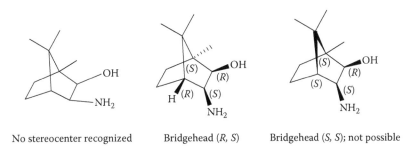

FIGURE 14.6 Computationally correct and incorrect representations of chiral structures.

chemical structure database. This may require querying all possible tautomeric forms of a substructure because tautomer canonicalization will not always obey substructure associativity; applying canonicalization to a substructure may not yield the same result as the (topological) substructure of the canonical tautomer (target) structure.

Standardized and consistent representations of stereoisomers and stereoisomeric mixtures are similarly important for the unique representations of distinct compounds. Recent file formats such as SDF v3000 and ChemAxon Extended SMILES provide clear definition and representation of complex relative and absolute stereochemical configurations. In practice these are not widely used because many commercially available files are represented by established v2000 or SMILES formats and also because HTS compounds are mostly relatively simple low molecular weight structures.

A computational system lacks a chemist's ability to understand and infer a structure representation. Figure 14.6 shows a simple example of a pyranoside. The chiral flag is often used to denote a (global) absolute versus a relative configuration of a stereocenter. Figure 14.7 illustrates the intuitive (chemist's) interpretation versus the actual defined bridgehead configuration in a bicyclic structure. If stereochemical mixtures are present, the authors suggest representing them as accurately as possible as a single parent structure component. This also applies to geometric isomers. However, in cases such as regio-isomeric mixtures or non-isomeric compound mixtures, this may not be possible and such mixtures may have to be represented as two (or more) components. Mixture representations can influence the perceived quantity of a (presumably bioactive) component or components.

FIGURE 14.7 Implied versus defined configuration of bridgehead atoms in bicyclic structures.

The real molecular weight of a sample should reflect the ratio of active component in the sample. For example, representing a mixture as a single component implies that the sample concentration refers to the combined concentrations of all components, while representing the (parent) structure as multiple components translates to each component having the indicated sample concentration. When integrating mass spectrometry-based analytical instrumentation, the confirmation of chemical structures is simplified if each sample is represented by a single most important or abundant component or an appropriate representation of a mixture in the case of an isomer. Implementing consistent structure representations organization-wide requires clearly defined user business rules and processes and carefully considered and optimized rules with application-based error handling and user notification.

14.3.2.3   Error Corrections

A chemical registration component of an enterprise screening informatics environment must also allow for rapid and economical correction of erroneous structures. Processes to facilitate error correction must be developed during system design and deployment. Even if a structure is identified as incorrect (e.g., by HPLC-MS) after it is associated with screening data endpoints, easy correction must be possible. The situation becomes more complicated if the (corrected) structure already exists in a database under another ID. In that situation it may be impractical to change the structure ID in all data tables of the operational system and it is not necessary.

When aggregating screening data into a data warehouse, one can associate two structure IDs as synonyms with the same canonical structure or aggregate the data by unique structures independent from their original IDs. Similarly any future corrections can retrigger the aggregation process to maintain all historical data in a correct structural context. In case of compound decomposition over time, a structure and corresponding screening results can be annotated accordingly. A data structure such as shown in Figure 14.3 makes such annotations visible across a database.

14.3.2.4   Chemical Structure Database Integration

In an enterprise operational system, it is critical to store chemical structures directly in Oracle for seamless integration with other data types and to assure enterprise-level system performance. Chemical data cartridge technology is now widely available for this purpose. The cartridges allow a relational database system to understand and index chemical information in the same way that it natively understands text and numeric data types, thus allowing the system to deal with all aspects of data storage, searching, and management.

Commercially available cartridges include Daycart (Daylight Chemical Information Systems), MDL Direct (Symyx), JChem (ChemAxon), Accord Chemistry (Accelrys), CS Oracle (CambridgeSoft), InfoChem, and AUSPYX (Tripos), many of which form the foundations of commercial chemical registration solutions. "Out-of-the-box" chemical registration systems include MDL Registration based on the MDL Direct and Isentris technologies from Symyx; Corporate Registration based on Accord Enterprise from Accelrys, Registration Enterprise from CambridgeSoft, Modgraph Chemical Registry, and ChemCart Registration from DeltaSoft.

14.3.2.5   Registering Non-Small Molecules

In recent years, screening of RNAi and shRNA libraries has become very popular and dozens of publications describe different screens in model organisms and humans (McManus et al., 2002; Grunweller et al., 2005; MacKeigan et al., 2005). Recording and storing information about such libraries does not require a chemical registration system as described above, but rather utilizes a gene-centric database in which corresponding RNAis and shRNAs are associated with gene data including ID, symbol(s), name(s), function(s), etc. (http://www.ncbi.nlm.nih.gov/sites/entrez?db=gene). In such screens, it is customary to include multiple oligos corresponding to different locations of the gene and/or mRNA to achieve higher confidence during screening. This arose because of the difficulty of obtaining consistent and reproducible results from such assays.

When screening RNAi and shRNA, 2D or 3D chemical structure-derived characteristics of these oligonucleotides are not informative. The primary sequence and corresponding location on the gene and/or mRNA are critical features for inclusion in a database. Several open source gene-centric and genome databases can be modified easily to include RNAi and shRNAi data. Such community efforts include GMOD (http://gmod.org/wiki/Main_Page), Ensembl (http://www.ensembl.org/index.html), and UCSC genomes (http://genome.ucsc.edu/). In the past 10 to 15 years, many organizations have also built internal gene-centric and genome databases to support their bioinformatics programs. Options are available to include information about RNAis and shRNAs in a database.

When screening small peptides (Marasco et al., 2008), the chemical structures are relevant for later data analysis and small peptides can be detailed in a compound registry. However, chemical structures of peptides are encoded in the sequences and it is therefore not essential to capture the structures in a chemical registry if the sequences are registered with IDs and other relevant data.

14.3.3 INVENTORY LOGISTICS AND SAMPLE TRACKING SYSTEM AND INTEGRATION

14.3.3.1 Function of Inventory Logistics System

An inventory logistics informatics system tracks all physical samples of an institutional compound collection in their various formats, associated containers, quantities, locations, identifiers, descriptions, genealogies, corresponding chemical structures, and (compound) batch data. It can also track analytical data, purity, date of sample preparation, storage conditions, solvent, freeze-and-thaw cycles, sample availability, etc. The key functions of an inventory logistics system are integrating and synchronizing sample storage and management operations with the inventory database and integrating the inventory database into HTS processes. The inventory system provides real-time enterprise-wide access to sample and associated property data; an integrated system also facilitates organization-wide compound requests and deliveries.

To support plate-based screening, a sample logistics system also creates and manages plate map information and tracks all plate and sample manipulations such as reformatting (Figure 14.8), copying, dilution, cherry picking, etc. An inventory logistics system integrates with laboratory automation instrumentation to perform such manipulations.

As shown in Figure 14.3, for subsequent data analysis, the inventory logistics system must also seamlessly integrate with the chemical compound and HTS databases. Because the inventory system tracks all plates and samples from the original source compounds all the way to screening plates and all intermediate transformations, the database may contain hundreds of millions of records and must therefore be robust and scalable with an efficient data model and architecture to allow real-time data access and process plate and sample manipulations. Any client interaction with the database must be transactional to avoid data corruption. Plate and sample logistics operations can become very complex and the system therefore must manage objects and relationships by which the samples and plates are organized including projects, locations, container templates, layouts, naming sequences, sample lists, sample map patterns, property maps, requests, etc.

By synchronizing and integrating sample inventory and transactions with operational screening informatics software, the inventory system enables the association of screening endpoints with compound structures and compound batch data (for example, commercial source, synthetic reference, etc.).

14.3.3.2 Integration with Operational Informatics Systems

The inventory system can be integrated with the other components of the operational HTS informatics infrastructure (HTS database and chemical structure registration system) at the database level or through the middle (business logic) layer (Figure 14.9). Integration via the middle layer of a system built in three-tier architecture is generally more flexible and expandable, although direct database integration may be faster if the schemas reside in the same Oracle instance.

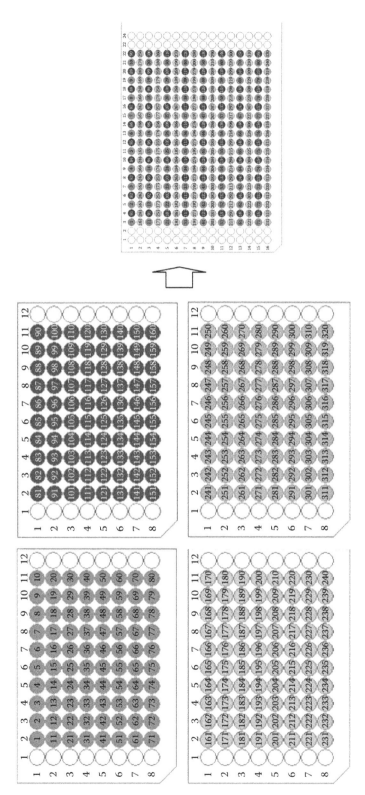

FIGURE 14.8 (See color insert following page 114.) Simple plate remapping operation: 4 × 96 (1, 12 empty) → 384 (1, 2, 23, 24 empty).

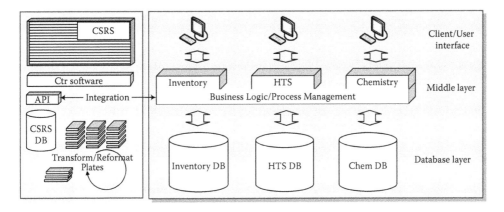

FIGURE 14.9 Integration of inventory system with other components of operational informatics environment (right) and with compound storage and retrieval system (CSRS) and plate and sample manipulation instrumentation (left).

Out-of-the-box HTS informatics software solutions with integrated functionality of all three operational components (HTS data, plate and sample tracking, chemical structures) include ActivityBase from IDBS, Assay Explorer from Symyx (former MDL), BioAssay Enterprise from CambridgeSoft, Accord Enterprise (Accelrys), ChemInnovation (CBIS), and others. However, we recommend careful evaluation of functionality, usability, architecture, and availability of a published application programming interface (API) if the system is to be integrated with other informatics components. Because the commercial solutions have different strengths, one answer is to combine different components or build certain functionalities in-house.

Three-tier architecture with a published API makes integration easier. Furthermore, any HTS informatics system must be customized to specific HTS and compound management workflows and robotics. Although a purchase usually includes initial set-up, extensive consulting may be required for further integration if the system is not open enough to allow in-house integration.

In the important decision to build a system in-house or buy a commercial solution, one should consider criteria such as functionality, initial (perpetual or annual) license fees, maintenance, consulting expenses, architecture and design for integration with HTS instrumentation, robustness and scalability, flexibility, ease of use, and user familiarity with the software. While developing screening informatics software in-house can provide a custom solution optimized to a specific organizational structure, HTS workflow, and user group, keep in mind that requirements and users change and maintaining an enterprise informatics system requires considerable resources for software engineering and user support in addition to the initial development investment.

It is often more economical to acquire a commercial system or individual components maintained by a vendor if they can be customized and integrated into the organizational workflow. The annual maintenance license fees of commercial software systems usually cover system and some level of user support, software updates addressing compatibility and dependency, and perhaps evolving functionalities. A similar consideration is the underlying database system. Most commercial software systems are built on Oracle, which provides commercial support. In many cases, the Oracle license can be bundled if an organization does not already have one. A custom system may be built on an open source relational database, for example, MySQL or PostgreSQL—both are robust but lack corporate support. In practice, lead discovery informatics requires a balance of in-house development and licensing of commercial solutions. Whether a commercial solution is chosen or a system is built partially or entirely in-house, it is an advantage to maintain in-house expertise to perform a certain level of integration and customization and provide user support if the organization can afford a small development and support team.

Data pipelining via enterprise server applications such as Pipeline Pilot (Accelrys/Scitegic), Inforsense, or the Talend (http://www.talend.com/) and KNIME (http://www.knime.org/) open source tools represent a powerful approach to integrate applications and develop custom functionality. As a specific application, Pipeline Pilot 7.0 includes a plate analytics collection for the development of complex plate-based data analysis and visualization protocols.

For small organizations or academic groups with limited informatics and IT resources, hosting their data outside the company may be an option. For example CDD (Collaborative Drug Discovery, http://www.collaborativedrug.com/) offers a cost-efficient solution for storing, accessing, and sharing chemical compound and screening data although it does not yet provide the full functionality of an industry-standard screening informatics solution.

14.3.3.3   Integration with Compound Storage and HTS Robotics Instrumentation

Integration of instrumentation with the operational informatics infrastructure is a critical requirement to keep HTS workflow efficient and error-free. Different proprietary vendor software and platform incompatibilities can complicate system integration and in most cases, some software development effort and/or consulting services from the instrumentation and software vendors will be required. Interoperability and ease of integration should also be considered when designing an HTS operation.

Robotics instrumentation is most effectively integrated via the instrumentation software's API into the business logic middle layer of an enterprise informatics system (Figure 14.9). Via the API, the compound storage and retrieval system (CSRS) compound inventory database (or inventory files) can be made available throughout the enterprise HTS informatics environment. Requests can be sent from the inventory system to the instrumentation control software to access certain samples and/or plates or perform certain manipulations such as plate reformatting (Figure 14.8). These manipulations are synchronized with the plate and sample management informatics system to track the transformation history (genealogy).

In the absence of an API, it may be possible to publish control software functionality as a Web service, for example, by using Microsoft's .NET framework to integrate via the inventory informatics system middle layer. It is also possible to achieve a certain level of integration by expanding the functionality of the plate logistics application (via its API) to create and send control files to the instrumentation software while performing the same manipulations in the plate inventory database. We recommend performing QC steps and verifying requested plate and sample manipulations against the instrumentation log files. Analytical instrumentation, e.g., HPLC MS, may be integrated in the same manner; purity information may be recorded in the inventory database. Integration of the plate and sample inventory system with compound registry and genealogy tracking facilitates real-time QC of screening results based on exact molecular mass.

14.3.4   HTS Assay Data Capture and Processing

14.3.4.1   Functions of HTS Assay Data Management System

An HTS data management system (Figures 14.2 and 14.3) facilitates and enforces the entire business process from capturing raw data assay reader files to releasing the final quality controlled (biological) endpoints to the end user or institutional data warehouse. The assay data systems must read instrumentation-specific reader file formats, capture and apply data quality criteria and rules for invalidating results, capture and perform various well-, plate-, and run-based calculations, and manage data release. All manipulations should be carried out in a user- and role-based secure manner to ensure that data and instructions data cannot be changed intentionally or inadvertently, thus comprising institutional data integrity.

The assay data system should also manage and capture assay metadata including biological and pharmacological categories, technical parameters related to assay method and endpoints, project-related information, and other data relevant to each assay. Metadata vocabulary should be managed by

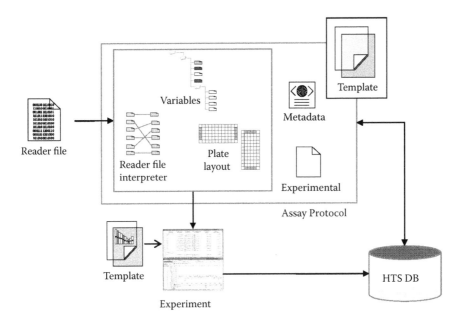

Reader file

Variables

Reader file interpreter

Plate layout

Metadata

Template

Experimental

Assay Protocol

Template

Experiment

HTS DB

FIGURE 14.10 (See color insert following page 114.) Components of assay data management system.

dictionaries so that metadata can be expanded in a controlled manner via an administrative mechanism. The system can also capture experimental details or integrate with an electronic laboratory notebook (ELN). Before building or deploying an assay data management system and optimizing the assay database, the organization and data structures of all assay-related information must be clearly defined along with HTS processes and information flow.

14.3.4.2 Components of Assay Data System

Unless very specific software with limited (and optimized) functionality is built in-house, an assay data management system should include the components shown in Figure 14.10. The variables are the basic components to describe an assay in the HTS data management system; they include all common data types such as real values, integers, text, Boolean, and can also include images and files or hyperlinks. Common types include input variables (from the instrumentation reader file), set variables as defined by assays, dimension variables, calculated variables that define all calculations performed during data processing, and also groups and tables to organize the variables and map them to the database schema.

The reader file interpreter is a set of instructions for parsing and mapping the rows and columns of the reader file to input variables. Plate layouts define the physical grid layout of an experiment such as positions of control, screening compound, and background wells. Although these components of the assay protocol appear complex to the end user, the flexible organization allows processing of assay data independent of the specific physical layout or reader file format.

After defining a protocol including all variables, instrumentation-specific reader file formats, and layouts, assay data can be processed automatically into the assembled end product containing the data of one screening experiment (Figure 14.10). All data is written to the HTS database where it can be accessed via client software to analyze and visualize results, perform quality control steps, invalidate erroneous data, and finally release the data of a screening run.

14.3.4.3 Managing Metadata

Other functions of an HTS data management system are capturing and organizing assay-related metadata that should be well structured and organized as ontologies. A well built ontology will facilitate data integration with external data sources such as gene-, protein-, pathway-, cell line- or licensed SAR databases; it also facilitates internal cross assay data analysis. No assay ontology is currently universally accepted. Many organizations develop their own optimized solutions.

As an example, Schuffenhauer et al. (2002) developed a biological target-based ontology. Major assay detection technologies in the context of different assay types (such an *in vitro* biochemical, cell-based reporter, and phenotypic assays) were reviewed by Inglese et al. (2007). The MIACA (Minimum Information about a Cellular Assay) project is an open source initiative developing information guidelines and a modular cellular assay object model (CA-OM) to capture the range of possible cellular assays and provide a foundation for efficient information exchange and data integration (Wiemann et al., 2007; http://miaca.sourceforge.net/).

14.3.4.4   Integration with Instrumentation

The assay data management system receives reader files generated by the plate reader either by monitoring the directory to which the files are written or by integration with the plate reader software via an API. This means screening progress can be monitored in real time and staff may be alerted automatically if assay statistics fall outside pre-defined stability criteria. If the screening robotics control software is accessible via an API, the assay management system can also be directly integrated with a robotics system (similar to the system shown in Figure 14.9) to start and control screening from the informatics system.

Via integration of the data management system, plate logistics system, and chemical structure registry (Figures 14.3 and 14.9), the chemical structure and batch IDs are linked into the assay data tables during processing of reader files. The screening data endpoints are thus associated with the original chemical structures and batch samples.

14.4   DATA ANALYSIS AND DISCOVERY INFORMATICS

14.4.1   SCREENING DATA ANALYSIS

HTS produces large numbers of individual measurements with inherent variability and error reflected in confirmation rates often significantly below 100%. Statistical and pattern recognition methods to analyze HTS data and optimize assay parameters are routinely used. (Padmanabha, Cook, and Gill, 2005). Various error sources can influence the variability of HTS data, leading to false positive and false negative results (Parker and Bajorath, 2006; Makarenkov et al., 2007).

Operational errors include compound handling, plate manipulations, compound annotations, quality control procedures, etc. Note that one of the key functions of an operational HTS informatics system is elimination of such errors. Strategic errors relate to the design of an assay campaign, for example, comprehensive or iterative screening, methods to identify compounds for confirmatory screening, or screening mixtures compared to individual samples. Measurement errors can be biological or technical, for example, expression level variations or plate reader errors.

Small systematic errors introduced into HTS results include imperfect pipetting, temperature gradients, uneven evaporation, concentration differences, variable growth patterns, time sequence differences, etc. Errors can also be caused by intrinsic compound properties such as stability, solubility, aggregation, etc., or relate to variations in the biological system upon which the assay is based. For single concentration assays, measurement errors are not distributed equally; they are typically higher for compounds exhibiting half maximum (50%) responses (as measured by the assay) close to the screening concentration.

Statistical data analysis methods have made it possible to identify and address HTS measurement errors (Zhang, Chung, and Oldenburg, 1999; Malo et al., 2006). Within-plate and assay-wide controls are required to monitor quality by plate and stability over an entire screening run. Terminology for controls can be context-dependent and must be unambiguously defined throughout the organization and tracked by the operational informatics system to avoid mistakes in HTS data analysis (Ling, 2008). Plate, well, and run statistics are typically used for validation and QC of HTS results.

Simple common statistical parameters include Z, Z′, signal to background (S/B), signal to noise (S/N), and coefficient of variance (%CV) (Table 14.1; Zhang, Chung, and Oldenberg, 1999). Plate-based normalization is usually based on controls. In the absence of controls, the Z score (Table 14.1) may be used to normalize HTS data by plate. To ensure that statistics can be applied accurately to HTS hit selection and data analyses, pre-processing of raw data has been suggested. For example, a logarithmic transformation prior to normalization can be performed if the HTS data reveal a Gaussian distribution because subsequent statistical analyses assume a normal error distribution (Kevorkov and Makarenkov, 2005). This transformation renders variation more independent of absolute magnitude and makes normalized data additive.

Shortfalls of the common control-based HTS data analysis approach have been suggested (Gribbon et al., 2005; Malo et al., 2006). For example plate-based normalization using controls (Table 14.1) intrinsically assumes a random error distribution for all wells in a plate because of the positional bias of the controls dictated by the formats of common screening libraries. Edge effects are particularly relevant in cell-based assays and vary from plate to plate (Malo et al., 2006). Classical HTS analysis relies on non-robust statistics; means and standard deviations are greatly influenced by outliers.

The B score (Brideau et al., 2003) is a robust analog of the Z score after median polish; it is more resistant to outliers and also more robust to row- and column-position related systematic errors (Table 14.1). The iterative median polish procedure followed by a smoothing algorithm over nearby plates is used to compute estimates for row and column (in addition to plate) effects that are subtracted from the measured value and then divided by the median absolute deviation (MAD) of the corrected measures to robustly standardize for the plate-to-plate variability of random noise. A similar approach uses a robust linear model to obtain robust estimates of row and column effects. After adjustment, the corrected measures are standardized by the scale estimate of the robust linear model fit to generate a Z statistic referred to as the R score (Wu, Liu, and Sui, 2008). In a related approach to detect and eliminate systematic position-dependent errors, the distribution of Z score-normalized data for each well position over a screening run or subset is fitted to a statistical model as a function of the plate; the resulting trend is used to correct the data (Makarenkov et al., 2007).

Various types of visualizations are also commonly used to evaluate quality, variability, and systematic errors in a screening run. These include scatter plots of raw and normalized data for different well types (data, controls, blanks), scatter plots of plate-based statistics, cityscape or bar chart visualizations of well position-based statistics, histograms of raw, normalized, and (if applicable) corrected HTS activity, plate grid heat map visualizations of HTS raw and normalized activity, etc. Beyond robust statistics and visualization to analyze large data sets and detect outliers, sophisticated pattern recognition algorithms are adopted to detect and correct systematic errors in complex HTS data sets with overlapping effects; commercial tools include Genedata and Partek (Gribbon et al., 2005; Wu et al., 2008).

Commercial statistics and visualization software packages include among many others S-Plus (http://www.insightful.com/), Systat (http://www.systat.com/), SAS (http://www.sas.com/), Matlab (http://www.mathworks.com/), TIBCO Spotfire (http://spotfire.tibco.com/), Partek (http://www.partek.com/), Graphpad (http://www.graphpad.com), and Minitab (http://www.minitab.com). Basic statistics and visualization functionality is usually built into operational HTS enterprise systems. Another powerful HTS analysis software is Genedata Screener (http://www.genedata.com/). R-project (http://www.r-project.org/) is the most common open source statistics and visualization environment integrated into many commercial solutions.

Depending on the specific control- or statistics-based normalization and error correction procedures to analyze single point concentration experiments, different samples are identified as hits and the methods differ significantly in their ability to efficiently identifying true actives (Wu et al., 2008).

TABLE 14.1
Common Equations Related to HTS and Statistical HTS Data Analysis

Parameter	Equation	Definition
Normalized percent control	$NPC = \dfrac{x_i - \bar{c}_-}{\bar{c}_+ - \bar{c}_-} \times 100$ $= \left(1 - \dfrac{\bar{c}_+ - x_i}{\bar{c}_+ - \bar{c}_-}\right) \times 100$	Control-based normalization of raw value; X_i ith raw value, $\bar{c}_+$ plate mean of positive control, $\bar{c}_-$ plate mean of negative control
Normalized percent inhibition	$NPI = \dfrac{\bar{c}_+ - x_i}{\bar{c}_+ - \bar{c}_-} \times 100$ $= \left(1 - \dfrac{x_i - \bar{c}_-}{\bar{c}_+ - \bar{c}_-}\right) \times 100$	Control-based normalization of raw value; X_i ith raw value, $\bar{c}_+$ plate mean of positive control, $\bar{c}_-$ plate mean of negative control
Z score	$Z = \dfrac{x_i - \bar{x}}{s_x}$	Control-independent normalization as distance between raw value x_i and plate mean $\bar{x}$ in units of plate raw value standard deviation s_x
Coefficient of variation	$CV = \dfrac{s_x}{\bar{x}} \times 100$	Measure of signal dispersion; $\bar{x}$ signal mean, s_x signal standard deviation
Signal to background	$S/B = \dfrac{\bar{c}_+}{\bar{c}_-}$	Measure of dynamic range of assay; $\bar{c}_+$, $\bar{c}_-$ mean of positive and negative control (background)
Signal to noise	$S/N = \dfrac{\bar{c}_+ - \bar{c}_-}{s_{c-}}$	Measure of signal strength; also expressed as $\dfrac{\bar{c}_+ - \bar{c}_-}{\sqrt{s_{c+}^2 + s_{c-}^2}} s_{c-}$ standard deviation of negative control (background)
Signal window	$SW = \dfrac{\bar{c}_+ - \bar{c}_- - 3(s_{c+} + s_{c-})}{s_{c+}}$	Significant signal between positive and negative controls; assay dynamic range
Z factor	$Z = 1 - \dfrac{3(s_x + s_{c-})}{\bar{x} - \bar{c}_-}$	Describes assay dynamic range based on range and data variation
Z′ factor	$Z' = 1 - \dfrac{3(s_{c+} + s_{c-})}{\bar{c}_+ - \bar{c}_-}$	Describes assay dynamic range based on range and data variation
Minimum significant ratio	$MSR = 10 \wedge (2s_{dp})$	Reproducibility test for potency (smallest statistically significant potency ratio between two measurements); s_{dp} standard deviation of difference in log potency
Minimum significant difference	$MSD = 2s_{de}$	Reproducibility of efficacy; s_{de} standard deviation of difference in efficacy
Hill-slope model	$y = y_{min} + \dfrac{(y_{max} - y_{min}) \cdot c^n}{IC_{50}{}^n + c^n}$ $= y_{min} + \dfrac{(y_{max} - y_{min}) \cdot c^n}{1 + (IC_{50}/c)^n}$	Concentration-response curve fitting of activity value y to obtain IC_{50} (or EC_{50}); y_{max} and y_{min} are fitted maximum and minimum values of data; if data do not define maximum or minimum asymptote, y_{max} can be fixed to 100% and y_{min} to 0% respectively; n is hill-slope coefficient and c is inhibitor (or activator) concentration

(continued)

TABLE 14.1 (CONTINUED)
Common Equations Related to HTS and Statistical HTS Data Analysis

Parameter	Equation	Definition
Michaelis–Menten	$$v = \frac{[S] \cdot V_{max}}{[S] + K_m}$$ $$K_m = [S] \cdot \frac{V_{max} - v}{v}$$	Enzyme kinetics; K_m, Michaelis–Menten constant is substrate concentration $[S]$ that produces half maximum enzyme velocity; v enzyme velocity, V_{max} maximum enzyme velocity
Cheng–Prusoff	$$K_i = \frac{IC_{50}}{1 + [S]/K_m}$$	Relates IC_{50} to K_i under conditions of competitive inhibition; K_i equilibrium enzyme inhibitor dissociation constant; K_m Michaelis–Menton constant, $[S]$ substrate concentration.
Chem–Prusoff ligand binding	$$K_i = \frac{IC_{50}}{1 + [L]/K_d}$$	Relates IC_{50} to K_i under conditions of competitive binding; K_d equilibrium dissociation constant of (labeled) ligand; $[L]$ ligand concentration
Median absolute deviation	$$MAD = median\{|\, r_{ijp} - median(r_{ijp})\,|\}$$	Robust estimate of spread of r_{ijp} values
Bscore	$$B = \frac{r_{ijp}}{MAD_p}$$ $$r_{ijp} = y_{ijp} - \hat{y}_{ijp} = y_{ijp} - (\hat{\mu} + \hat{R}_{ip} + \hat{C}_{jp})$$	Residual r_{ijp} is difference between observed result y_{ijp} and fitted value $\hat{y}_{ijp}$ defined as sum of estimated average of plate μ_p and estimated systematic measurement offsets for row i and column j on plate p, $\hat{R}_{ip}$ and $\hat{C}_{jp}$ respectively, obtained by two-way median polish and smoothing; B score is adjusted value scaled to plate median absolute deviation above

Typically the activity cut-off value to advance a number of samples to confirmation or concentration response screening is chosen by three times the standard deviation of the normalized activity of all samples of a run; it can also be dictated by robotic capacity. Statistically calculated expected confirmation rates as a function of the cut-off value can guide the optimal activity cut-off. In one example, the empirically estimated null distribution (Efron, 2004) was used to estimate the false discovery rate using R scores. The resulting estimated confirmation rates correlated closely to experimental confirmation rates (Wu et al., 2008). False positives are easily identified and eliminated via confirmatory screening. False negatives on the other hand cannot be eliminated without using improved statistical HTS data processing as described above if chemical structure information are not utilized.

Replicate measurements can be used to estimate real error rates and improve precision of activity measurements. Statistical approaches to handle replicate data have been described (Malo et al., 2006; Wu et al., 2008). However, replicate measurements are rare in practice because of cost constraints. Quantitative high-throughput screening (qHTS) developed at the NIH Chemical Genomics Center (NCGC) is an approach that generates concentration response curves for large libraries in a single experiment (Inglese et al., 2006). Concentration response data can in principle be normalized using the same statistical methods described above while accounting for the specific concentration-based plate layout.

The potency of compounds derived from concentration response assays is expressed most commonly as IC_{50} or EC_{50} defined as the compound concentration that produces half maximum response. A common model is the four-parameter Hill-slope equation (Table 14.1). A three-parameter model can be used if a maximum or minimum asymptote is not available because compound potency falls outside the concentration range. One recommendation is to fit the logarithm (log10) of IC_{50} or EC_{50} instead of the untransformed concentration because the concentration response errors are normally distributed in log space. Because the uncertainty of IC_{50} or EC_{50} is not symmetrical in regular space, the standard error derived from fitting untransformed IC_{50} or EC_{50} values is less useful than the standard error in log space that translates to more realistic (asymmetric) confidence intervals in regular space. Under competitive binding conditions, the Cheng–Prusoff equation (Table 14.1) relates the binding constant K_i to the IC_{50}. K_i is the equilibrium inhibitor dissociation constant, the concentration of a competing inhibitor that results in binding to half the enzyme sites at equilibrium in the absence of a substrate or other competitor. Binding constants can provide more meaningful comparisons because they depend less on assay conditions.

A variety of curve fitting models and algorithms are available in various statistics software packages including commercial HTS systems. Depending on the curve fitting model and specific settings of the regression algorithms, IC_{50} or EC_{50} values can vary; it is therefore important to standardize the curve fitting method throughout an organization to produce more easily comparable and reproducible data. It is also important to set standardized criteria for reporting data falling outside the tested concentration range and reporting samples that plateau below maximum efficacy (partial inhibitors or activators). The minimum significant ratio and minimum significant difference have been used as statistical parameters to characterize the reproducibility of potency and efficacy estimates in concentration response assays (Eastwood et al., 2006).

A key functionality of an operational HTS data system is the ability to interactively visualize concentration response curves, invalidate outliers to manually optimize curve fitting, and annotate results. Automated curve fitting for large numbers of concentration response curves requires efficient algorithms and must consider all possible response profiles. Manual review and notation are generally required at least for a small number of concentration response curves. Genedata Screener implements high-throughput curve fitting. Large-scale curve fitting, curve classification algorithms, and freely available software have been developed by the NCGC (Inglese et al., 2006). Also, NCGC's *Assay Guide* is a valuable resource that explains various aspects of HTS data quality and related statistical parameters and tests including assay stability, reproducibility, replicate experiments, curve fitting, and variability.

14.4.2 Application of Cheminformatics to HTS

14.4.2.1 Basic Cheminformatics Principles Related to HTS

The goal of a screening campaign is to identify the best structural series that may be optimized into lead compounds of a desired profile (Figure 14.1). Cheminformatics methods are required to analyze HTS and uHTS screening results in the context of chemical structures and effectively follow-up results of a primary screen or screening campaign. Cheminformatics should organize and visualize data based on chemical structures, suggest explanations of the data to develop hypotheses, and identify possible false positives or false negatives by finding similarity relationships among compounds that may relate to similar mechanisms.

Cheminformatics data analysis and modeling generally focus on exploration of molecular similarities under the fundamental principle that structural similarity is related to similarity of biological activity and—more generally—physical properties (Martin, Kofron, and Traphagen, 2002; Bender and Glen, 2004). This similar property principle is a useful assumption for HTS data because HTS data sets are large and usually applicable to statistical analysis. Various types and implementations of molecular descriptors have been developed (Todeschini et al., 2000) to quantify similarity relationships, cluster chemical structures, and develop statistical predictors. For HTS analysis, the most frequently applied descriptors include 2D topological fingerprints, structure keys, fragment-based descriptors, and numeric properties. Common topological fingerprints include path-based fingerprints developed by Daylight and circular fingerprints such as Scitegic extended connectivity fingerprints. Structure keys are predefined substructures, e.g., MDL Maccs keys.

Fragment- or substructure-based descriptors include maximum common substructures (MSC) or combinations of scaffolds or subscaffolds. Simple numeric descriptors include calculated LogP, polar surface area, molecular weight, and number of hydrogen bond donors and/or acceptors that are well known descriptors for characterizing drug likeness (Lipinski et al, 2001; Oprea et al., 2007a). Depending on the descriptors and the specific objective, various similarity metrics can be applied. For fingerprint and structure key descriptors, the Tanimoto metric is the most widely used. In addition to screening data analysis, visualization, and statistical modeling, another important application of cheminformatics is the design and diversity analysis of screening libraries.

14.4.2.2 HTS Data Mining and Modeling

Cheminformatics methods play a critical role in the statistical analysis of HTS data from primary assays and many HTS data analysis and mining approaches have been developed (Parker and Schreyer, 2004; Davies, Glick, and Jenkins, 2006; Harper and Pickett, 2006). Although primary screening may cover a million compounds, the capacity of concentration response and secondary screening and logistical considerations limit the number of compounds that can be carried forward to usually a few thousand. The goals of primary screening data analysis are to maximize the confirmation rate, identify series that exhibit SAR, eliminate artifacts, recover false negatives, and expand hit lists and chemotypes.

Structure-based clustering is used to group related compounds for the purpose of HTS data analysis, identification of SAR series, and detection of potential outliers (Engels et al., 2002). In one example, researchers at GNF reported a statistical approach to dynamically score each scaffold family (obtained by prior clustering of screened structures) based on family members' HTS activities. This method identifies compounds that share structural similarities and similarly high HTS activities; it yielded greatly improved confirmation rates compared to using a static (scaffold-independent) activity cut-off (Yan et al., 2005).

Various statistical methods have been successfully applied to HTS data mining in combination with 2D topological descriptors. Recent examples include Laplacian-modified naive Bayesian (Rogers, Brown, and Hahn, 2005), reverse partitioning (RP; Rusinko et al., 1999), and support vector machines (SVM; Ma et al., 2008). An advantage of these methods is their relative resistance to noise that makes them applicable to HTS data (Glick et al., 2006). One effective method to analyze HTS data and identify structurally homogeneous classes with high average activities is a variation of RP that uses simulated annealing (SA). RP/SA produces a regression tree of biological activity in which each node is characterized by a specific combination of features; the terminal nodes correspond approximately to different structural classes (Blower et al., 2004). It is implemented in the Leadscope software package (http://www.leadscope.com/).

Statistical models can be used to predict the likelihood of activities of compounds that are present in the screening set (for example, to recover false negatives). Ligand- and structure-based molecular similarity analysis and virtual screening approaches are also routinely applied to follow up screening results and expand hit lists or to identify potentially new scaffolds (scaffold hopping; Klebe, 2006; Muegge and Oloff, 2006; Eckert and Bajorath, 2007). With well established business processes, the complementary nature of in silico testing and experimental HTS can be aligned in the lead discovery process and the synergy of the approaches can improve the efficiency of a campaign (Bajorath, 2002; Bleicher et al., 2003; Karnachi and Brown, 2004; Davies, Glick, and Jenkins, 2006; Parker and Bajorath, 2006).

14.4.2.3 HTS Structure–Activity Relationship (SAR)
Analysis, Clustering, and Visualization

From a chemist's perspective, it is necessary to interactively analyze and visualize screening results by structural scaffolds to identify chemical series that can be synthetically optimized. Depending on the business process, interactive HTS analysis from a medicinal chemistry perspective can be performed at any stage—after primary, concentration response, or secondary screening (Figure 14.1).

Visualizing large data sets is a powerful way to explore and analyze data. Dynamic visualization techniques to interactively query, explore, and analyze very large data sets are increasingly used across disciplines (Shneiderman, 2008). One of the most established interactive data visualization tools is Spotfire (http://spotfire.tibco.com); it is also used to analyze HTS data sets (Ahlberg, 1999). Visualization is commonly used to survey and analyze structure–activity relationships (SARs) in HTS data—usually after a dimensionality reduction (Gedeck and Willett, 2001).

To visualize and analyze SARs in large data sets, chemically meaningful scaffolds and in many cases their hierarchical relationships must be generated and represented in the context of quantitative screening results. The exact definition of "scaffold" may be context-dependent. From a chemist's perspective, a scaffold is a common core structure characterizing a group of molecules that may share a common synthetic pathway. More generally a scaffold may be considered a representation of a group of structures that share common characteristics such as topological or pharmacophoric features that are related to activity, but they may not necessarily share the exact same core structure.

Different methods are used to cluster chemical structures into scaffolds and visualize them for the purpose of analyzing SARs. Among commercial tools, the desktop HTS DataMiner from Tripos (http://www.tripos.com/) clusters structures using 2D unity fingerprints and visualizes the structural similarities after dimensionality reduction as 2D SAR landscape scatter plots and bull's eye representations. DataMiner can also generate heat maps and SAR tables, perform substructure analysis, and includes an integrated data modeling component (HQSAR). Leadscope (http://www.leadscope.com/) is a powerful Oracle-backed client–server application for interactive and visual analysis of large sets of chemical compounds, their properties, and biological activities. Chemical structures are organized in an expandable hierarchy of about 27,000 substructures motivated by those typically found in small molecule drug candidates (Roberts et al., 2000). The software comprises a number of options to visualize numerical data sets in the context of these chemical features and identify statistical correlations of structural features and data. It includes different algorithms to cluster chemical structures and identify chemical series and feature combinations related to biological activity. Cluster results and chemical structures can be represented in matrix views facilitating its interactivity. Leadscope can also generate SAR tables and maximum common substructure scaffolds and has a data mining component to build quantitative statistical models from SAR data. The software facilitates a typical workflow for analysis of HTS data sets that involve three major phases (after loading structures and data): (1) removing undesired compounds based on calculated descriptors, reactive functionalities, toxicity, or other criteria, (2) identifying local structural neighborhoods around active compounds, and (3) analyzing in detail the local SARs (Blower et al., 2004).

Organizing compounds by maximum common substructure (MCS) or molecular framework is an alternative to fingerprint- or feature-based clustering. In HierS (Wilkens, Jones, and Su, 2005), researchers at GNF adopted the concepts of molecular frameworks (Bemis and Murcko, 1996) to construct a hierarchical relationship of ring features to intuitively (and uniquely) organize chemical structures according to their highest priority scaffold. HierS is useful to interactively analyze and navigate HTS data as an easy-to-interpret scaffold hierarchy. A related approach systematically prunes side chains and rings using prioritization rules to generate a hierarchy of chemically well defined canonical frameworks in which each scaffold in a tree has only one parent (Schuffenhauer et al., 2007). In another report, a hierarchy of maximum common frameworks (MCFs) is generated and combined with an interactive tool to visualize the tree-like organization of structures in the context of biological activity (Cho and Sun, 2008). Molecules can be assigned to multiple MCFs and a user can interactively prune undesired branching nodes.

SARvision from Altoris (http://www.chemapps.com/) is a commercial desktop application targeted at medicinal chemists. It analyzes screening data based on an automatically generated and user-expandable scaffold hierarchy; it can also generate R tables (Reichard, 2008).

To cluster very large chemical databases while preserving the chemically intuitive nature of MCS, fast fingerprint-based clustering can be combined with MCS (Stahl and Mauser, 2005). One shortcoming of MCS in detecting similarities among pairs of structures in which the same features are connected by linkers has been addressed by a clustering algorithm based on finding the largest set of disconnected fragments that two structures share in combination with their relative orientation (Stahl et al., 2005). Biologically meaningful relationships among compounds of different chemical classes may be detected.

In a recent study, network representations were used to visualize the similarity relationships of scaffolds of compound libraries (Shelat and Guy, 2007). Using Murcko assemblies (Bemis and Murcko, 1996) and atom-type path-based (Daylight) fingerprints to calculate their similarities, the resulting networks of scaffold nodes and similarity-scaled edges illustrate both inter- and intra-cluster distributions. Onto such a network one could overlay meaningfully aggregated screening data that can be visualized in Cytoscape (http://cytoscape.org/) as node properties to construct a bioactivity scaffold map.

A powerful suite of applications and Java-based tools for manipulating and processing chemical structures including database integration, very fast structure searching, structure drawing and visualization, generating topological and pharmacophoric fingerprints, calculating various descriptors and properties, MCS- and descriptor-based clustering, virtual screening, and structure handling and property calculations in local databases and Microsoft Excel is available from ChemAxon (http://www.chemaxon.com/products.html).

14.4.2.4 Design, Selection, and Diversity Analysis of Screening Libraries

Cheminformatics methods are required to design screening libraries, enrich existing libraries, select diverse or focused subsets, and pre- or post-filter undesired or reactive compounds. At present more than 10 million unique compounds are commercially available for screening (http://www.emolecules.com/, http://www.chemnavigator.com) and the theoretical number of synthetically tractable compounds even with lead-like properties is enormous (Bleicher et al., 2003; Fink, Bruggesser, and Reymond, 2005).

Various criteria must be considered for designing an HTS collection or enriching an existing screening library. The fundamental focus should be on finding lead-like compounds using simple numerical Lipinski criteria (Oprea et al., 2001, 2007a). General filters have been developed to identify compounds with reactive functional groups or other undesired characteristics from a medicinal chemistry view that should not be part of an HTS collection (Rishton, 1997; Hann et al., 1999). A practical tool to filter compound collections with many built-in definitions is Filter from Open Eye (http://www.eyesopen.com/). Other considerations of an HTS library include aggregators for *in vitro* assays (Seidler et al., 2003), frequent hitters in cell-based reporter gene assays (Crisman et al., 2007), and fluorescent compounds (Simeonov et al., 2008). These criteria and filters may depend on the specific biology, assay format, and screening goal and should be considered in the context of the HTS campaign.

Design efforts should differentiate focused and diverse libraries (Bleicher et al., 2003; Shelat and Guy, 2007). Focused libraries are usually optimized for one target class and based on similarities of known actives or other target data such as receptor structures. Ligand- and structure-based molecular similarity analysis and virtual screening approaches can be applied for designing or selecting target-focused libraries (Klebe, 2006; Muegge and Oloff, 2006; Eckert and Bajorath, 2007).

General purpose screening libraries should be diverse and biologically relevant. However, the specific meaning and implementation of structural diversity and biological relevance depend on the perspective and objective. From a pharmacological view, similarity is a context-dependent parameter and thus diversity (or dissimilarity) can be defined meaningfully only in a relevant context (Bleicher et al., 2003). General methods for designing a diverse library apply algorithms that maximize dissimilarities in a chemical descriptor space. Diversity selection algorithms using high dimensional fingerprints in combination with the Tanimoto (T) metric have been used successfully and included in a number of commercial software packages including Leadscope and Pipeline Pilot. When using these methods, note that high dimensional fingerprints and the T metric have been developed to identify and quantify similarity, not necessarily dissimilarity. Also, 1-T is not a formal measure for distance because it does not obey the triangle inequality.

Burden, CAS, and University of Texas (BCUT) descriptors are well suited and widely used to describe diversity of a chemical population in a low dimensional Euclidian space and they allow for fast cell-based diversity selection algorithms (Pearlman and Smith, 1998). The DiverseSolutions software developed by Pearlman and commercialized by Tripos (http://www.tripos.com) includes tools and algorithms to generate BCUT descriptors, select and optimize the descriptor space for a given population, and perform various diversity-related and library design tasks.

In practice, an HTS collection of microclusters around diverse scaffolds provides a certain redundancy in noisy primary HTS data and facilitates the data analysis methods described above to identify active series. Recent library design efforts have shifted away from the optimal diversity approach and focus more on scaffold representation, high quality lead-like (Hann and Oprea, 2004) compounds, synthetic feasibility, and smart focused libraries to provide the best starting points for medicinal chemistry (Davies, Glick, and Jenkins 2006; Schnur, 2008).

14.5   INFORMATICS FOR HIGH CONTENT IMAGE-BASED SCREENING

14.5.1   HIGH CONTENT SCREENING (HCS)

During the past few years, high content image-based phenotypic screening has emerged as a very powerful, transforming technology for early drug discovery (Korn and Krausz, 2007; Bullen, 2008). High content screening (HCS) combines automated microscopy with image analysis, producing multiparametric readouts that (quantitatively) characterize phenotypic responses in intact cells. Small molecule HCS thus enables the simultaneous measurement of multiple phenotypic features related to therapeutic and toxic activities of compounds (Eggert and Mitchison, 2006; Korn and Krausz, 2007). Advances and miniaturization of HCS instrumentation make it possible today to run image-based assays on a scale of 100,000 compounds and more (Carpenter, 2007; Korn and Krausz, 2007).

14.5.2   OPERATIONAL HCS INFORMATICS CHALLENGES

HCS poses significant informatics requirements in addition to the operational and data analysis informatics systems and business processes needed for an effective HTS operation described earlier (Dunlay, Czekalski, and Collins, 2007). The HCS operational data handling informatics environment requires a solution to store large amounts of image-based data, a database-backed system to manage the images with their metadata, a solution to integrate the image management system with the other operational screening informatics components, integrated image analysis algorithms and software, and also solutions to integrate with instrumentation. HCS generates large amounts of data as high resolution images. Dealing with a file size of ~2 MB per image, multiple images per well, and counting controls, an HCS campaign of 100,000 compounds can readily require 1 TB of storage—exceeding the capacities of typical HTS data centers. If images must be transferred or translated into non-proprietary formats for integration, additional space is needed.

Commercial HCS instrumentation typically includes software for image analysis, visual review, and in some cases image data management (e.g., Cellomics' Store or INCell Miner HCM). There is no open standard and vendors use different proprietary data formats that complicate systems integration and reduce flexibility (Carpenter, 2007). The Open Microscopy Environment (OME; http://www.openmicroscopy.org/) is a promising multisite collaboration to address image data management in an open source approach, facilitating integration with operational software systems. OME also integrates with statistics software (Matlab) to perform image analysis. In addition to managing images, the quantity of derived (calculated) data after quantitative image analysis can be substantial (up to gigabytes per run of a few thousand samples) and the required relational database storage must be considered. Because relational storage is far more expensive than flat file storage, it may not be practical to permanently store all raw data after image analysis in a relational format; raw data can easily be archived as files that can be linked to relational data tables of the assay endpoints or the most relevant results.

An integrated informatics solution should also provide interoperability between the HCS platform and other transactional informatics components such as a compound database, sample inventory system, and assay data management system. It must also integrate HCS image data and image analysis results with other screening endpoints in a compound-centric data warehouse. No commercial or open software solution for such an integrated data environment is available. To develop a custom solution meeting requirements described for HTS, significant development resources and/ or consulting services of vendors should be budgeted.

14.5.3 HCS Image Analysis and Integration

Quantitative high content screening requires assay-specific image analysis and involves computation of many parameters for each individual cell recorded in images corresponding to sample wells at a given time point. The resulting distribution of each parameter must be statistically analyzed to identify the specific population subset of interest and (if desired) compute a population summary value. Both the perception and quantification of image features into numeric values and their statistical analysis are areas of ongoing research (Smellie, Wilson, and Ng, 2006; Carpenter, 2007; Denner, Schmalowsky, and Prechtl, 2008; Dorn, Danuser, and Yang, 2008; Stacey and Hitomi, 2008). Software including various algorithms for common image analysis tasks is usually bundled with HTS instrumentation (e.g., INCell Investigator) and is also available from third parties (e.g., Scitegic Pipeline Pilot Imaging Component). CellProfiler (http://www.cellprofiler.org/) is an open source software project for automated image analysis of numerous phenotypes (Carpenter et al., 2006).

While current software systems work well for typical HCS campaigns in which image analysis is optimized in parallel with assay development, miniaturization, automation, and validation, it is challenging due to the lack of integration of the HCS informatics components and significant computational requirements to re-analyze hundreds of thousands of images from different campaigns to generate additional data and knowledge.

After image perception and statistical feature population analysis, the generated multiparametric assay readouts must be processed in an assay data management system (as part of the operational screening informatics infrastructure) such as described for HTS, but suitable to process many layers of readouts.

14.5.4 HCS Data Analysis

The most important parameters to analyze in an HCS assay are typically well understood, based on the biological event monitored by the assay. In such cases, with an optimized population summary value, one can practically apply the methods for data normalization and assess data quality as described for statistical analysis of HTS data. However, it is desirable to take greater advantage of information-rich image-derived data sets. In many cases, certain parameters generated in image analysis carry no obvious biological meaning, although they may report useful data about compound activity. For these large multidimensional data sets, the standard HTS statistics described earlier are less useful and multivariate statistical methods are required to detect correlations and reduce dimensionality with the goal to maximize information content in as few parameters (or parameter combinations) as necessary and define activity and hits.

Along with statistical techniques, data visualization is effective for exploring and analyzing HTC data (Smellie, Wilson, and Ng, 2006; Anstett, 2007). Spotfire (http://spotfire.tibco.com)—an industry standard for interactive data visualization and analysis—is particularly useful in the area of high content data analysis to gain insight into complex multidimensional data sets, identify trends, or develop hypotheses (Anstett, 2007). It also has been integrated with image analysis software tools (e.g., INCell Investigator and Cellomics' vHCS Discovery ToolBox). Another commercial solution is the high content module of Genedata's Screener (http://www.genedata.com/).

In one recent example, factor analysis was used for data reduction of a compound profiling assay that generated 36 cytological features (after primary image analysis) related to cell cycle states (Young et al., 2008). The factor loadings could be used to infer the underlying phenotypic attributes associated with that factor. Based on the six-factor model obtained, regression was used to score each factor on a cell-by-cell basis followed by summarizing each compound treatment (well average). Hits were then defined independently of the exact phenotype after computing the Euclidian distance between each compound and the average control (untreated) phenotype for a composite vector of the factor scores of that compound. The biological activity of the hit compounds was then profiled using unsupervised hierarchical clustering of the factor scores revealing seven primary phenotype clusters.

Cheminformatics methods, such as those described earlier for mining and visualizing SARs of HTS data, can also be applied to high content readouts resulting in phenotypic statistical models and phenotypic SARs. In the same example (Young et al., 2008), the hit compounds were clustered and chemical similarity was found to be statistically significantly correlated with phenotypic similarity. Using Bayesian "target-fishing" models (Nidhi et al., 2006), the top five most probable targets of the hit compounds were found to also be statistically significantly correlated with the primary phenotypes, presumably linking phenotypic effects to mechanisms of action via chemical features.

14.6   SUMMARY AND OUTLOOK

HTS has evolved into an industrialized process in which hundreds of thousands to a million compounds can be screened in an automated fashion with the goal to identify novel entry points into drug discovery programs. A sophisticated HTS informatics infrastructure is required to manage the operational aspects of an HTS campaign and process the large data sets generated.

The components of this infrastructure must be robust and interoperable and the informatics systems must be fully integrated with the HTS instrumentation. Commercial HTS informatics systems address many of these challenges and are valuable starting points in implementing an HTS informatics environment. However, because of the lack of industry standards among the HTS instrumentation vendors and the lack of data representation standards, informatics system and instrumentation integration is not trivial and usually requires additional expertise to deploy a robust operational HTS informatics environment. Organizational business processes and standardized operating procedures must be in place for these informatics systems to work effectively and successfully perform an HTS campaign. Because of their information-driven nature, implementing and optimizing these processes should be guided by informatics expertise.

While an integrated operational informatics infrastructure is required to run a screening campaign (acquiring, processing, and associating very large data sets), HTS discovery informatics including statistical methods, visualization tools, and cheminformatics is concerned with analyzing data and effectively following up screening results. Statistical methods and visualization are important to sift through the large amounts of HTS data generated, assess screening quality and stability, and correct systematic errors. A number of statistical methods have been described for this purpose, but no common standard exists for HTS data analysis.

Further advances in statistical modeling of HTS data can be expected to provide more objective benchmarks against which to compare experimental results and will contribute to standardizing the hit identification process. Cheminformatics relates HTS results to chemical structures. A number of cheminformatics methodologies in combination with statistical data analysis have been developed to mine primary screening data, identify and evaluate hit series, effectively follow up screening results, and facilitate decision making for lead optimization. Cheminformatics approaches are also useful for developing a screening strategy including library design and "smart" (sequential) screening involving experimental and virtual screenings.

HTS discovery informatics should be seen as part of a global discovery informatics environment encompassing various computational disciplines (Stahl, Guba, and Kansy, 2006). An integrated discovery informatics environment facilitates exploration and computational modeling of heterogeneous data in a distributed environment. Integration of HTS with available cheminformatics methods and tools facilitates synergy of experimental and computational screening approaches. Better tools are required to enable scientists with little informatics expertise to query and explore very large heterogeneous data sets and apply computational prioritization protocols. Interactive visualization is a powerful trend in this direction.

Community standards for data representation that span industry and academia will further facilitate effective integration of discovery informatics systems and various public and proprietary data sources for functional genomics, proteomics, HTS, and high content image-based phenotypic screening. Today, no universally accepted standards exist due to the multiplicity and increased complexity of the assays. However, projects such as MIACA (Wiemann et al., 2007; http://miaca.sourceforge.net/) and the recent MIAHA initiative related to high content assays are important developments in this direction and underline the recognition in the community for the need of standards to communicate and integrate assay data. With the NIH Molecular Libraries Initiative making HTS resources available in academia and increasing amounts of HTS data in the public domain (PubChem), existing standards will develop further and gain more acceptance as informatics systems mature. Some HTS reporting guidelines have been proposed (Inglese, Shamu, and Guy, 2007). In addition to academic and pharmaceutical industry efforts, the Pistoia project is an industry collaboration to develop an open foundation of data standards, ontologies, and Web services (http://sourceforge.net/projects/pistoia/).

High content image-based phenotypic screening poses additional informatics requirements, both for operational data processing and data analysis. Although HCS is a fast evolving field, instrumentation and image analysis for common phenotypes are already well developed. Methods and informatics systems for downstream data processing and mining of multiparametric biological results are less established. The development of metadata schemas to describe HTS and HCS assays will make it possible for informatics integration to seamlessly explore and mine the phenotypic effects of small molecules in the context of protein targets, gene expression, regulatory networks, and metabolic pathways. It may also facilitate the departure from the "one-indication–one-target–one-drug" paradigm.

Improved integration and usability of drug discovery informatics systems will also further facilitate incorporating later stage clinical data into the early discovery process. With emerging system chemical biology (Oprea et al., 2007b), the various computational disciplines of drug discovery will evolve more toward translational drug informatics to improve the overall process of identifying tractable and novel entry points for drug discovery programs with increased probability of clinical success.

ACKNOWLEDGMENTS

We wish to acknowledge financial support from National Institutes of Health Molecular Libraries Screening Center Network and Molecular Libraries Probe Production Center Network grants U54 MH074404-01 and U54MH084512-01, respectively, and from the Scripps Florida Funding Corporation. We wish to acknowledge support from the University of Miami Center for Computational Sciences (CCS). This is CCS publication number 156. We thank Caty Chung, Chris Mader, Mark Southern, Peter Hodder, Patrick Griffin, Hugh Rosen, and an unknown reviewer for critical suggestions and helpful discussion.

REFERENCES

Ahlberg, C. 1999. Visual exploration of HTS databases: bridging the gap between chemistry and biology. *Drug Disc. Today* 4, 370–376.

Anstett, M.J. 2007. Visualization of high content screening data. *Meth. Mol. Biol.* 356, 301–317.

Augen, J. 2002. The evolving role of information technology in the drug discovery process. *Drug Disc. Today* 7, 315–323.

Austin, C.P. et al. 2004. NIH Molecular Libraries Initiative. *Science* 306, 1138–1139.

Bajorath, J. 2002. Integration of virtual and high-throughput screening. *Nat. Rev. Drug Disc.* 1, 882–894.

Baker, M. 2006. Open access chemistry databases evolving slowly but not surely. *Nat. Rev. Drug Disc.* 5, 707–708.

Bemis, G.W. and Murcko, M.A. 1996. Properties of known drugs 1: molecular frameworks. *J. Med. Chem.* 39, 2887–2893.

Bender, A. et al. 2008. Which aspects of HTS are empirically correlated with downstream success? *Curr. Opin. Drug Disc.Devel.* 11, 327–337.

Bender, A. and Glen, R.C. 2004. Molecular similarity: key technique in molecular informatics. *Org. Biomol. Chem.* 2, 3204–3218.

Bleicher, K.H. et al. 2003. Hit and lead generation: beyond high-throughput screening. *Nat. Rev. Drug Disc.* 2, 369–378.

Blower, P.E., Jr. et al. 2004. Systematic analysis of large screening sets in drug discovery. *Curr. Drug Disc. Technol.* 1, 37–47.

Brideau, C. et al. 2003. Improved statistical methods for hit selection in high-throughput screening. *J. Biomol. Screen.* 8, 634–647.

Bullen, A. 2008. Microscopic imaging techniques for drug discovery. *Nat. Rev. Drug Discov.* 7, 54–67.

Carpenter, A.E. 2007. Image-based chemical screening. *Nat. Chem. Biol.* 3, 461–465.

Carpenter, A.E. et al. 2006. CellProfiler: image analysis software for identifying and quantifying cell phenotypes. *Genome Biol.* 7, R100.

Cho, S.J. and Sun, Y. 2008. Visual exploration of structure–activity relationship using maximum common framework. *J. Comput. Aided Mol. Des.* 22, 571–578.

Claus, B.L. and Underwood, D.J. 2002. Discovery informatics: its evolving role in drug discovery. *Drug Disc. Today* 7, 957–966.

Crisman, T.J. et al. 2007. Understanding false positives in reporter gene assays: in silico chemogenomics approaches to prioritize cell-based HTS data. *J. Chem. Inf. Model.* 47, 1319–1327.

Davies, J.W., Glick, M., and Jenkins, J.L. 2006. Streamlining lead discovery by aligning in silico and high-throughput screening. *Curr. Opin. Chem. Biol.* 10, 343–351.

Denner, P., Schmalowsky, J., and Prechtl, S. 2008. High-content analysis in preclinical drug discovery. *Comb. Chem. High-Throughput Screen.* 11, 216–230.

Dorn, J.F., Danuser, G., and Yang, G. 2008. Computational processing and analysis of dynamic fluorescence image data. *Meth. Cell Biol.* 85, 497–538.

Dunlay, R.T., Czekalski, W.J., and Collins, M.A. 2007. Overview of informatics for high content screening. *Meth. Mol Biol*, 356, 269–280.

Eastwood, B.J. et al. 2006. Minimum significant ratio: a statistical parameter to characterize the reproducibility of potency estimates from concentration response assays and estimation by replicate experiment studies. *J. Biomol. Screen.* 11, 253–261.

Eckert, H. and Bajorath, J. 2007. Molecular similarity analysis in virtual screening: foundations, limitations and novel approaches. *Drug Disc. Today* 12, 225–233.

Efron, B. 2004. Large-scale simultaneous hypothesis testing: choice of a null hypothesis. *J. Am. Stat. Assn.* 99, 96–104.

Eggert, U.S. and Mitchison, T.J. 2006. Small molecule screening by imaging. *Curr. Opin. Chem. Biol.* 10, 232–237.

Engels, M.F. et al. 2002. Outlier mining in high-throughput screening experiments. *J. Biomol. Screen.* 7, 341–351.

Fink, T., Bruggesser, H., and Reymond, J.L. 2005. Virtual exploration of the small-molecule chemical universe below 160 Da. *Angew. Chem. Int.* 44, 1504–1508.

Gedeck, P. and Willett, P. 2001. Visual and computational analysis of structure–activity relationships in high-throughput screening data. *Curr. Opin. Chem. Biol.* 5, 389–395.

Glick, M. et al. 2006. Enrichment of high-throughput screening data with increasing levels of noise using support vector machines, recursive partitioning, and Laplacian-modified naive Bayesian classifiers. *J. Chem. Inf. Model.* 46, 193–200.

Gribbon, P. et al. 2005. Evaluating real-life high-throughput screening data. *J. Biomol. Screen.* 10, 99–107.

Gribbon, P. and Sewing, A. 2005. High-throughput drug discovery: what can we expect from HTS? *Drug Disc. Today* 10, 17–22.

Grunweller, A. et al. 2005. RNA interference as a gene-specific approach for molecular medicine: past and future perspectives of synthetic peptide libraries. *Curr. Med. Chem.* 12, 3143–3161.

Hann, M. et al. 1999. Strategic pooling of compounds for high-throughput screening. *J. Chem. Inf. Comput. Sci.* 39, 897–902.

Hann, M.M. and Oprea, T.I. 2004. Pursuing the lead likeness concept in pharmaceutical research. *Curr. Opin. Chem. Biol.* 8, 255–263.

Harper, G. and Pickett, S.D. 2006. Methods for mining HTS data. *Drug Disc. Today* 11, 694–699.

Hertzberg, R.P. and Pope, A.J. 2000. High-throughput screening: new technology for the 21st century. *Curr. Opin. Chem. Biol.* 4, 445–451.

Houston, J.G. et al. 2008. Case study: impact of technology investment on lead discovery at Bristol-Myers Squibb, 1998–2006. *Drug Disc. Today* 13, 44–51.

Inglese, J. et al. 2006. Quantitative high-throughput screening: a titration-based approach that efficiently identifies biological activities in large chemical libraries. *Proc. Natl. Acad. Sci. USA* 103, 11473–11478.

Inglese, J. et al. 2007a. High-throughput screening assays for the identification of chemical probes. *Nat. Chem. Biol.* 3, 466–479.

Inglese, J., Shamu, C.E., and Guy, R.K. 2007b. Reporting data from high-throughput screening of small-molecule libraries. *Nat. Chem. Biol.* 3, 438–441.

Karnachi, P.S. and Brown, F.K. 2004. Practical approaches to efficient screening: information-rich screening protocol. *J. Biomol. Screen.* 9, 678–686.

Kevorkov, D. and Makarenkov, V. 2005. Statistical analysis of systematic errors in high-throughput screening. *J. Biomol. Screen.* 10, 557–567.

Klebe, G. 2006. Virtual ligand screening: strategies, perspectives and limitations. *Drug Disc. Today* 11, 580–594.

Korn, K. and Krausz, E. 2007. Cell-based high-content screening of small-molecule libraries. *Curr Opin Chem Biol*, 11, 503–510.

Lazo, J.S. 2006. Roadmap or roadkill: a pharmacologist's analysis of the NIH Molecular Libraries Initiative. *Mol. Interv.* 6, 240–243.

Ling, X.B. 2008. High-throughput screening informatics. *Comb. Chem. High-Throughput Screen.* 11, 249–257.

Lipinski, C.A. et al. 2001. Experimental and computational approaches to estimate solubility and permeability in drug discovery and development settings. *Adv. Drug Delivery Rev.* 46, 3–26.

Ma, X.H. et al. 2008. Evaluation of virtual screening performance of support vector machines trained by sparsely distributed active compounds. *J. Chem. Inf. Model.* 48, 1227–1237.

MacKeigan, J.P. et al. 2005. Sensitized RNAi screen of human kinases and phosphatases identifies new regulators of apoptosis and chemoresistance for RNA interference as a gene-specific approach for molecular medicine. *Nat. Cell Biol.* 7, 591–600.

Makarenkov, V. et al. 2007. An efficient method for the detection and elimination of systematic error in high-throughput screening. *Bioinformatics* 23, 1648–1657.

Malo, N. et al. 2006. Statistical practice in high-throughput screening data analysis. *Nat. Biotechnol.* 24, 167–175.

Marasco, D. et al. 2008. Past and future perspectives of synthetic peptide libraries. *Curr. Protein Pept. Sci.* 9, 447–467.

Martin, Y.C., Kofron, J.L., and Traphagen, L.M. 2002. Do structurally similar molecules have similar biological activity? *J. Med. Chem.* 45, 4350–4358.

McManus, M.T. et al. 2002. Gene silencing in mammals by small interfering RNAs: Sensitized RNAi screen of human kinases and phosphatases identifies new regulators of apoptosis and chemoresistance. *Nat. Rev. Genet.* 3, 737–747.

Muegge, I. and Oloff, S. 2006. Advances in virtual screening, *Drug Disc. Today* 3, 405–411.

Nidhi, M. et al. 2006. Prediction of biological targets for compounds using multiple-category Bayesian models trained on chemogenomics databases. *J. Chem. Inf. Model.* 46, 1124–1133.

Oprea, T.I. et al. 2001. Is there a difference between leads and drugs? A historical perspective. *J. Chem. Inf. Comput. Sci.* 41, 1308–1315.

Oprea, T.I. et al. 2007a. Lead-like, drug-like or pub-like: how different are they? *J. Comput. Aided Mol. Des.* 21, 113–119.

Oprea, T.I. et al. 2007b. Systems chemical biology. *Nat. Chem. Biol.* 3, 447–450.

Padmanabha, R., Cook, L., and Gill, J. 2005. HTS quality control and data analysis: a process to maximize information from a high-throughput screen. *Comb. Chem. High-Throughput Screen.* 8, 521–527.

Parker, C.N. and Bajorath, J. 2006. Toward unified compound screening strategies: critical evaluation of error sources in experimental and virtual high-throughput screening. *QSAR Comb. Sci.* 25, 1153–1161.

Parker, C.N. and Schreyer, S.K. 2004. Application of chemoinformatics to high-throughput screening: practical considerations. *Meth. Mol. Biol.* 275, 85–110.

Pearlman, R.S. and Smith, K.M. 1998. Novel software tools for chemical diversity. *Persp. Drug Disc. Des.* 9–11, 339–353.

Pereira, D.A. and Williams, J.A. 2007. Origin and evolution of high-throughput screening. *Br. J. Pharmacol.* 152, 53–61.

Posner, B.A. 2005. High-throughput screening-driven lead discovery: meeting the challenges of finding new therapeutics. *Curr. Opin. Drug Disc. Dev.* 8, 487–494.

Reichard, G.A. 2008. SARVision Plus. *J. Chem. Inf. Model* 48, 1287–1288.

Rishton, G.M. 1997. Reactive compounds and *in vitro* false positives in HTS. *Drug Disc. Today* 2, 382–384.

Roberts, G. et al. 2000. LeadScope software for exploring large sets of screening data. *J. Chem. Inf. Comput. Sci.* 40, 1302–1314.

Rogers, D., Brown, R.D., and Hahn, M. 2005. Using extended connectivity fingerprints with Laplacian-modified Bayesian analysis in high-throughput screening follow-up. *J. Biomol. Screen.* 10, 682–686.

Rusinko, A. et al. 1999. Analysis of a large structure/biological activity data set using recursive partitioning. *J. Chem. Inf. Comput. Sci.* 39, 1017–1026.

Sayle, R. 1999. *Canonicalization and Enumeration of Tautomers*. http://www.daylight.com/meetings/emug99/Delany/taut_html/index.htm.

Schnur, D.M. 2008. Recent trends in library design: rational design revisited. *Curr. Opin. Drug Disc. Dev.* 11, 375–380.

Schuffenhauer, A. et al. 2007. Scaffold tree visualization of the scaffold universe by hierarchical scaffold classification. *J. Chem. Inf. Model.* 47, 47–58.

Schuffenhauer, A. et al. 2002. An ontology for pharmaceutical ligands and its application for in silico screening and library design. *J. Chem. Inf. Comput. Sci.* 42, 947–955.

Seidler, J. et al. 2003. Identification and prediction of promiscuous aggregating inhibitors among known drugs. *J. Med. Chem.* 46, 4477–4486.

Shelat, A.A. and Guy, R.K. 2007. Scaffold composition and biological relevance of screening libraries. *Nat. Chem. Biol.* 3, 442–446.

Shneiderman, B. 2008. Extreme visualization: squeezing a billion records into a million pixels. In *Proceedings of the 2008 ACM SIGMOD International Conference on Management of Data.* Vancouver, Canada: ACM, 3–12.

Simeonov, A. et al. 2008. Fluorescence spectroscopic profiling of compound libraries. *J. Med. Chem.* 51, 2363–2371.

Smellie, A., Wilson, C.J., and Ng, S.C. 2006. Visualization and interpretation of high content screening data. *J. Chem. Inf. Model.* 46, 201–207.

Stacey, D.W. and Hitomi, M. 2008. Cell cycle studies based upon quantitative image analysis. *Cytometry A* 73, 270–278.

Stahl, M., Guba, W., and Kansy, M. 2006. Integrating molecular design resources within modern drug discovery research: the Roche experience. *Drug Disc. Today* 11, 326–333.

Stahl, M. and Mauser, H. 2005. Database clustering with a combination of fingerprint and maximum common substructure methods. *J. Chem. Inf. Model.* 45, 542–548.

Stahl, M. et al. 2005. Robust clustering method for chemical structures. *J. Med. Chem.* 48, 4358–4366.

Szegezdi, J. and Csizmadia, F. 2007. Tautomer generation. pKa based dominance conditions for generating dominant tautomers. In *American Chemical Society Fall National Meeting*, Boston.

Todeschini, R. et al., Eds. 2000. *Handbook of Molecular Descriptors*. New York: Wiley-VCH.

Weininger, D. 1988. SMILES, a chemical language and information system 1: Introduction to methodology and encoding rules. *J. Chem. Inf. Comput. Sci.*, 28, 31–36.

Wiemann, S. et al. 2007. MIACA: minimum information about a cellular assay, and the cellular assay object model. http://www.nature.com/nbt/consult/pdf/Wiemann.pdf.

Wilkens, S.J., Janes, J., and Su, A.I. 2005. HierS: hierarchical scaffold clustering using topological chemical graphs. *J. Med. Chem.* 48, 3182–3193.

Wu, Z., Liu, D., and Sui, Y. 2008. Quantitative assessment of hit detection and confirmation in single and duplicate high-throughput screenings. *J. Biomol. Screen.* 13, 159–167.

Yan, S.F. et al. 2005. Novel statistical approach for primary high-throughput screening hit selection. *J. Chem. Inf. Model.* 45, 1784–1790.

Young, D.W. et al. 2008. Integrating high-content screening and ligand target prediction to identify mechanism of action. *Nat. Chem. Biol.* 4, 59–68.

Zhang, J.H., Chung, T.D., and Oldenburg, K.R. 1999. Simple statistical parameter for use in evaluation and validation of high-throughput screening assays. *J. Biomol. Screen.* 4, 67–73.

WEB SITE REFERENCES

ChemAxon Extended SMILES, SMARTS:
http://www.chemaxon.com/marvin/help/formats/cxsmiles-doc.html#cxsmiles.
MDL's Enhanced Stereochemical representation: http://www.mdli.com/solutions/white_papers/stereochemistry-white-paper.jsp.
MDL CTfile formats:
http://www.mdli.com/solutions/white_papers/ctfile_formats.jsp.
SMILES simplified molecular input line entry system: http://www.daylight.com/smiles/index.html.
National Institutes of Health *Assay Guidance Manual* Version 5.0: http://www.ncgc.nih.gov/guidance/manual_toc.html.

Index

9 780367 384708